T0249745

CONTROL AND DYNAMIC SYSTEMS

*Advances in Theory
and Applications*

Volume 68

CONTRIBUTORS TO THIS VOLUME

ALBERT BENVENISTE
JAMES B. BURR
D. HATZINAKOS
SIMON HAYKIN
CORNELIUS T. LEONDES
WEIPING LI
KEN MARTIN
MUKUND PADMANABHAN
ALLEN M. PETERSON
PETER A. STUBBERUD
MICHAIL K. TSATSANIS

CONTROL AND DYNAMIC SYSTEMS

ADVANCES IN THEORY AND APPLICATIONS

Edited by
C. T. LEONDES

School of Engineering and Applied Science
University of California, San Diego
La Jolla, California

VOLUME 68: DIGITAL SIGNAL PROCESSING SYSTEMS: IMPLEMENTATION TECHNIQUES

ACADEMIC PRESS

San Diego   New York   Boston
London   Sydney   Tokyo   Toronto

This book is printed on acid-free paper. ∞

Copyright © 1995 by ACADEMIC PRESS, INC.

All Rights Reserved.
No part of this publication may be reproduced or transmitted in any form or by any
means, electronic or mechanical, including photocopy, recording, or any information
storage and retrieval system, without permission in writing from the publisher.

Academic Press, Inc.
A Division of Harcourt Brace & Company
525 B Street, Suite 1900, San Diego, California 92101-4495

United Kingdom Edition published by
Academic Press Limited
24-28 Oval Road, London NW1 7DX

International Standard Serial Number: 0090-5267

International Standard Book Number: 0-12-012768-7

Printed and bound in the United Kingdom
Transferred to Digital Printing, 2011

CONTENTS

CONTENTS

CONTRIBUTORS

Numbers in parentheses indicate the pages on which the authors' contributions begin.

Albert Benveniste (121), *Institut de Recherche en Informatique et Systemes Aleatoires, Campus Universitaire de Beaulieu, F-35042 Rennes, France*

James B. Burr (1), *Electrical Engineering Department, Stanford University, Stanford, California 94305*

D. Hatzinakos (279), *Department of Electrical and Computer Engineering, University of Toronto, Toronto, Ontario, Canada M5S 1A4*

Simon Haykin (89), *Communications Research Laboratory, McMaster University, Hamilton, Ontario, Canada L8S 4K1*

Cornelius T. Leondes (163, 255), *School of Engineering, University of California, San Diego, La Jolla, California 92037*

Weiping Li (1), *Electrical Engineering and Computer Science Department, Lehigh University, Bethlehem, Pennsylvania 18015*

Ken Martin (197), *Electrical Engineering Department, University of Toronto, Toronto, Ontario, Canada M5S 1A4*

Mukund Padmanabhan (197), *IBM T. J. Watson Research Center, Yorktown Heights, New York 10598*

Allen M. Peterson (1), *Electrical Engineering Department, Stanford University, Stanford, California 94305*

Peter A. Stubberud (163, 255), *Department of Electrical and Computer Engineering, University of Nevada, Las Vegas, Las Vegas, Nevada 89154*

Michail K. Tsatsanis, *(333), University of Virginia, Charlottesville, Virginia 22903*

PREFACE

From about the mid-1950s to the early 1960s, the field of digital filtering, which was based on processing data from various sources on a mainframe computer, played a key role in the processing of telemetry data. During this time period the processing of airborne radar data was based on analog computer technology. In this application area, an airborne radar used in tactical aircraft could detect the radar return from another low-flying aircraft in the environment of competing radar return from the ground. This was accomplished by the processing and filtering of the radar signal by means of analog circuitry in order to take advantage of the Doppler frequency shift due to the velocity of the observed aircraft. This analog implementation was lacking in the flexibility and capability inherent in programmable digital signal processor technology, which was just coming onto the technological scene.

Developments and powerful technological advances in integrated digital electronics coalesced soon after the early 1960s to lay the foundations for modern digital signal processing. Continuing developments in techniques and supporting technology, particularly very large scale integrated digital electronics circuitry, have resulted in significant advances in many areas. These areas include consumer products, medical products, automotive systems, aerospace systems, geophysical systems, and defense-related systems. Hence, this is a particularly appropriate time for *Control and Dynamic Systems* to address the area of "Digital Signal Processing Systems: Implementation Techniques," the theme for this volume.

The first contribution to this volume is "VLSI Signal Processing," by James B. Burr, Weiping Li, and Allen M. Peterson. This contribution gives an overview of digital implementations and techniques for high-performance digital signal processing. The authors are to be most highly complimented for producing a self contained treatment with a high degree of clarity of this major and broad topic which underpins the implementation of digital signal processing systems. Needless to say, this is a most appropriate contribution with which to begin this volume.

The next contribution is "Recurrent Neural Networks for Adaptive Filtering," by Simon Haykin. Adaptive filtering, which is an inherently nonlinear process, arises in such major application areas as identification, equalization (inverse modeling), prediction, and noise cancellation. This contribution presents rather powerfully effective neural network techniques which are ideally suited to adaptive filtering applications. As such this is also a most important contribution to this volume.

The next contribution is "Multiscale Signal Processing: From QMF to Wavelets," by Albert Benveniste. It is noted in this contribution that multirate filtering, multiscale signal analysis, and multiresolution are, in fact, closely related to each other. The techniques of Quadrature Mirror Filters (QMF) and orthonormal wavelet transforms are presented for dealing with these important problems.

The next contribution is "The Design of Frequency Sampling Filters," by Peter A. Stubberud and Cornelius T. Leondes. Many digital signal processing systems require linear phase filtering. Digital linear phase filters designed by either the window design or the optimal filter design method are generally implemented by direct convolution, which uses the filter's impulse response as filter coefficients. In this contribution, frequency sampling filters use frequency samples, which are specific values from the filter's frequency response, as coefficients in the filter's implementation, and are presented as an effective alternative means for linear phase filtering implementation.

The next contribution is "Low-Complexity Filter-Banks for Adaptive and Other Applications," by Mukund Padmanabhan and Ken Martin. Filter-banks, single-input/multiple-output structures, find use in a wide variety of applications such as subband coding, frequency domain adaptive filtering, communication systems, frequency estimation, and transform computations. While a great deal of attention has been focused in the literature on their design and properties, it is only recently that the issues of their implementation have started to receive treatment in the literature. This contribution is an in-depth treatment of implementation techniques with a number of important illustrative examples.

The next contribution is "A Discrete Time Nonrecursive Linear Phase Transport Processor Design Technique," by Peter A. Stubberud and Cornelius T. Leondes. A discrete time transport processor is a discrete time system that is composed only of delays, adds, and subtracts; such that more complex operations including multiplication are not required in their implementation. This contribution develops the details of the implementation of such signal processors and illustrates their applications. For instance, such processors are well suited for the design and implementation of frequency-selective linear phase discrete time transport processors.

The next contribution is "Blind Deconvolution: Channel Identification and Equalization," by D. Hatzinakos. Blind deconvolution refers to the problem of separating the two convolved signals $\{f(n)\}$, discrete filter impulse response, and $\{x(n)\}$, discrete time input signal to the filter, when both signals are unknown or partially known. This is an important problem in seismic data analysis, transmission monitoring, deblurring of images in digital image processing, multipoint network communications, echo cancellation in wireless telephony, digital radio links over fading channels, and other applications when there is either limited knowledge of the signals due to practical constraints or a sudden change in the properties of the signals. This contribution presents an in-depth treatment of the major approaches to the problem of blind deconvolution with numerous illustrative examples.

The final contribution to this volume is "Time-Varying System Identification and Channel Equalization Using Wavelets and Higher-Order Statistics," by Michail K. Tsatsanis. This contribution is an in-depth treatment of basis expansion ideas to identify time-varying (TV) systems and equalize rapidly fading channels. A number of illustrative examples make evident the great potential of basis expansion tools for addressing challenging questions regarding adaptive and blind estimation of these TV channels. The powerful results presented in this contribution are also, of course, importantly applicable to the other significant problems already mentioned. As such this is a most appropriate contribution with which to conclude this volume.

This volume on implementation techniques in digital signal processing systems clearly reveals the significance and power of the techniques that are available, and, with further development, the essential role they will play as applied to a wide variety of areas. The authors are all to be highly commended for their splendid contributions to this volume, which will provide a significant and unique international reference source for students, research workers, practicing engineers, and others for years to come.

VLSI Signal Processing

James B. Burr
Electrical Engineering Department
Stanford University, Stanford, CA 94305

Weiping Li
Electrical Engineering and Computer Science Department
Lehigh University, Bethlehem, PA 18015

Allen M. Peterson
Electrical Engineering Department
Stanford University, Stanford, CA 94305

1 Introduction

This chapter gives an overview of digital VLSI implementations and discusses techniques for high performance digital signal processing. It presents some basic digital VLSI building blocks useful for digital signal processing and a set of techniques for estimating chip area, performance, and power consumption in the early stages of design to facilitate architectural exploration. It also shows how technology scaling rules can be included in the estimation process. It then uses the estimation techniques to predict capacity and performance of a variety of digital architectures.

The assumption about the readership of this chapter is that the reader knows about signal processing algorithms very well, has little knowledge about VLSI design, and would like to understand how to implement signal processing algorithms using VLSI. We hope to put enough relevant materials for the reader to understand the opportunities and problems in VLSI signal processing.

Copyright © 1995 by Academic Press, Inc.
All rights of reproduction in any form reserved.

2   Basic CMOS digital circuits

This section is not intended to be a thorough presentation of digital logic design, as there are many excellent sources for this [MC80, WF82, WE85, HP89]. Rather, we highlight specific structures which are especially useful in designing digital elements for signal processing. Each element has a behavioral specification which defines its funtionality, a graphical symbol which is used in a circuit diagram, and circuit description(s) at the transistor level.

There are many logic design styles to choose from in implementing CMOS circuits (see Table 1). Logic design styles achieve different tradeoffs in speed, power, and area. The highest speed logic families also tend to consume the most power. The most compact tend to be slow.

year	what	who	description
-	Static	-	Fully static CMOS
1982	Domino	[KLL82]	Domino logic
1987	DCVSL	[CP87]	Differential cascode voltage switch logic
1987	DPTL	[PSS87]	Differential pass transistor logic
1990	CPL	[YYN$^+$90]	Complementary pass transistor logic
1991	L-DPTL	[LR91]	Latched differential pass transistor logic

Table 1: Logic design styles.

Fully static logic offers high speed, reasonably small area, and low power. Domino logic [KLL82] offers less area but higher power, and is dynamic. Differential cascode voltage switch logic (DCVSL) [CP87] is popular in asynchronous design because each complex logic gate generates its own completion signal. It is quite high power, however, since every output toggles on every cycle. Complementary pass-transistor logic (CPL) offers modest performance, is compact, and low power.

For very low voltage operation, fully static works better than dynamic or pass transistor logics since dynamic nodes leak too fast and pass transistor logics normally require N-transistors to pass ones. The best logic style for very low voltage operation reported to date is latched differential pass transistor logic (L-DPTL) [LR91], which uses N-transistors to pull down one side or the other of a cross coupled inverter. This style is very similar to CPL or DPTL [PSS87], but the use of inverters rather than cross coupled P-transistors turns the output stage into a differential sense amplifier. A static latch can be implemented with only two additional pass transistors, turning the output stage into a 6-transistor SRAM cell. L-DPTL is especially effective at low voltage when the threshold voltage of the pass devices is reduced to around 200mV.

Some basic CMOS digital circuits are presented in this section. Because these circuits are used in VLSI designs frequently, the most efficient design in terms of size and speed makes a big difference in system performance. The most interesting circuit is the exclusive or (XOR) gate which plays a crucial role in many digital systems. Unless specified, we mainly use static logic style.

1. Transmission Gate

Behavior of a transmission gate is specified as follows:

$$Out = \begin{cases} In & \text{if} \quad C = 1 \\ HI & \text{if} \quad C = 0 \end{cases}$$

where HI means high impedance. The symbol and a circuit diagram of a CMOS transmission gate is shown in Figure 1.

Figure 1: Transmission gate

When the control signal "C" is high, the transmission gate passes the input "In" to the output "Out". When "C" is low, "In" and "Out" are isolated from each other. Therefore, a transmission gate is actually a CMOS switch. A reason to use a transmission gate, which consists of a pair of N and P transistors, as a switch instead of a single N or P transistor is to prevent threshold drop. To illustrate this point, a single N-transistor is shown in Figure 2 as a switch.

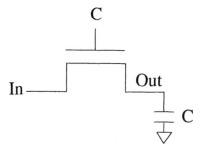

Figure 2: Pass transistor switch

When the control signal "C" is high (5V), the input signal "In" should be passed to the output "Out". If "In" is low (0V), the capacitor discharges through the switch transistor so that "Out" becomes low

(0V). However, if "In" is originally high (5V) and "Out" is originally low (0V), the capacitor is charged upto 5V-V_{Nth} through the switch transistor, where V_{Nth} is the threshold voltage of the N-transistor. The output becomes high with a threshold drop (5V-V_{Nth}). Therefore an N-transistor can pass a good "0" but not a good "1". On the other hand, a P transistor can pass a good "1" but not a good "0". To prevent threshold drop for both "1" and "0", a pair of N and P transistors are used in the transmission gate.

2. Inverter

An inverter is specified as follows:

$$Out = \overline{In}$$

The symbol and a circuit diagram are shown in Figure 3.

Figure 3: Inverter

When the input signal "In" is high, the N-transistor "pulls down" the output node "Out". When "In" is low, the P-transistor "pulls up" the output node "Out".

3. Nand Gate

The logic specification of a nand gate is as follows:

$$Out = \overline{In1 \cdot In2}$$

The symbol and a circuit diagram are shown in Figure 4. The two N-transistors connected in series pull down the output node when both inputs are high. The output node is pulled up when one of the two inputs is low.

4. Nor Gate

A nor gate is specified as follows:

$$Out = \overline{In1 + In2}$$

Figure 5 shows the symbol and a circuit diagram of a nor gate. In a nor gate, the two pull-down N-transistos are connected in parallel and the two pull-up P-transistors are in series.

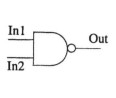

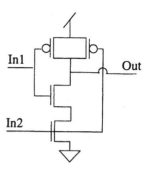

Figure 4: NAND gate

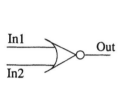

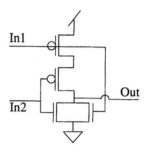

Figure 5: NOR gate

5. Tri-State Inverter

A tri-state inverter is specified as follows:

$$\text{Out} = \begin{cases} \overline{\text{In}} & \text{if} \quad C = 1 \\ \text{HI} & \text{if} \quad C = 0 \end{cases}$$

The name of this circuit comes from the fact that the output node "Out" can be either "1" or "0" or high impedance. The symbol and a circuit diagram are shown in Figure 6. A tri-state inverter can also be implemented by cascading an inverter with a transmission gate as shown in Figure 7. An interesting question is whether we can change the order of cascading the inverter and the transmission gate? The answer is no because, with the inverter at the output, there are only two unpredictable states "1" or "0" instead of three. Since the output is unpredictable when the control signal "C" is low, such a circuit should be avoided.

Figure 6: Tri-state inverter

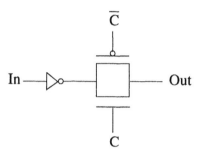

Figure 7: An alternative tri-state inverter circuit

6. Multiplexer

Multiplexer is a circuit to selectively pass one of two inputs to the output depending on a control signal. The behavior of a multiplexer can be written as follows:

$$Out = \begin{cases} \text{In1} & \text{if} \quad C = 1 \\ \text{In2} & \text{if} \quad C = 0 \end{cases}$$

The symbol and two multiplexer circuits are shown in Figure 8. The first multiplexer circuit consists of two transmission gates with complementary controls. The second circuit uses three nand gates plus an inverter and thus requires 14 transistors versus 6 transistors needed in the first circuit (an inverter is needed to generate the $\overline{C}$ signal). The second circuit with a higher transistor count provides better driving capability to the output because the resistance between the output node and either Vdd or Gnd is limited to the transistors in the output nand gate. On the other hand, the driving capability of the first circuit depends on the input signals. The resistance between the output node and a power source (either Vdd or Gnd) can be very large if many transmission gates are cascaded in series.

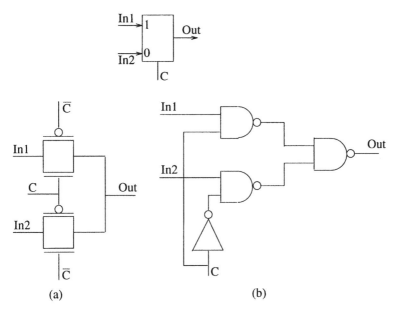

Figure 8: Multiplexer

7. Demultiplexer

A demultiplexer performs the opposite function of a multiplexer, i.e., it selectively passes the input to one of two outputs depending on a control signal. The behavior of a demultiplexer is usually specified as follows:

$$\text{Out1} = \text{In} \quad \text{if} \quad C = 1$$
$$\text{Out2} = \text{In} \quad \text{if} \quad C = 0$$

The symbol and two demultiplexer circuits are shown in Figure 9. It is interesting to note that the above behavior specification of the demultiplexer is actually incomplete. This is why the two circuits both satisfy the demultiplexer behavior specification but they are not equivalent to each other. In the first circuit, when the input is passed to one of the outputs, the other output node is high impedance. On the other hand, it is "0" in the second circuit. This is a good example of how an incomplete specification may lead to some unexpected behavior of a design.

8. Exclusive OR Gate

Exclusive OR (XOR) gate is an interesting unit to be designed using CMOS circuits. Its behavior specification can be given as follows:

$$\text{Out} = \begin{cases} 1 & \text{if} \quad \text{In1} \neq \text{In2} \\ 0 & \text{if} \quad \text{In1} = \text{In2} \end{cases}$$

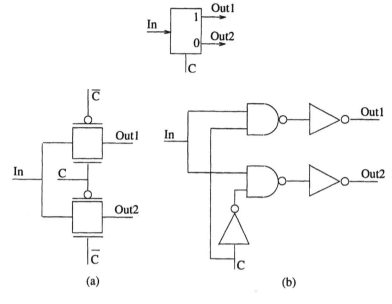

Figure 9: Demultiplexer

The symbol for an XOR gate is shown in Figure 10. There are many different ways to design an XOR gate. We discuss 10 of them with transistor counts range from 22 to 3.

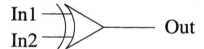

Figure 10: XOR gate symbol

(a) XOR-A

From the above behavior specification, we can obtain a logic relationship of the output with the inputs as follows:

$$\text{Out} = \text{In1} \cdot \overline{\text{In2}} + \overline{\text{In1}} \cdot \text{In2}$$

A direct implementation of this logic relationship requires 22 transistors as shown in Figure 11. Because logic "and" and logic "or" are implemented in CMOS based on inverting nand and nor outputs respectively, the transistor count is high in such a direct design.

(b) XOR-B

Since "nand" and "nor" require less transistors than "and" and "or" in CMOS design, one can write the XOR logic relationship based on "nand" and "nor" as follows:

$$\text{Out} = \overline{\overline{\text{In1} \cdot \overline{\text{In2}}} \cdot \overline{\overline{\text{In1}} \cdot \text{In2}}}$$

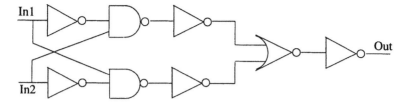

Figure 11: XOR gate circuit-a (22T)

An implementation of this logic relationship requires 16 transistors as shown in Figure 12. This would be the most efficient XOR circuit in terms of transistor count if "nand" gate, "nor" gate, and inverter were the lowest level units and all other logic circuits had to be built upon them.

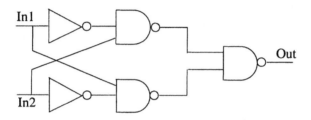

Figure 12: XOR gate circuit-b (16T)

(c) XOR-C

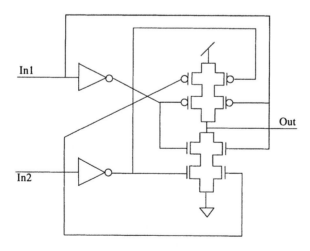

Figure 13: XOR gate circuit-c (12T)

A smaller transistor count can be obtained for an XOR gate as shown in Figure 13. In this XOR circuit, the two inverters simply generate the complements of the two input signals. The four pairs of transistors determine the output value according to the four possible input cases with a one-to-one correspondance. For example, in the case of In1=1 and In2=0, the upper left pair of P-transistors pull up the output node and the rest three paths to Vdd and Gnd are off. This circuit requires 12 transistors with 4 for the two inverters and 8 for the four pull-up and pull-down paths.

(d) XOR-D

Comparing the XOR-B circuit with the second multiplexer circuit in Figure 8, one can notice that an XOR gate is almost the same as a multiplexer. The multiplexer circuit can be changed to an XOR circuit by adding one more inverter so that the original "In1" node takes the complement of the "In2" signal. Conceptually, an XOR gate can be considered as a multiplexer with one input selectively controls the passage of the other input signal or its complement. Using this concept, one can reduce the transistor count of an XOR circuit by using the first multiplexer circuit shown in Figure 8. A circuit diagram of such an XOR gate design is shown in Figure 14 with a transistor count of 8. As shown here, "In1" signal and its complement control the passage of "In2" or its complement. When In1=0 and In2 is passed to the output, the driving capability of the output depends on the "In2" signal.

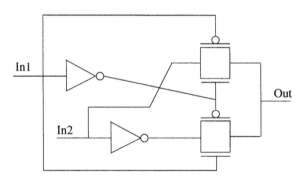

Figure 14: XOR gate circuit-d (8T)

(e) XOR-E

To push further the concept of an XOR gate being equivalent to a multiplexer, one may think 4 transistors would be enough for an XOR gate circuit as shown in Figure 15. The thought is that the inverter generates the complement of In2 and In1 controls the two transistor switches to pass either In2 or its complement. However, as discussed earlier, this circuit has the threshold drop problem in two input cases, namely, when In1=1 and In2=0, the output may be 5V-V_{Nth}, and when In1=0 and In2=0, the output may be V_{Pth}.

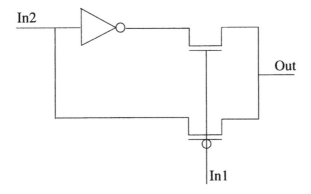

Figure 15: XOR gate circuit-e (4T)

(f) XOR-F

One way to solve the threshold drop problem is to use an inverter at the output to restore the voltage level as shown in Figure 16. The input inverter is changed to the lower path because the logic relationship has to be maintained. Although there are still threshold drops at the output of the pass-transistors in two input cases, the output inverter can function correctly so that the final output voltage is either 5V or 0V provided by the Vdd or Gnd in the output inverter. However, in the two input cases that cause threshold drop, the output inverter becomes slower than usual because its input voltage doesn't go to 5V or 0V.

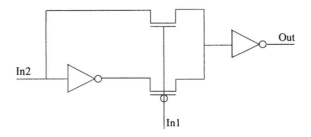

Figure 16: XOR gate circuit-f (6T)

(g) XOR-G

Another way to solve the threshold drop problem is to add a transmission gate as shown in Figure 17. As discussed in XOR-E, the threshold drop happens when In2=0. The added transmission gate passes In1 when In2=0 and thus the output doesn't have a threshold drop anymore. Note that the logic value of the XOR output is supposed to be the same as In1 when In2=0. This circuit doesn't provide as a strong driving capability as XOR-F which has an inverter at the output. Therefore, this circuit can be slow when many of them are cascaded in series.

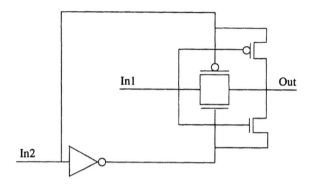

Figure 17: XOR gate circuit-g (6T)

(h) XOR-H

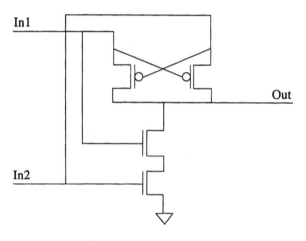

Figure 18: XOR gate circuit-h (4T)

Figure 18 shows another possibility for a 4-transistor XOR gate. The two N-transistors in series pull down the output node when both inputs are high. When one of the inputs is low and the other is high, the cross-coupled P-transistor pair passes the high input to the output. When both inputs are low, both P-transistors pass the low signal to the output. However, there is a threshold drop in last case (In1=In2=0). This is better than XOR-E which has threshold drop in two input cases.

(i) XOR-I

To solve the threshold drop problem in XOR-H, an inverter can be used to restore the output voltage level in a similar way to that in XOR-F. Because the added inverter changes the logic value, the original 4 transistors have to be rearranged as shown in Figure 19. This circuit uses the same number of transistors (6) as that in XOR-F but has only one input case causing a slow operation of the output inverter versus two cases in XOR-F.

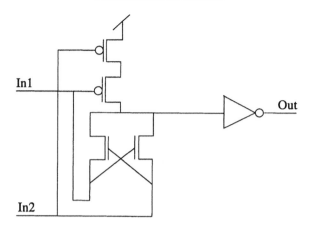

Figure 19: XOR gate circuit-i (6T)

(j) XOR-J

The XOR gate circuits discussed so far are all static CMOS circuits. If dynamic logic is allowed in a design, it is possible to implement an XOR gate using only 3 transistors as shown in Figure 20. Compared with the circuit in XOR-H, this circuit uses a single N-transistor instead of two and the N-transistor is controlled by a clock signal instead of the inputs. When clock is high, the circuit is in precharging phase and the output is pulled down. When clock goes low, the circuit goes into evaluating phase and the output logic value is determined by the inputs. If both inputs are high, the two P-transistors are both off and the output stays low as in the precharging phase. If both inputs are low, the P-transistors would not be on either because the output node was precharged to low and therefore the output node stays low. Note that this is the threshold drop case in XOR-H. If one of the inputs is low and the other is high, the cross-coupled P-transistor pair passes the high signal to the output.

In summary, XOR gate is an interesting logic function with many different ways to implement it using CMOS circuits. From the 10 circuits discussed above, 4 and 3 are the minimum transistor counts for static and dynamic logic XOR gate respectively. However, the two 4-transistor static XOR circuits (XOR-E and XOR-H) suffer from the threshold drop at the output which may not be tolerable. Among the 6-transistor XOR circuits (XOR-F, XOR-G, and XOR-I), XOR-I performs the best in terms of speed, especially when many XOR gates are cascaded in series. In [HL89], simulation results on various XOR circuits are reported. All the XOR circuits are simulated using LSIM, SPICE [Nag75] and IRSIM [Ter82]. Figure 21 shows the setup for simulating the XOR circuits. The input signals are chosen to make sure that all possible combinations of transitions are simulated.

The most interesting result of this study is the comparison of XOR-G

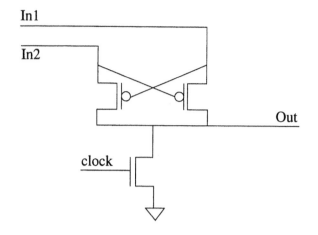

Figure 20: Dynamic XOR gate circuit (3T)

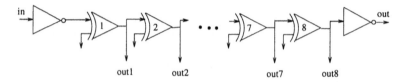

Figure 21: Setup for simulating XOR gate circuits

and XOR-I circuits. Both circuits have the same number of transistors (six). The single-stage worst case performance of XOR-I circuit is a little better than that of XOR-G circuit. Further analysis indicates that XOR-I circuit is much faster than XOR-G circuit in applications where many cascaded XOR gates are used in a chain. Figure 22 plots the delay time vs. number of cascaded XOR stages for both circuits.

This simulation shows that XOR-I circuit is indeed much faster than XOR-G circuit. The reason is that the threshold drop problem of XOR-I circuit is isolated from stage to stage because of the restoring inverter. This is why the delay of XOR-I circuit chain is growing linearly with the number of stages as shown in Figure 22. For XOR-G circuit chain, the delay is much worse than the linear relationship because each stage introduces both series resistance and parallel capacitance. The delay of XOR-G circuit chain is approximately proportional to square of the number of stages as shown in and Figure 22.

9. Latch (Register)

Latches play an important role in any digital design, especially if the design is pipelined. The behavior of a latch can be written as follows:

$$Out(n) = In(n-1)$$

where n is the time index. The right latch to use depends on the

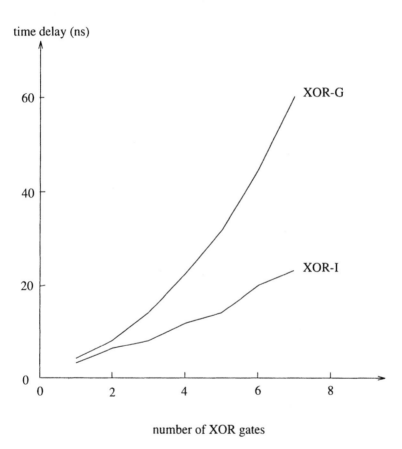

number of XOR gates

Figure 22: Delay Time vs. Number of Cascaded XOR Stages

clocking discipline and the desired performance. Some clocking styles
are safer than others. We have been using nonoverlapping two-phase
clocks in most of our chips to date. Most of our designs use either
the fully static or pseudo-static latch shown in Figure 23. There is
a good section on latches in Weste and Eshraghian's book [WE85].
There are also many clocking styles to choose from in implementing
CMOS circuits (see Table 2). Each style achieves different tradeoffs in
speed, power, area, and operating margin. The highest performance
clocking schemes tend to have the narrowest operating margins. Wide
operating margins generally cost area and power.

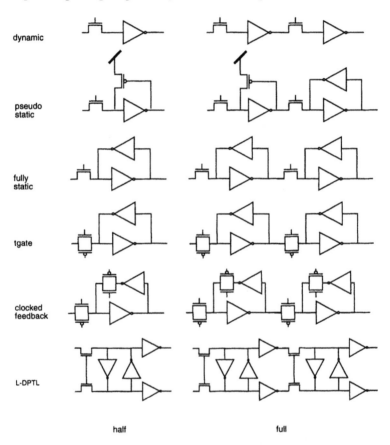

dynamic

pseudo
static

fully
static

tgate

clocked
feedback

L-DPTL

half   full

Figure 23: latches

Clocked CMOS (C²MOS) [SOA73] is a dynamic scheme relying on
the availability of two clocks which may overlap. NORA [GM83] is
another dynamic scheme which embeds the clocks with the logic in
alternating pullup and pulldown stacks. True single phase clocking
(TSPC) is discussed in [JKS87, Kar88, YS89]. This is a dynamic
clocking strategy with good performance. Its principal advantage is
its robustness. Although the transistor count of a TSPC latch is
relatively high (see Figure 24), the layout is quite compact. TSPC-1

year	what	who	description
1973	C^{2}MOS	[SOA73]	Clocked CMOS
1980	2phase	[MC80]	Nonoverlapping 2-phase clocking
1983	NORA	[GM83]	Race-free dynamic CMOS
1987	TSPC-1	[JKS87]	True single-phase clocking, version 1
1988	SSPC	[Lu88a]	Safe single-phase clocking
1989	TSPC-2	[YS89]	True single-phase clocking, version 2

Table 2: Clocking styles.

has the disadvantage that the internal node is precharged during clkL, resulting in excess power dissipation. Latch type and clocking style are closely related. C^{2}MOS, NORA, and TSPC all have their own latch circuits (see Figure 24).

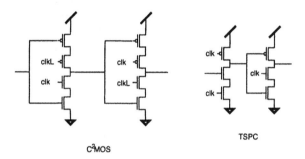

Figure 24: C^{2}MOS and TSPC dynamic latches.

Nonoverlapping two phase clocking is discussed in Mead and Conway [MC80]. This scheme has wide operating margins but can be quite slow at high frequencies due to the mandatory gaps, and is fairly high power. The non-overlap requirement can be specified using the following equation:

$$\phi_1(t) \cdot \phi_2(t) = 0 \quad \text{for all} \quad t$$

Figure 25 shows a typical clock cycle of the two-phase non-overlap clocking scheme. There are actually four time intervals in a cycle. The first one is ϕ_1 high and ϕ_2 low. During this time interval, an input signal is clocked into the latch. The second time interval is both ϕ_1 and ϕ_2 low, which is referred to as gap($\phi_1 \rightarrow \phi_2$). This gap is needed to ensure the non-overlap requirement. The third time interval is ϕ_1 low and ϕ_2 high. This is the time for the latch to clock out the signal stored in it. After this time interval, both ϕ_1 and ϕ_2 are low again to form gap($\phi_2 \rightarrow \phi_1$). The two gaps in a clock period ensures that a latch is either clocking a signal in or clocking a signal out but not

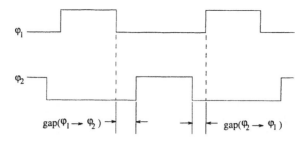

Figure 25: Non-overlap two-phase clock waveforms

both. Obviously, the advantage of this kind of clocking scheme is its clear timing definition for avoiding race and hazard. The price paid for achieving this timing safety is the extra delay introduced by the gap. To illustrate how the gaps affect the speed, a simple system is shown in Figure 26. Assume that the combinational logic between the two latches has a maximum delay of D_{max} seconds. The time from ϕ_2 going high to ϕ_1 coming low has to be greater than or equal to D_{max}. Thus the gap from ϕ_1 coming low to ϕ_2 going high is wasted time. The minimum clock period for the system to work is

$$T_{min} = D_{max} + \text{gap}(\phi_1 \rightarrow \phi_2)$$

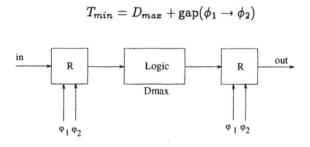

Figure 26: A simple latched system

In the circuit diagram of the static latch shown in Figure 23, the transistor sizes of the inverters are very important. The two feedforward inverters should be stronger than the two feedback inverters. We have been using 16:2 for the feedforward inverters and 3:8 for the feedback inverters. The feedback inverters have two major functions. One is to make the latch static by feeding back the correct logic value to the input of the feedforward inverters and the other is to speed up the latch by overcoming the threshold drop of the two pass-transistor switches.

Another issue in using two-phase non-overlap clocking scheme is clock skew caused by qualified ϕ_1. Figure 27 shows how a latch can be used with a qualified ϕ_1 clock. Qualified clock is a way to control the register transfer activities using a control signal. When the control signal En is low, the ϕ_1 signal is blocked from turning the input switch on and thus the register transfer is stopped. When En is high, the transfer is resumed. A problem of using such a qualified ϕ_1 clock is the delay caused by the logic circuit between the ϕ_1 signal and the $Q\phi_1$ signal supplied to the register input switch. This delay may cause an overlap of the clock signal supplied to the first pass-transistor and ϕ_2 signal. However, qualified clocks can result in significant power savings over techniques which use a separate load signal and clock latches on every cycle.

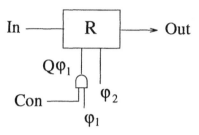

Figure 27: Qualified ϕ_1 clock

Using qualified ϕ_1 signals, a multiplexer can be merged with a register as shown in Figure 28. When the control signal is high, in1 is clocked into the register by ϕ_1. Otherwise, in2 is clocked into the register.

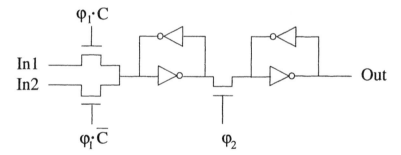

Figure 28: Two input register

Clock distribution at the chip level is extremely important, especially in high speed, low power systems using qualified clocks. It is becoming more difficult as chips grow larger and feature sizes become smaller. In our designs, we qualify all clocks with identical circuits, so that both qualified and unqualified clocks have the same latency. We then tune the transistor sizes to compensate for differences in loading.

10. Memory

Storage density is an important issue in large digital systems, so memory optimization is important. One-transistor (1T) dynamic random access memories (DRAMs) have the highest density. Six-transistor (6T) static memories (SRAMs) consume the least power. A typical 1T DRAM cell measures $6\lambda \times 11\lambda$ ($66\lambda^2$) [KBD+90]. A typical 4T SRAM measures $12\lambda \times 20\lambda$ ($240\lambda^2$) [AOA+90]. Shift registers can often be implemented with either SRAM or DRAM, saving substantial amounts of area and power.

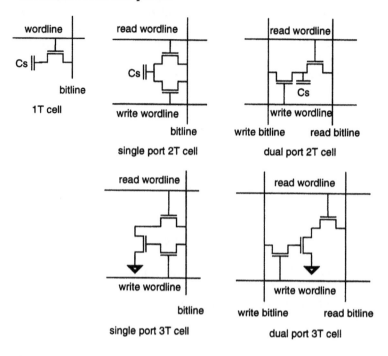

Figure 29: dynamic memory cells

DRAM

DRAM cells must be refreshed due to leakage current [CTTF79], and therefore consume more power than SRAMs. Normally the refresh power is a small fraction of the operating power, but could be significant in very large systems. Figure 29 shows a variety of DRAM circuits. One-transistor (1T) DRAMs are the most compact but the most difficult to sense and control.

Commercial DRAM processes offer storage densities limited by the metal pitch. The storage node is buried near the orthogonal intersection of a wordline and bitline. An 8λ metal pitch would give a $64\lambda^2$ cell area, comparable to reported values. DRAMs are high energy because reads are destructive: every time a cell is read it has to be rewritten, and writing requires swinging the bitlines to at least a threshold drop below V_{dd}. This is much higher energy than SRAM, which only has to swing the bitlines 100mV or so.

SRAM

Figure 30 shows a variety of SRAM circuits. Commercial SRAMs are normally implemented using a high resistance poly pullup, and can achieve densities only a factor of 4 worse than 1T DRAM. The 6T SRAM is lower power because it has a CMOS pullup which shuts off. SRAMs can be very low energy because reads and writes only require

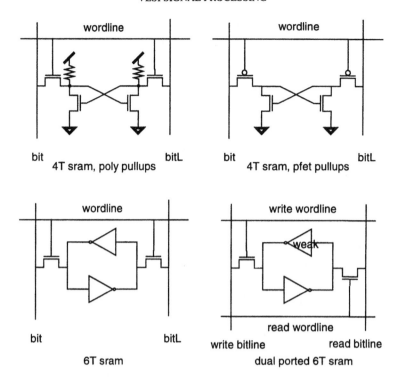

Figure 30: static memory cells

small changes in bitline voltage. This small difference can be amplified by a cross-coupled inverter sense amplifier without disturbing the bitlines.

FLASH

FLASH EPROM memory is as dense or denser than 1T DRAM because the data is stored on a floating gate at the intersection of the word and bitline, and there is no need for buried storage cells. FLASH can be used to store analog or digital values. Single transistor FLASH is normally read at reduced voltage and written at elevated voltage, so read times are limited by the reduced voltage swings, and writes take longer and consume more power than other types of memory.

A pair of FLASH cells can be used to form a differential pair to reduce read energy and read times. A disadvantage of FLASH is the high write energy (20pJ/bit) and long write times (10μsec). Write energy can be reduced in analog applications by only making incremental changes.

Decoders

Decoders form an important part of any memory system. Much of the power consumed in accessing memory is consumed in the decoder. Decoders are usually implemented in several stages. Decoding before the wordline is called "x-decoding". Decoding after the bitline is called "y-decoding". X-decoding is often broken into 3 stages: pre-decode, decode, and postdecode. Predecoders determine which of a potentially hierarchical set of memory blocks contains the data, and recode address bits to reduce the fanout to the wordline decoders of a single block. One or more wordline decoders will respond to an address; the postdecoder selects a single wordline by selectively enabling its power supply.

During a read operation, an entire row of data is read out of the memory block. The desired piece of the row is then multiplexed onto the data bus through the y-decoder. SRAMs can share sense amps among multiple bits; DRAMs cannot, since DRAM reads are destructive.

During a write operation, a row of data is written to the memory block. Again, SRAMs can perform partial writes by only biasing the bitlines of active bits. DRAMs must write the entire row, since a cell is corrupted as soon as its access transistor is turned on.

X-decoders can be implemented using a tree multiplexer. This structure is more compact but slower than a gate decoder. If the memory is pipelined, the decoder latency can be hidden.

If memory access is predominantly sequential, decoder energy can be reduced substantially by bit-reversing the address presented to a block. In sequential access, the least significant bit toggles at the highest frequency. If adjacent addresses are stored in adjacent words in the memory block, the lsb must gate $2^N/2$ transistors, where N is the number of bits in the block address. If the address is bit-reversed, then the lsb only toggles 1 transistor, and the msb toggles $2^N/2$ transistors, but only once every $2^N/2$ sequential accesses. So, conventional addressing toggles $2^N/2 + \frac{1}{2}2^N/4 + ... + 2^{N-2i+1}$ transistors every cycle, whereas bit-reversed addressing toggles $1/1 + 2/2 + 4/4 + ...$ For $N = 5$, conventional addressing toggles 21 transistors per cycle; bit-reversed addressing toggles 5. Bit reversal can be combined with gray coding to reduce decoder energy by another factor of 2, but the added complexity of the gray coder may more than offset the incremental reduction in energy.

Memory access energy can be reduced further using low voltage differential signals for long lines, especially address and data buses at higher levels of the hierarchy. In a large memory subsystem, the power dissipated in a single 100MHz 10cm wire swinging 5 volts is 25mW. The same wire swinging 50mV around 0.25V dissipates 12.5μW.

Wordline drivers are difficult to pitchmatch to DRAM. The layout becomes very inefficient when the cell pitch is less than 15λ. In DRAMs, decoders are often placed on both sides of the memory array and the wordlines interleaved to meet the pitch. Wordline series resistance can be significantly reduced by strapping the poly with metal2. It can be further cut in half by placing the decoder in the middle of the memory array and driving the wordlines in both directions.

The number of transistors in a gate decoder is about $2N\log_2 N$ for the gates and $2N$ for the wordline drivers, where N is the number of rows in a block. The number of transistors in a tree decoder is $2N - 2$ for the tree and also $2N$ for the wordline drivers. Wires dominate decoder area since there are N horizontal wires for the wordlines and $2log_2 N$ vertical wires for the address lines in a tree decoder, and $N log_2 N$ horizontal wires for wordlines in a gate decoder.

Sense amplifiers

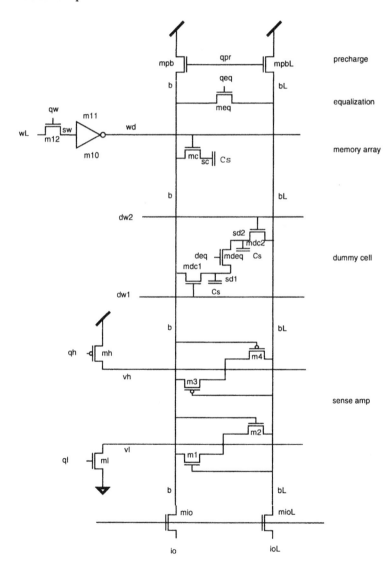

Figure 31: differential sense amplifier

Figure 31 shows a high performance differential dense amplifier reported in [DLHP88], and recommended for use with 1T DRAMs.

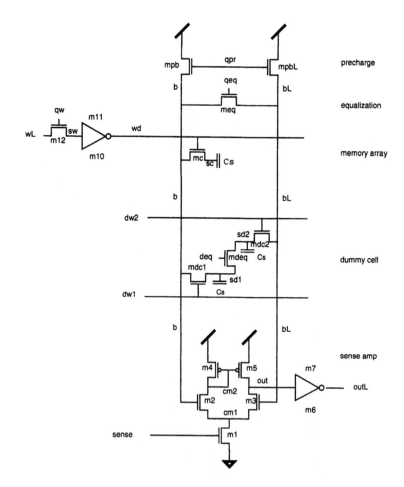

Figure 32: current mirror sense amplifier

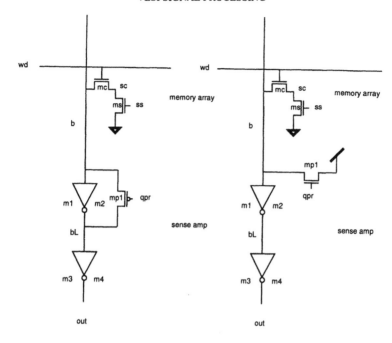

Figure 33: single ended sense amplifier

This sense amp has the nice property that the crosscoupled inverters automatically perform a refresh write after read. Figure 32 shows a current mirror sense amplifier similar to one reported in [AOA+90]. Figure 33 shows a single ended sense amp which can be used with ROMs, 3T DRAMs, and single-ended SRAMs.

Pipelined memory

Memory can be pipelined to increase throughput and reduce storage energy. The wordline driver power supplies can be clocked, as can the sense amps, leaving an entire cycle to equalize the bitlines, clock the wordlines, and spin up the sense amps.

Caches

For inner product architectures, memory accesses are predictable and sequential. The memory can be interleaved by fetching a number of words in parallel from the block into a cache and then accessing those words from the cache. N-way interleaving permits a single memory access to take N cycles. A k-port memory can be implemented with k caches and N-way interleaving as long as the memory access time is less than N/k cycles.

11. Power supply sizing

Power supplies need to be sized to avoid excessive resistive drops along the rails and to stay within metal migration limits.

The metal migration limit is a limit on the average current which can flow through a wire without dislodging the metal atoms and eventually breaking the wire. The metal migration limit for metal in a 2.0μ process is about $0.7\text{mA}/\mu$. w_{min}, the minimum width of the supply rail, is given by

$$w_{min} = \frac{I_{on}}{(ld/a)J_{max}}$$

where I_{on} is the current through an ON device in milliamps, ld is the logic depth in the design, a is the activity, and J_{max} is the metal migration limit in milliamps/micron. For example, given a multiplier in 2.0μ CMOS, $I_{on} = 1\text{mA}$, $ld = 10$, $a = 0.4$, and $J_{max} = 0.7\text{mA}$, so $w_{min} = 0.057\mu$ per gate. Then a 4 micron wide power line could support 70 gates.

There is also a resistive drop limit. To keep the voltage drop on a power bus under 100mV at 5V while a single 6:2 device is switching, we need to keep $0.1/r_s > 4.9/r_d$, where r_s is the resistance in the supply and r_d is the resistance of the device. For example, if $r_d = 3000\Omega$, then $r_s = 61.2\Omega$. Metal resistance is about $40\text{m}\Omega/\square$, so a 61.2Ω line would be $1530\square$.

Older designs tend to place wide supply buses at either side of the datapath. If the supply rails are run parallel to the dataflow then the peak current per rail will be less than if the rails are run perpendicular to the dataflow. A newer approach is to run the rails horizontally in metal1, strapping them vertically in metal2 to form a power mesh. This works better because the resistive drop is proportional to the square of the wirelength, since the width depends on the number of active devices as well as the length of the wire.

For example, if we have 1280 sense amps, 320 per block, each drawing $500\mu\text{A}$. That's about $10\text{K}\Omega$ per sense amp ($5\text{V}/0.0005\text{A} = 10\text{K}\Omega$). But we have 320 of these in parallel, so $r_d = 10000/320 = 31.25\Omega$. Then we need $r_s = 0.6378\Omega$. At $40\text{ m}\Omega/\square$ that's $0.6378/0.040 = 15.9\square$. If we have a single bus 4000λ long it has to be 252λ wide. If we can feed the bus from both ends, it only has to be $1/4$ as wide (twice the resistance in each half and half the length) or 63λ. If we place a metal2 strap in the middle, the bus only needs to be 16λ wide. If we place one every 1000λ, the bus only needs to be 4λ wide.

In general, the strapping distance l_s required for a bus of width w is given by

$$l_s = 2\sqrt{(Rp/n)vw/r}$$

where R is individual device resistance in Ω, p is the cell pitch in λ, n is the number of active devices in a cell, $v = dV/(V_{dd} - dV)$ is the allowed voltage degradation, and r is the resistivity of the supply rail in $\Omega/\square$. For the sense amp example,

$$l_s = 2\sqrt{(10000 \times 500/40) \times 0.1/(5 - 0.1) \times 4/0.040} = 1010\lambda$$

Since the dataflow is vertical, the required current can be supplied from adjacent rails in the mesh through the vertical straps.

For metal migration, we have 20 devices each drawing $500\mu A$ between straps for 7ns out of 44ns or 16% of the time. Then the average current is 1.6mA, requiring $1.6/0.7 = 2.28$ microns or 3.8λ of metal. In this case, the metal migration limit also permits minimum width rails.

3 Arithmetic elements

Floating point is becoming widespread because it makes life easy for the end user. However, the area, performance, and energy penalty for using IEEE standard implementations is substantial. Taking advantage of just enough precision is key to compact, low energy implementations of digital signal processing. In the discussions of this section, we assume normalized 2's complement fixed point arithmetic.

$$x = x_0 \cdot x_1 x_2 \cdots x_{B-1} = \sum_{i=0}^{B-1} x_i 2^{-i}$$

where B is the number of bits to represent the binary number x, x_i is either 1 or 0 for $i > 0$, and x_0 is the sign bit with a value of either -1 or 0.

3.1 Adders

In an adder design, overflow is the first problem we have to consider. Let's look at the following examples. ($B = 4$ bits)

Example 1:

Let a $= -0.5 = 1.100$ and b $= 0.75 = 0.110$. A binary addition of 1.100 and 0.110 produces 10.010. If we discard the highest bit (1 in this case), we obtain $0.010 = 0.25$, which is the correct answer.

Example 2:

Let a $= 0.5 = 0.100$ and b $= 0.75 = 0.110$. A binary addition of 0.100 and 0.110 produces 01.010. If we discard the highest bit again, we have $1.010 = -0.75$, which is obviously a wrong answer. This is called an overflow problem in binary addition because the answer of a+b is larger than 1 which cannot be represented in the number system although both a and b are in the number system. A conventional way to avoid this problem is to scale the inputs to the binary adder by a factor of 0.5 so that the result is guaranteed to be in the number system. This can be called input-scaling because the inputs to the binary adder are scaled. A problem associated with such an input scaling scheme is the lose of precision. If several additions are performed in a row, the final result may suffer from underflow. The following example shows an extreme case.

Example 3:

Let a $= 0.001$ and b $= 0.001$. If the input scaling scheme were used in this addition, 0.000 and 0.000 would be the inputs to the 4-bit adder and the result would be 0.000. A non-zero result would become a zero.

A better way to handle the overflow problem is to use output scaling. In the previous example, a+b without input scaling should be 00.010. Scaling the output by a factor of 0.5 generates 0.001, which is a better result. Application of this scheme to Example 2 provides the correct result. However, this scheme seems to give a wrong answer for the case in Example 1. The reason is that the sign bit is treated in the same way as the other bits although the symbol "1" at the sign bit represents a value -1 while the symbol "1" at other locations represents a value 1. Therefore, with a special design of the sign bit adder, this problem can be solved.

a	b	cin	cout	sum
0	0	0	0	0
0	0	1	0	1
0	1	0	0	1
0	1	1	1	0
1	0	0	0	1
1	0	1	1	0
1	1	0	1	0
1	1	1	1	1

Table 3: Truth table for a regular full adder

The truth table for a regular full adder is shown in Table 3. From this truth table, the logic relationship of inputs and outputs can be derived as follows:

$$\text{sum} \quad = \quad \text{parity-checking}(a,b,cin)$$

$$\text{cout} \quad = \quad \text{majority-voting}(a,b,cin)$$

where parity-checking(a,b,cin) can be implemented using (a xor b xor cin) and majority-voting(a,b,cin) can be implemented using ((a and b) or (b and cin) or (a and cin)). For the sign bit in output scaling, the truth table is as follows, keeping in mind that a symbol "1" means a value -1 for inputs a and b, a value -2 for output cout, and a value 1 for input cin and output sum.

a	b	cin	cout	sum
0	0	0	0	0
0	0	1	0	1
0	1	0	1	1
0	1	1	0	0
1	0	0	1	1
1	0	1	0	0
1	1	0	1	0
1	1	1	1	1

Table 4: Truth table for sign bit full adder

From this truth table, the logic relationship of inputs and outputs can be derived as follows:

$$\text{sum} = \text{parity-checking(a,b,cin)}$$

$$\text{cout} = \text{majority-voting(a,b,}\overline{\text{cin}})$$

The logic for sum is the same as that in a regular full adder. The only difference is in the logic for cout where one more inverter is needed to generate $\overline{\text{cin}}$ for the majority-voting circuit. Figure 34 shows a regular full adder and a sign-bit full adder for output scaling.

Figure 35 shows bit-parallel implementations of input scaling and output scaling schemes. For input scaling, a_{B-1} and b_{B-1} are discarded before a and b go to the B-bit adder. a_0 and b_0 are sign-extended to the left-most full adder. The sum bit of the left-most full adder is the sign bit of the result and the cout bit is discarded. For output scaling, no bits are discarded at the input and there are no sign-extentions either. The sum bit of the right-most full adder is discarded and the cout bit of the left-most full adder, which has one more inverter than the regular full adder, is the sign bit of the result.

Figure 36 shows bit-serial implementations of input scaling and output scaling schemes. For input scaling, the control signal con1 determines if a "0" or the previous carry-out should be the input to the carry-in of the full adder. At the beginning cycle of each addition, con1 allows "0" to pass. During the rest of addition cycles, con1 allows the previous carry-out to pass. The control signal con2 provides a qualified ϕ_1 signal to the input registers to prevent the LSBs of the input from getting into the full adder so that sign-extention is automatically performed for the previous addition. A total of B cycles are needed for each addition and the output result is obtained with a two-cycle latency.

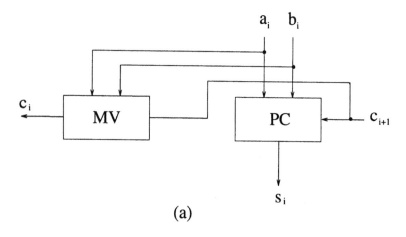

(a)

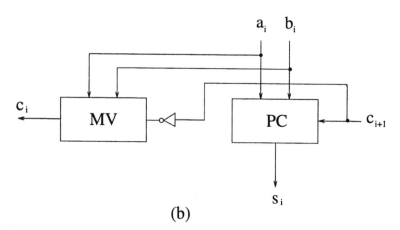

(b)

Figure 34: A regular full adder (a) and a sign-bit full adder (b) for output scaling

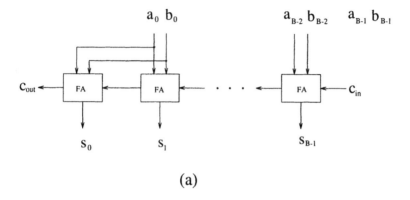

(a)

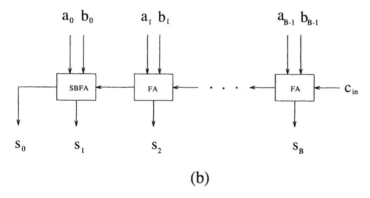

(b)

Figure 35: Parallel implementations of an adder with input scaling (a) and output scaling (b)

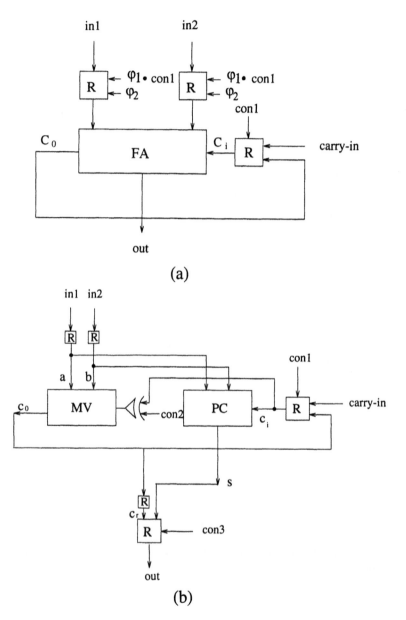

Figure 36: Bit-serial implementations of an adder with input scaling (a) and output scaling (b)

For output scaling, an extra XOR gate is needed as a conditional inverter to generate a complement signal for the majority-voting circuit when the sign bits arrive. The control signal con1 clocks a "0" into the register one cycle before the LSBs of the inputs arrive. The con1 signal also controls the conditional inverter (XOR gate) because, in a pipelined operation, the cycle before new data arrive is the one with sign-bits of the previous input data. The control signal con2 determines if the output of parity-checking or the output of majority-voting should be passed. At the time new data arrive, the carry-out of the previous addition should pass and during the rest of the cycles, sum should pass. While the con2 signal allows the previous carry-out to pass, it actually prevents the LSB of the new result from passing to the output. This is equivalent to discarding the LSB of the output, which is part of the output scaling scheme. Same as the input scaling implementation, it takes B cycles to perform a B-bit addition and the output is obtained with a two-cycle latency.

3.2 Biased Redundant Binary (4:2) Arithmetic

Carry propagation is an expensive operation in digital arithmetic. Several families of arithmetic have been developed to reduce the impact of carry propagation. Signed digit [Avi61] and various redundant binary methods [HNN+87] have been proposed. Biased redundant binary (4:2) arithmetic [SW78] interfaces cleanly to standard two's complement data format, implements an efficient and compact accumulator [LB87, LB, Li88, SH88, SH89, San89], and has optimal logic depth [BP91b].

Again, a normalized conventional 2's complement binary number is represented as follows:

$$X_{2c} = \sum_{n=0}^{N-1} x_{2c,n} 2^{-n}$$

where $x_{2c,n} \in \{0,1\}$, for $n = 1,2,3,\ldots,N-1$, and $x_{2c,0} \in \{0,-1\}$. A normalized signed-digit (SD) number is represented as follows.

$$X_{SD} = \sum_{n=0}^{N-1} x_{sd,n} 2^{-n}$$

where $x_{sd,n} \in \{-1,0,1\}$, for $n = 0,1,2,3,\ldots,N-1$, Because each digit $x_{sd,n}$ in an SD number may take three different values $\{-1, 0, 1\}$, two bits are needed to represent each digit. However, only three states of the two bits are allowed. A way to code the two bits for a digit is as follows:

$x_{sd+,n}$	$x_{sd-,n}$	$x_{sd,n}$
1	0	1
0	0	0
0	1	−1
1	1	not allowed

Because of this coding scheme, $x_{sd+,n}$ and $x_{sd-,n}$ are usually referred to as the positive bit and the negative bit of the nth digit respectively. Conversion of a 2's complement number to an SD number, i.e., introducing redundancy, can be a simple assignment operation as follows:

$x_{sd+,0} = 0$, $x_{sd-,0} = x_{2c,0}$,

$x_{sd+,n} = x_{2c,n}$, $x_{sd-,n} = 0$, for $n = 1, 2, 3, \ldots, N - 1$.

Conversion of an SD number to a 2's complement number, i.e., eliminating redundancy, is a binary subtraction operation as follows:

$$X_{2c} = \sum_{n=0}^{N-1} x_{sd+,n} 2^{-n} - \sum_{n=0}^{N-1} x_{sd-,n} 2^{-n}$$

Biased redundant binary (BRB) arithmetic is another type of redundant arithmetic to eliminate carry propagation. A normalized BRB number can be represented as follows:

$$X_{BRB} = \sum_{n=0}^{N-1} x_{brb,n} 2^{-n}$$

where $x_{brb,n} \in \{0, 1, 2\}$, for $n = 1, 2, 3, \ldots, N - 1$, and $x_{brb,0} \in \{0, -1, -2\}$. Each digit $x_{brb,n}$ in a BRB number may also take three values but the center value is not 0 anymore. It is "biased" relative to the SD number representation. $x_{brb,0}$ is the most significant digit which has a bias "–1", and other digits, $x_{brb,n}$, have a bias "1". To represent a digit in a BRB number, two bits are needed too. A coding table is as follows:

$x_{brb1,n}$	$x_{brb2,n}$	$x_{brb,n}$	$x_{brb1,0}$	$x_{brb2,0}$	$x_{brb,0}$
1	1	2	1	1	−2
1	0	1	1	0	−1
0	1	1	0	1	−1
0	0	0	0	0	0

A 2's complement number can be condisered as a half of a BRB number with the other half being zero, i.e.,

$x_{brb1,n} = x_{2c,n}$ $x_{brb2,n} = 0$,

for $n = 0, 1, 2, \ldots, N-1$. Conversion of a BRB number to a 2's complement number is an addition operation as follows:

$$X_{2c} = \sum_{n=0}^{N-1} x_{brb1,n} 2^{-n} + \sum_{n=0}^{N-1} x_{brb2,n} 2^{-n}$$

Although both BRB and SD number formats have the same property of eliminating carry propagation, they have different features. One feature of the SD number format is that the representation of zero is unique although the digit pattern of any nonzero number is not unique. Since the BRB number format does not have this feature, the SD number format is better than the BRB number format in operations such as comparison of two numbers or memory address decoding. On the other hand, the BRB number format has a feature that the SD number format does not have. The feature is that two 2's complement numbers can form one BRB number without needing any operations. This feature makes the BRB number format a more desirable choice than the SD number format for operations such as partial product summation in a multiplier. It also makes BRB adder design simpler than SD adder design.

Since the BRB and the SD number formats have different advantages and disadvantages for different operations, it is desirable to use them for different subsystems in a signal processing system. To have two different number representations in a system, we have to be able to convert one number representation to the other and vice versa. One way to do the BRB–SD conversion is through 2's complement format since, as mentioned above, both BRB and SD number formats can be converted to and from 2's complement format. But conversion of either BRB or SD number format to 2's complement format requires carry propagation addition or subtraction. A parallel algorithm for converting BRB numbers to and from SD numbers is very desirable.

The difference between a BRB digit and an SD digit is that the BRB digit is biased with 1 or -1 while the SD digit is unbiased. Because they both can have three consecutive values, we may convert one representation to the other by simply subtracting or adding the biases from or to the digits. To convert a BRB number X_{BRB} to an SD number, we first eliminate the biases in the digits of X_{BRB} as follows:

$$X'_{SD} = (x_{brb,0} + 1)2^0 + \sum_{n=1}^{N-1}(x_{brb,n} - 1)2^{-n}$$

$$= x_{brb,0}2^0 + \sum_{n=1}^{N-1} x_{brb,n}2^{-n} + 1 - \sum_{n=1}^{N-1}2^{-n}$$

$$= X_{BRB} + 2^{-(N-1)}$$

This shows that the process of subtracting the biases actually generates an SD number that is the orginal BRB number plus $2^{-(N-1)}$. In order to get the equivalent SD number of the same algebraic value, a subtraction of $2^{-(N-1)}$ from X'_{SD} is needed. The logic for subtracting the bias 1 in the nth digit $(n > 0)$ is as follows:

$$x'_{sd+,n} = x_{brb1,n}x_{brb2,n}$$

$$x'_{sd-,n} = \overline{x_{brb1,n} + x_{brb2,n}}$$

This is derived from the following truth table:

$x_{brb1,n}$	$x_{brb2,n}$	$x'_{sd+,n}$	$x'_{sd-,n}$
1	1	1	0
1	0	0	0
0	1	0	0
0	0	0	1

The logic for subtracting the bias -1 from the most significant digit is as follows:

$$x'_{sd+,0} = \overline{x_{brb1,0} + x_{brb2,0}}$$

$$x'_{sd-,0} = x_{brb1,0}x_{brb2,0}$$

This is exactly the same as that for subtracting the bias 1 except renaming the output bits. Logic circuits for subtracting the biases from a two bit BRB digit are shown in Figure 37.

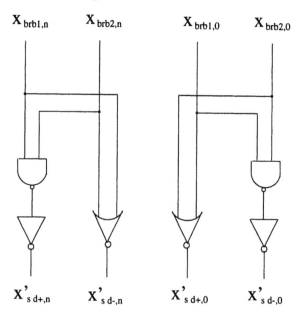

Figure 37: Logic circuits for subtracting biases for BRB to SD conversion

After subtracting the biases from the digits of X_{BRB}, an SD number X'_{SD} is obtained. Subtraction of $2^{-(N-1)}$ from X'_{SD} is needed to obtain X_{SD}. This can be performed by the logic circuit shown in Figure 38.

From Figure 37 and Figure 38, we can see that conversion of X_{BRB} to X_{SD} can be done in parallel because there is no serial signal propagation in either elimination of biases or subtraction of $2^{-(N-1)}$.

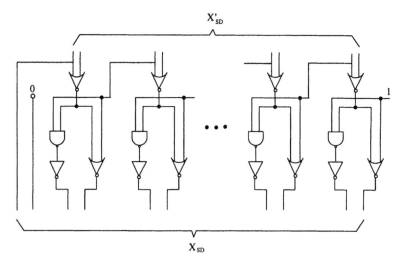

Figure 38: Logic circuit for subtracting $2^{-(N-1)}$ from X'_{SD}

To convert an SD number X_{SD} to a BRB number, we need to do the reverse, i.e., adding the biases first as follows:

$$X'_{BRB} = (x_{sd,0} - 1)2^0 + \sum_{n=1}^{N-1} (x_{sd,n} + 1)2^{-n}$$

$$= x_{sd,0}2^0 + \sum_{n=1}^{N-1} x_{sd,n}2^{-n} - 1 + \sum_{n=1}^{N-1} 2^{-n}$$

$$= X_{SD} - 2^{-(N-1)}$$

This shows that the process of adding the biases actually generates a BRB number that is the orginal SD number minus $2^{-(N-1)}$. In order to get the equivalent BRB number of the same algebraic value, $2^{-(N-1)}$ has to be added to X'_{BRB}. The logic for adding the bias 1 to the nth digit $(n > 0)$ is as follows:

$$x'_{brb1,n} = x_{sd+,n}$$

$$x'_{brb2,n} = \overline{x}_{sd-,n}$$

This is derived from the following truth table:

$x_{sd+,n}$	$x_{sd-,n}$	$x'_{brb1,n}$	$x'_{brb2,n}$
1	0	1	1
0	0	0	1
0	1	0	0
1	1	any	thing

The logic for adding the bias -1 to the most significant digit is as follows:

$$x'_{brb1,0} = \overline{x}_{sd+,0}$$

$$x'_{brb2,0} = x_{sd-,0}$$

Again, this is the same logic but inputs are renamed. Logic circuits for adding the biases to a two bit SD digit are shown in Figure 39.

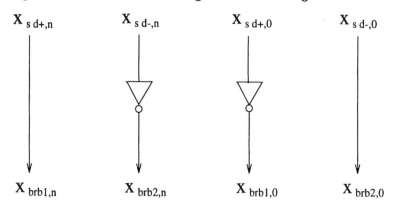

Figure 39: Logic circuits for adding biases for SD to BRB conversion

Only one inverter per digit is needed for adding the biases. Logic in Figure 39 is simpler than that in Figure 37, because $(x_{sd+,n}, x_{sd-,n}) = (1,1)$ is a "never happen" ("don't care") state. After adding biases to the digits of X_{SD}, a BRB number X'_{BRB} is obtained. Addition of $2^{-(N-1)}$ to X'_{BRB} is needed to obtain X_{BRB}. This can be done by N half adders as shown in Figure 40.

From Figure 39 and Figure 40, we can see that conversion of X_{SD} to X_{BRB} is a parallel process too, because there is no serial signal propagation in either adding biases or addition of $2^{-(N-1)}$. Thus we have shown that conversion of a BRB number to or from an SD number is a parallel process.

Although a BRBA can be implemented using two full adders, we have used a "direct logic" implementation [LB87] that reduces the number of xors in series from four to three, increasing the speed by 33% (see Figure 41).

3.3   Fast comparator

Unloaded manchester carry chains make very fast, efficient comparators. They are generally much faster than propagating adders because there are no sum terms to load down the carry chain (see Figure 42).

3.4   Multiplier

Multiplication can be divided into two parts, namely, partial product generation (PPG) and partial product summation (PPS). As an example, let

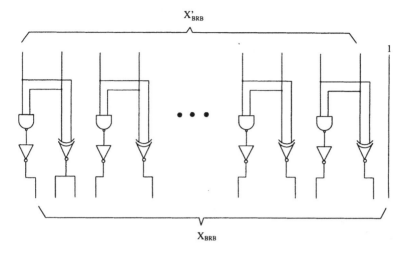

Figure 40: Addition of $2^{-(N-1)}$ to X'_{BRB} by half adders

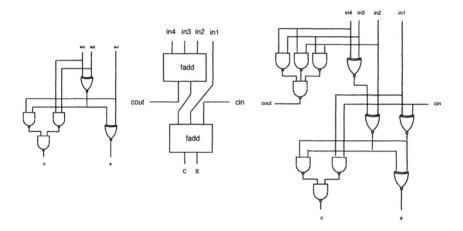

Figure 41: BRB adders. Left is a full adder. Its critical path is 2 xors in series. Center is a BRB adder using 2 full adders. Its critical path is 4 xors in series. Right is a direct logic implementation of a BRB adder. Its critical path is 3 xors in series.

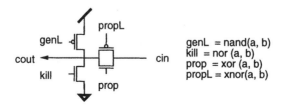

Figure 42: Unloaded manchester carry chain.

$X = x_0 \cdot x_1 x_2 x_3$ and $Y = y_0 \cdot y_1 y_2 y_3$ be two numbers to be multiplied with each other. The following procedure is usually involved in the multiplication:

	x_0			x_0	x_1	x_2	x_3
y_3	$p_{3,0}$	$p_{3,0}$	$p_{3,0}$	$p_{3,0}$	$p_{3,1}$	$p_{3,2}$	$p_{3,3}$
y_2	$p_{2,0}$	$p_{2,0}$	$p_{2,0}$	$p_{2,1}$	$p_{2,2}$	$p_{2,3}$	
y_1	$p_{1,0}$	$p_{1,0}$	$p_{1,1}$	$p_{1,2}$	$p_{1,3}$		
y_0	$p_{0,0}$	$p_{0,1}$	$p_{0,2}$	$p_{0,3}$			
				1			
	z_0	z_1	z_2	z_3	z_4	z_5	z_6

where $p_{i,0}p_{i,1}p_{i,2}p_{i,3}$ is called the ith partial product and $z_0 z_1 z_2 z_3 z_4 z_5 z_6$ is the final result. PPG is the process of gerenating the partial product array and PPS is the process to add up the partial products for the final result of the multiplication. Figure 43 shows how PPG is performed using NAND and AND gates.

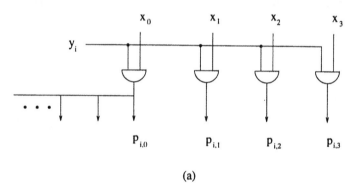

(a)

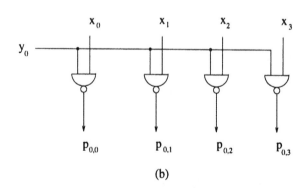

(b)

Figure 43: Partial product generation

As shown in Figure 43, sign extension in two's complement multiplication is normally a high fanout operation, since the msb of the least significant partial product must fan out as far as the most significant bit of the most significant partial product. This can be circumvented by making use of the fact that

$$ssss = 1111 + \bar{s}$$
$$1 + \bar{s} = s + 2\bar{s}$$

where s is either 0 or 1 and $\bar{s}$ is its logical complement. Then the partial product array becomes

				x_0	x_1	x_2	x_3
y_3			$\overline{p_{3,0}}$	$p_{3,0}$	$p_{3,1}$	$p_{3,2}$	$p_{3,3}$
y_2			$\overline{p_{2,0}}$	$p_{2,1}$	$p_{2,2}$	$p_{2,3}$	
y_1		$\overline{p_{1,0}}$	$p_{1,1}$	$p_{1,2}$	$p_{1,3}$		
y_0	$\overline{p_{0,0}}$	$p_{0,1}$	$p_{0,2}$	$p_{0,3}$			
				1			
	z_0	z_1	z_2	z_3	z_4	z_5	z_6

Figure 44 shows the logic circuit for PPG without sign extention.

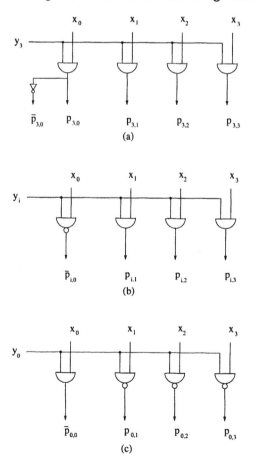

(a)

(b)

(c)

Figure 44: Partial product generation without sign extention

Modified Booth encoding is a technique to reduce the number of partial

y_{i-1}	y_i	y_{i+1}	P_i	explanation
0	0	0	0	a string of 0's
0	0	1	X	high end of a string of 1's
0	1	0	X	a single 1 at y_i
0	1	1	2X	high end of a string of 1's
1	0	0	-2X	low end of a string of 1's
1	0	1	-X	low end of a string of 1's AND high end of a string of 1's
1	1	0	-X	low end of a string of 1's
1	1	1	0	middle of a string of 1's

Table 5: Truth table of the modified Booth encoder

products from N to $N/2$ where N is the number of bits in Y. Table 5 shows the truth table for designing a modified Booth encoder.

After partial products are generated, PPS can be considered as simply adding up several numbers together. Figure 45 shows the most straight-forward way of using carry-propagate adders (CPAs) to perform PPS. In practice, this structure is not used because CPAs are too slow. Figure 46 shows how to use a full adder array to perform PPS. Every stage of full adders performs an addition of three numbers and generates two numbers. Therefore, except the first stage which takes three partial products as input, every stage of full adders takes as input only one partial product. Thus a total of $N - 2$ stages of full adders are needed to reduce N partial products to two which are then added by a final CPA. Excluding the final CPA which is usually replaced by a carry-lookahead adder (CLA), the delay time to reduce N partial products to two in an array multiplier is linearly related to the number of partial products. Wallace tree is another way to perform PPS as shown in Figure 47. The time to recuce the partial products from N to 2 is proportional to $\log_{1.5} \frac{N}{2}$. The base 1.5 is from the fact that Wallace tree has a 3:2 reduction of partial products per stage. Signed-digit redundant binary (SDRB) adders have also been used for PPS. Figure 48 illustrates an SDRB adder tree for PPS. The B-to-SDRB conversion box introduces redundancy so that the number of lines at the output of the box is twice as many as that at the input of the box. $\log_2 N$ stages of SDRB adders are needed to reduce N SD numbers to one SD number which is then converted to regular 2's complement number by a subtractor. As mentioned ealier, BRB adders are better than SDRB adders for PPS because two 2's complement numbers can be put together to form a BRB number. Therefore, there is no need to have any conversion box at the top to introduce redundancy. Figure 49 shows how to construct a BRB adder tree for PPS. At the top, four 2's complement numbers from PPG are taken as two BRB numbers. $\log_2 \frac{N}{2}$ stages of BRB adders are needed to reduce $\frac{N}{2}$ BRB numbers (or equivalently N 2's complement numbers) to one BRB number (or equivalently two 2's complement numbers). The final stage of converting the BRB number to a 2's complement number is a regular adder.

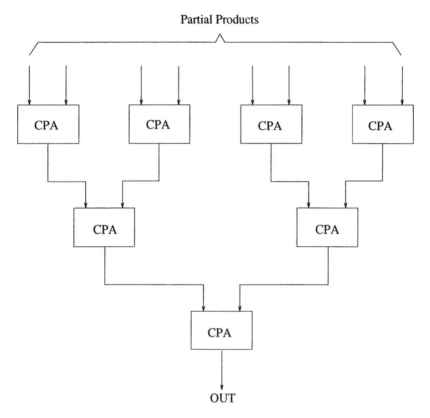

Figure 45: CPA tree for PPS

BRB adder also provide a good solution to a problem in multiplier accumulator. Figure 50 shows a conventional multiplier accumulator. Ususally, the multiplier is pipelined to increase the throughput. However, such an effort would be useless if the CPA in the accumulation loop is much slower than a single pipeline stage in the multiplier. A natural question is whether the CPA in the accumulation loop can also be pipelined. The answer is unfortunately no because the extra cycles in the accumulator would delay the result to be accumulated to the new input and the final result would be wrong. Although this problem can be solved by interleaving the input, a better solution is to use a BRB adder as an accumulator as shown in Figure 51. The multiplier in this figure generates as output a BRB number. One more BRB adder in the middle performs accumulation in BRB format. Because the BRB accumulator has the same delay time as a pipeline stage in the multiplier, the throughput bottleneck is eliminated. The output of the BRB accumulator is then converted to a 2's complement number using a CPA which can be pipelined without causing timing problems.

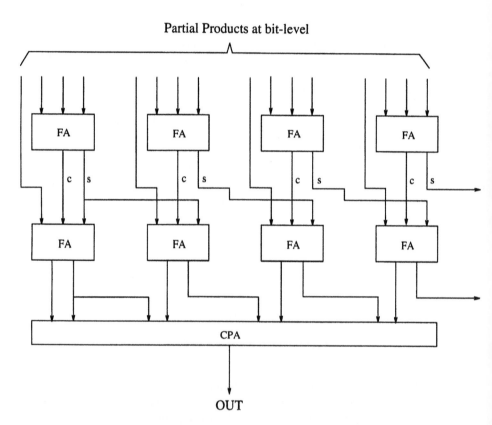

Figure 46: Full adder array for PPS

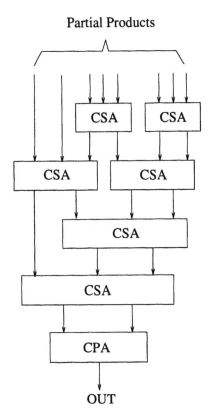

Figure 47: Wallace tree for PPS

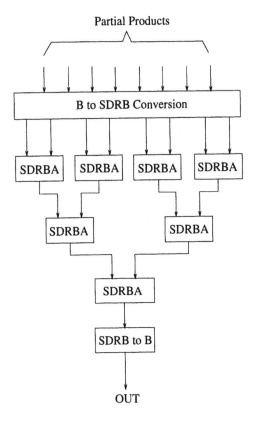

Figure 48: Signed-digit redundant binary adder tree for PPS

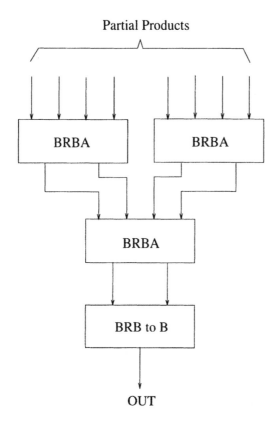

Figure 49: Biased redundant binary adder tree for PPS

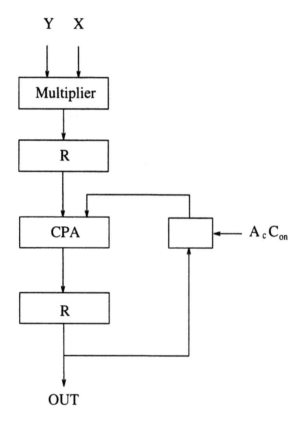

Figure 50: Multiplier accumulator using CPA

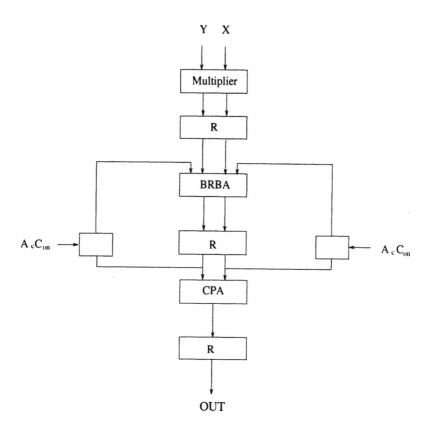

Figure 51: Multiplier accumulator using BRB adder

4   Mapping algorithms to architectures

An algorithm is a set of tasks to be applied to data in a specified order to
transform inputs and internal state to desired outputs. An architecture is a
set of resources and interconnections. Mapping algorithms to architectures
involves assigning tasks to resources. Optimizing an architecture to execute
a class of algorithms involves an iterative process in which both the algorithm and the architecture may be modified to improve overall efficiency.

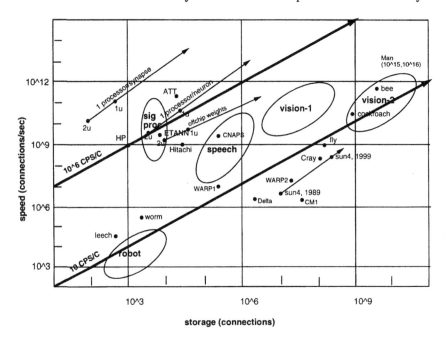

Figure 52: DARPA study application capacity and performance requirements. The biological trendline lies at 10 connections per second per connection (CPS/C); for signal processors, at 10^6 CPS/C.

In many types of systems, a first order concern is the relative area dedicated to program store, data store, and processor resources. In digital
neural net design, for example, this can be expressed as the number of connections per processor (CPP). The DARPA neural net study [Wei88] lists
a variety of applications and their requirements in connections (C) and
connections per second (CPS). These are reproduced in Figure 52. The
optimum CPP is related to the C and CPS of an application by CPP =
(CPS/C)/(CPS/P), where CPS/P, the number of connections per second
per processor, is determined mainly by the technology. An optimal architecture is therefore a function of both the application and the technology.

This equation gives the optimal architecture for a family of applications
which lie along a trendline of constant CPS/C. CPS/C is equivalent to patterns per second whenever all the weights in the network must be accessed
on every pattern presentation.

The biological nets in Figure 52 fall along the trendline of 10 CPS/C.

Conventional computers lie closer to 1 CPS/C. Neurochips implemented so far lie between 10^4 CPS/C and 10^7 CPS/C.

A single digital VLSI multiplier can support 10^8 CPS. Therefore, a biologically motivated architecture capable of 10 CPS/C could have 10^7 connections per processor (CPP). An accelerated learning engine capable of 1000 CPS/C could have 10^5 CPP. An architecture matched to the signal processing applications in the DAPRA report (10^6 CPS/C) could have 100 connections per processor.

So far, all the neurochips (analog or digital) which have been implemented are better suited to signal processing applications than to the other DARPA applications because they feature a relatively small synaptic store for each processor. The Adaptive Solutions CNAPS [Ham90], for example, has 64 processors and 4K 8-bit weights per processor. Each processor can execute 25 million CPS, resulting in about 10^4 CPS/C.

Processors for large networks will need to be optimized for minimum decision energy, not minimum silicon area, since power dissipation will be a dominant constraint. Processor area optimization is much more important in signal processing networks, where the memory/processor area ratio is closer to 1:10.

0.5μ technology supports about 10^6 connections/cm^2, so we might expect about 1 processor for every 10cm^2 at 10 CPS/C, or one fractional processor capable of 10^7 CPS for every cm^2. 0.2μ technology will support about 10^7 connections/cm^2, and therefore 1 processor/cm^2 at 10 CPS/C.

who	bits	type	t_c	C	CPS	CPS/C	W	J
[CTK$^+$89]	16M	DRAM	60ns	2M	17M	8	425mW	25nJ
commercial91	4M	DRAM	40ns	512K	25M	49	-	-
[KBD$^+$90]	16M	DRAM	10ns	2M	100M	50	-	-
M5M44C256	1M	DRAM	60ns	128K	8M	63	300mW	38nJ
[LCH$^+$88]	512K	DRAM	12ns	64K	80M	1250	1.7W	21nJ
[Ram92]	4M	RDRAM	2ns	512K	500M	953	-	-
[HKT$^+$90]	4M	SRAM	23ns	512K	43M	84	350mW	8nJ
[AOA$^+$90]	4M	SRAM	15ns	512K	67M	131	650mW	10nJ
[SIY$^+$90]	4M	SRAM	9ns	512K	111M	217	970mW	9nJ
Paradigm91	1M	SRAM	17ns	128K	59M	449	-	-

Table 6: Maximum pattern presentation rates at full capacity for various memory chips. CPS/C is connections per second (CPS) per connection (C) and is equal to the maximum pattern presentation rate if the chip is loaded to its capacity. In each case, t_c is the fastest cycle time reported. DRAM cycle under 60ns and SRAM under 10ns usually implies page mode access.

The trendline along which an architecture can fall depends heavily on the available memory bandwidth, and whether a single weight fetch can be applied to multiple patterns. If weights are stored in off-chip memory, then the CPS/C of the system is determined by the CPS/C of a single memory chip and the depth of the on-chip pattern cache. A single 16Mbit DRAM chip with 200ns cycle time organized as 4M×4 can supply 2.5MCPS and has a capacity of 2MC, assuming 8-bit weights. This is only 1.25 CPS/C. 10CPS/C would require 25ns access at 16Mbits or a depth 8 on-chip pattern cache. Table 6 shows a variety of memory chips which can provide in excess of 10 CPS/C. However, if we wish to build a neural net capable of accelerated learning at 100×10 CPS/C, Table 6 suggests that conventional memory chips do not yet have the desired bandwidth without a large (1K-pattern) on-chip pattern cache.

Rambus Inc. [Ram92] has used a DRAM architecture with greatly enhanced throughput. They organize a DRAM core so 1024 bits can be latched at a time into an on-chip SRAM cache. The cache contents can be read or written over a byte-wide interface in packets up to 128 bytes long. Data is transferred on both edges of a 250MHz clock for a throughput of 500MBytes per second. The SRAM cache is double buffered so the next line in the DRAM can be read while the current line is being transferred.

The Rambus architecture is ideally suited to block oriented data transfers and applications where the data accesses are to successive words, such as video scanline operations or line replacement in microprocessor caches. It is not well suited to random access when each word is in a separate block.

In fully connected networks, the Rambus architecture can efficiently support $\Sigma w_{ij}x_j$. Note that if w_{ij} is efficient, w_{ji} is not. This can be a significant problem in back propagation. It can be solved by keeping two copies of the weights, one for feedforward and one for feedback computations, and exchanging their roles following each weight update.

In sparsely connected nets, the throughput of the Rambus architecture depends strongly on how the weights are packed in memory. In change-driven computation, the throughput depends strongly on the patterns of activity.

An alternative to the Rambus approach which permits high throughput and random access is pipelining. With pipelined memory, each memory transaction may take several cycles to complete, but a new transaction can be initiated on each cycle. Pipelined memory is well suited to applications which require high throughput random access, but tolerate high latency.

Bandwidth to weight memory can be reduced by storing multiple patterns on-chip and applying each incoming weight to all the cached patterns before fetching the next weight. This can be done during feedforward computation provided the extra latency can be tolerated. To maintain a uniform data rate in patterns and weights, reducing weight bandwidth by a factor of k while keeping the pattern presentation rate constant requires a $2k$-element pattern cache and $2k$ times the latency from pattern input to pattern output. Weight memory bandwidth can be reduced during learning provided k weight changes can be accumulated between weight updates. The impact of batch updating is application dependent.

Although partitioning the system into separate processor and memory chips makes sense for traditional architectures, it increases the memory access energy since the memory is removed from the processor. During training, all the weights tend to be modified incrementally all the time, so caches do not reduce memory access energy. In this case memory access energy can be substantially reduced by distributing processors to the memory.

The optimum architecture balances the power dissipated accessing memory with the power dissipated communicating between processors.

A disadvantage of combining processors and memory on the same chip is that logic processes are not optimized for the smallest memory cell, and memory processes are not optimized for the fastest logic circuits.

In the foregoing discussion, the word "processor" refers to an entity capable of computing a connection. The issue of SIMD vs MIMD is not addressed. For example, the SPERT architecture [Waw92] has a single scalar processor which controls a datapath that has 8 multiply-accumulators. In terms of the previous discussion, this is an 8-processor system, since there are 8 entities capable of computing connections. Its advantage is that it can amortize the energy and resources of one scalar control processor over 8 datapath processors, at some sacrifice in flexibility.

5 Architectural weapons

The performance of digital VLSI systems can be enhanced using a variety of techniques. These are: pipelining, precision, iteration, concurrency, regularity, and locality. Of these, precision may be particularly useful in signal processing applications, and iteration may become more widespread as technology scales down. "Architectural weapons" are design techniques which can increase performance, reduce area, or decrease energy. In this section we focus on several techniques which are generally useful in performance driven signal processor design.

5.1 Pipelining

	const V	scale V
devices/area	S^2	S^2
speed	S^2	S
ops/sec/area	S^4	S^3
energy/op	$1/S$	$1/S^3$
power/area	S^3	1

Table 7: Energy scaling

For many problems in signal processing, performance can be maximized by maximizing pipelining and parallelism. This is an excellent technique for small and medium scale systems with generous power budgets. But as systems get very large, the level of parallelism achievable either by pipelining or replication will be limited by power considerations. Table 7 suggests that scaled voltage is the way to go for implementing very large systems.

In highly pipelined systems, performance is maximized by placing pipe stages so delay is the same in every stage. This requires good timing analysis and optimization tools.

As technology scales, the clock rates achievable by highly pipelined structures will be difficult to distribute globally. Self-timed iterative circuits can save area by performing high speed local computation.

Signal processing applications are well suited to deep pipelining because the latency of an individual unit is not nearly as important as the computation and communication throughput of the system.

If a system is too heavily piped, too much area is taken in latches, and the load on the clock is too large, increasing clock skew and reducing the net performance gain.

A system clock based on the propagation delay of a BRBA provides a good balance between area and perforamnce. This is typically 1/4 of a RISC processor clock period in the same technology.

Pipelining has been used extensively to improve performance. It can also be used to lower power at the same performance, if the supply voltage is reduced to maintain constant throughput.

For a given resource, there is an optimum logic depth which minimizes energy at constant throughput. At shallower logic depths, latch energy dominates. At greater logic depths, logic energy dominates. We have found the optimum logic depth for a 32×32 bit multiplier to be just about equal to the propagation delay through a BRBA [BP91b]. At this logic depth, the area penalty is 37%. We have obtained similar results for both array and tree multipliers ranging in size from 4×4 bits to 256×256 bits, and various types of adders, including ripple, carry select, and carry lookahead.

5.2 Precision

Precision can have a significant impact on area, power, and performance. The area of a multiplier is proportional to the square of the number of bits. A 32 × 32 bit parallel multiplier is 16 times the area of an 8 × 8 bit multiplier. At 8 or fewer bits, a multiplier is just an adder. A Booth encoder can cut the number of partial products in half and a BRBA can accept up to 4 inputs.

Precision should be leveraged where it is available in the system. For example, an $N \times N$ bit multiply generates a $2N$ bit result. A Hebbian style learning algorithm could compute a weight adjustment to higher precision than is available and choose a weight value probabilistically.

Low precision arithmetic has its own special problems, primarily having to do with keeping intermediate results within the available dynamic range, and minimizing systematic errors. Range control techniques include using overflow and underflow detection together with saturation logic. Error statistics will be affected by the type of quantization method chosen. These include truncation, rounding, and jamming. Jamming places a one in the lsb of the result if any of the lower order bits being discarded are ones. This approach has the advantage that it is zero mean [HH91].

5.3   Iteration

Iteration involves using an area-efficient arithmetic element and a local high speed clock to compute a single result in several clock cycles. Iteration reduces logic area by time multiplexing resources at a higher frequency than can be managed globally.

The conventional way to achieve high computational throughput is to implement a parallel arithmetic element clocked at the system clock rate. An alternate approach is to generate a local clock which is at least some multiple of the system clock rate, and then to implement only a fraction of the arithmetic element.

The basic idea behind iteration can be illustrated by the following example. Suppose we want to build a multiplier which can accept two inputs and output a product on every clock cycle. Using conventional techniques, we would have to use N^2 full adders and clock the data through the multiplier at the system clock rate. Suppose, however, that we could generate a local clock which was at least twice the frequency of the system clock. We could then implement the multiplier in half the area using $N^2/2$ fulladders. Iterative structures trade space for time. Iteration will become more widespread as technology scales down and local clock rates scale up.

5.4   Concurrency

Concurrency is a widely used technique for increasing performance through parallelism. The degree of parallelism will increasingly be limited more by power density than by area. Although there is a great deal of concurrency at the system level, the concurrency of a single chip depends on the CPS/C ratio of the application. In signal processing applications, the degree of parallelism in a single chip will increasingly be limited by power density.

5.5   Regularity

Regularity in an architecture or algorithm permits a greater level of system complexity to be expressed with less design effort. Many signal processing allgorithms are composed of large numbers of similar elements, which should translate directly to VLSI implementation.

5.6   Locality

Local connections are cheaper than global ones. The energy required to transport information in CMOS VLSI is proportional to the distance the information travels. The available communication bandwidth is inversely proportional to wirelength, since the total available wirelength per unit area in a given technology is constant.

The energy required to switch a node is CV^2, where C is the capacitance of the node and V is the change in voltage. C is proportional to the area

of the node, which for fixed width wires, is proportional to the length of a
node.

It is very important to maximize locality to minimize communication
energy. Wires cost about 1pF/cm, whether they are on a 0.2μ chip or a 24
inch circuit board. This translates directly to 25pJ/cm if signals switch 5V,
and 10fJ/cm if they switch 100mV. In either case, the shorter the wires,
the better, since communication energy is proportional to wirelength and
communication bandwidth is inversely proportional to wirelength. Massively parallel architectures need to be very careful about the number of
long range connections.

6   Area, power, and performance estimation

We have developed a simple area, performance, and power estimation technique which we use to construct spreadsheets in the early stages of architectural exploration, feasibility analysis, and optimization of a new design.
Area is computed by estimating the number of transistors required. Performance is estimated by building an RC timing model of the critical paths
into the spreadsheet. Power is estimated using CV^2f, where f is obtained
from the performance section of the spreadsheet.

Area, performance, and power are parametrized by technology. We have
a "technology section" of the spreadsheet where we build in technology
scaling rules to compute transistor transconductance and device parasitics.

6.1   Area estimation

We use a simple technique to estimate area of chips before we build them.
We identify the major resources on the chip, and estimate the number of
transistors for each resource. We then multiply the number of transistors
in each case by an area-per-transistor which depends on how regular and
compact we think we can make the layout.

1-transistor DRAM and ROM are about $100\lambda^2$/transistor. 3T DRAM
and 6T SRAM are about $200\lambda^2$/transistor. Tightly packed, carefully hand-
crafted logic is also about $200\lambda^2$/transistor. Loosely packed full custom
logic is about $300\lambda^2$/transistor. Standard cells are about $1000\lambda^2$/transistor.
In some cases (shifters, muxes, decoders), we also include wiring area in our
estimates.

Block routing takes about 30% of the chip area. Standard cell routing
takes 60% of the block. The pad frame reduces the die by about 1mm. For
example, the largest die available on a MOSIS 2μ run is 7.9×9.2mm. Of
this, 6.9×8.2mm, or 56.6mm², is available for logic and routing. Of this,
17mm² is routing, and 40mm² is logic. In 2μ CMOS, lambda = 1μ, so
there is room for 133,000 transistors at $300\lambda^2$/transistor.

The number of transistors required to implement a function can vary
significantly depending on the design style. For example, a full adder im-

plemented with gate logic requires about 30 transistors. However, it can be implemented in 15 transistors using pass-transistor logic. Which is best depends on desired performance and power dissipation, input drive and output load.

We maintain a list of leafcells, the number of transistors they require, and their area-per-transistor, which we reference in estimating requirements of new designs. We also have a set of "tiling functions" which we use to construct complex blocks. For example, an $N \times M$ bit multiplier requires roughly NM full adders, whether it is implemented as an array or a tree.

This technique is especially well suited to spreadsheet implementation, and is especially useful during the early stages of architectural exploration and feasibility analysis in an area-limited design.

item	area
memory	$200\lambda^2$/transistor
logic	$300\lambda^2$/transistor
control	$900\lambda^2$/transistor
block routing	+30%
std cell routing	+60%
padframe	+1mm

Table 8: Area estimation rules.

Table 8 summarizes the technique described above. Transistor counts of various logic elements and tiling functions are summarized in Table 9. This information can be supplemented by using block sizes and transistor counts of subsystems from already implemented chips.

6.2   Performance estimation

We estimate performance using a simple RC timing model based on the RSIM simulator [Ter82], in which transistors are calibrated to have an effective resistance charging or discharging a node capacitance.

We build the following equations into our spreadsheet models to allow the performance estimates to scale with technology. The symbology of these equations follows the development in Hodges and Jackson [HJ83].

function	transistor count
DRAM	1
SRAM	6
tgate	1
inv, mux	2
nand, nor	4
xor	6
latch	10
TSPC latch	12
halfadder	20
fulladder	30
3:2 adder	40
BRB adder	60
N-input mux	$4N/3$
N-bit shifter	$\log_2 N$
N-bit comparator	$20N$
N-bit counter	$14N$
N-bit ripple carry adder	NT_{fa}
N-bit carry select adder	$2N(T_{fa} + 2)$
$N \times M$-bit multiplier	NMT_{fa}

Table 9: Transistor count summary. T_{fa} is the number of transistors in a fulladder.

Effective resistance of transistors

```
Ilin = k/2(2 * (Vgs-Vt) * Vds - Vds^2) Vds <= Vgs - Vt
Isat = k/2(Vgs - Vt)^2                  Vds >= Vgs - Vt

Iav  = integrate(I * dt)/T .
reff = DV / Iav
     = const * k

kn   = un * cox         kp   = up * cox     A/V^2
rn   = 1/kn/(Vdd - Vt)  rp   = 1/kp/(Vdd - Vt) ohms/sq
Rn   = rn * 1 / w       Rp   = rp * 1 / w       ohms
```

Parasitic capacitance

```
cg   = eox / tox                            farad/m^2
cox  = cg                                   farad/m^2
xj   = (2*esi/q/NA*(V-Vt))^.5               meters
cj   = esi / xj                          farad/m^2
cjsw = 3 * cj * xj = 3 * esi                farad/m
ci   = eox / hi                          farad/m^2
cisw = ci * ti                              farad/m
```

Propagation delay

```
tpu  = Rp * (Cd + Ci + Cg)                  sec
tpd  = Rn * (Cd + Ci + Cg)                  sec
```

Table 10 summarizes the equations we use to compute propagation delay. Capacitance formulas were obtained from [ST83]. The effective resistance should be increased to reflect velocity saturation, especially in short channel devices. Velocity saturation occurs at around 4V in 2.0μ CMOS.

6.3   Power estimation

There are three principal components to power dissipation in most CMOS systems:

$$P_{dc} = V^2/R_{dc}$$

$$P_{sc} = I_{sc}V$$

$$P_{ac} = CV^2f$$

where

Transistor resistance

$$I_{ds} \quad = \quad k/2(V_{gs} - V_t)^2 \qquad V_{ds} \geq V_{gs} - V_t$$

$$
\begin{aligned}
k \quad &= \quad u_o c_{ox} & A/V^2 \\
r \quad &= \quad V_{dd}/I_{ds}(V_{dd}) & \Omega/\square \\
R \quad &= \quad rl/w & \Omega
\end{aligned}
$$

Parasitic capacitance

$$
\begin{aligned}
c_{ox} \quad &= \quad \epsilon_{ox}/t_{ox} & \text{farad/m}^2 \\
\phi \quad &= \quad V_T ln(N_A N_D/n_i^2) & \text{volts} \\
x_d \quad &= \quad \sqrt{2\epsilon_{si}(\phi - V)/q/N_A} & \text{meters} \\
c_j \quad &= \quad \epsilon_{si}/x_d & \text{farad/m}^2 \\
c_{jsw} \quad &= \quad 3c_j x_j & \text{farad/m} \\
c_i \quad &= \quad 1.15\epsilon_{ox}/h_i & \text{farad/m}^2 \\
c_{isw} \quad &= \quad 1.40\epsilon_{ox}(t_i/h_i)^{0.222} & \text{farad/m}
\end{aligned}
$$

Propagation delay

$$t_{pd} \quad = \quad R(C_d + C_i + C_g) \qquad \text{sec}$$

Table 10: Performance equations. Propagation delay is determined from the effective resistance of transistors and the parasitic capacitance of the nodes being switched. Current in short channel devices is reduced by velocity saturation.

V is the supply voltage

C is the total capacitance being switched

f is the clock frequency

R_{dc} is the total static pullup or pulldown resistance

I_{sc} is the short-circuit current

P_{dc} is the power dissipated at DC

P_{sc} is the power dissipated due to short-circuit current

P_{ac} is the power dissipated by switching capacitance.

P_{dc} can be designed out of a system, except leakage, which is usually on the order of a few microwatts [CTTF79, HKT+90], but P_{sc} and P_{ac} cannot. In a CMOS inverter, I_{sc} is the current which flows when both the N-transistor and P-transistor are on during switching. Powell and Chau [PC90] have reported that P_{sc} can account for up to half the total power.

We have done some investigations which suggest that the short circuit current can be significant if rise times are long and transistors are large, and that in most cases P_{sc} can be reduced to less than 10% of total power by sizing transistors. This implies that short circuit current can be more of a problem in gate array or standard cell design, where transistors are fixed sizes or are sized to drive large loads.

Consider a single CMOS inverter driving a purely capacitive load. Initially, assume the gate is at 0 volts, so the P-transistor is on, the N-transistor is off, and the output is at 5 volts. Now, switch the gate from 0 to 5 volts. Ideally, the P-transistor should turn off instantly, the N-transistor should turn on and drain the charge off the output (actually supply electrons to the output) until the output potential reaches 0 volts. The work done (or energy consumed) by the inverter is just QV where Q is the charge on the output and V is the initial potential difference between the output and GND. But $Q = CV$ so the work done is CV^2.

In practice, the input does not switch instantly, so both the N-transistor and the P-transistor are on for a short time, causing excess current to flow.

We measured the short circuit charge for a variety of transistor sizes, rise times, and output loads using spice [Nag75] on a typical 2 micron CMOS technology from MOSIS.

Figure 53 shows the current flowing through vdd and gnd supplies as the gate is switched first from 0 to 5 volts between 2ns and 7ns, and then from 5 to 0 volts between 40ns and 45ns. The short circuit current in each case is the smaller spike; it is the current flowing through the supply which should be off. The short circuit charge is the area under the short circuit current. In the figure the short circuit charge is about 15% of the charge initially on the output, so P_{sc} will be about 15% of P_{ac}.

Figure 54 shows short circuit charge as a percentage of output load vs input rise/fall time. Each graph has a pair of curves for each of 5 output loads: 0, 100, 200, 500, and 1000 fF. One curve in each pair is for a rising input, the other for a falling input. The short circuit charge for a given output load is nearly the same whether the input is rising or falling. The pair with no output load has the largest short circuit charge.

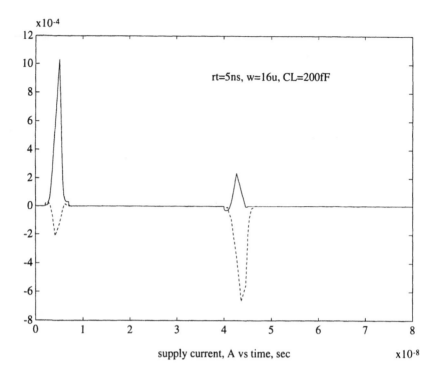

Figure 53: short circuit current

The short circuit charge is negative for fast rise times because the coupling capacitance between the gate and the drains pulls the output above 5 volts. This deposits additional charge on the output which must then be removed. So there is an energy cost associated with switching an input too fast, and a finite rise time at which the net short circuit charge is exactly zero.

In practice, we estimate power in a variety of ways. In some cases, when we have experience or know statistically what percentage of the nodes are switching, we use CV^2f. In other cases, we use Powell and Chau's power factor approximation (PFA) technique, which uses the energy of existing devices to predict new ones. In still other cases, when the devices have an analog behavior, such as sense amplifiers in memories and reduced voltage swing logic, we use spice to compute current and integrate to find charge dumped.

We have modified the RSIM timing simulator to accumulate the charge dumped as nodes switch during simulation. This is easy to do if the simulator is event driven. We compared RSIM's results on a signal processing chip we fabricated through MOSIS [B+87] with power measurements done on a performance tester and found agreement to within 20%.

We estimate power by estimating the capacitance switched on each clock cycle. We ignore short circuit current and DC leakage. In [BBP91],

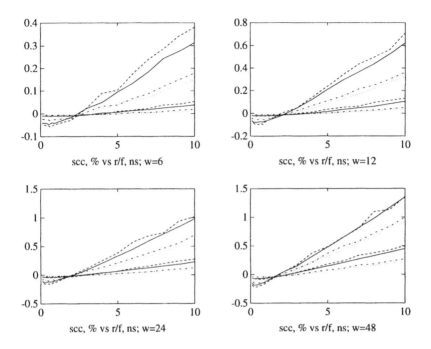

Figure 54: short circuit charge as a percentage of output load vs rise time for different size devices

we showed that short circuit current can be optimized out of the system, and becomes negligible at low voltage because the current at the switching threshold of a gate is only a few percent of the current when the gate is fully on due to the increase in V_t/V_{dd}. Leakage, on the other hand, becomes very important if the threshold voltage is reduced. We have discussed energy optimization at low voltage in [BP91b] and [BP91a].

The key element in our power estimation technique is obtaining an accurate estimate of circuit activity a. Results reported from event driven simulation suggest that a is around 10%. That is, around 10% of the nodes in the system switch at any given time. We have modified our RSIM simulator to gather activity statistics during simulation, and have found good agreement with published results. We have found that a as a measure of nodes switched is in most cases negligibly different from a as a measure of the fraction of capacitance switched.

We analyzed one of the signal processing chips we designed which has 19,918 nodes and 46,906 transistors. The chip has two 16-bit Brent-Kung adders, 3 32×16 4-port memories, and a 16×16 multiplier-accumulator. One 16-bit I/O bus was used in the simulation, and about 60 control pin inputs. There is an on-chip 2-phase clock generator.

We found activities of 33% in the adders, 22% in the multiplier, 5% in the memories, and 10.6% in the chip overall. The I/O pads had an activity of 13% even though they were in use on every clock cycle. The total energy of the chip was 17.3nJ per cycle. The clock circuits accounted for 25% of this total. Buses accounted for about 6%, and pads about 12% (the pads were not driving any external load).

Given the activity a and the number of transistors N_t in a circuit, we can compute energy by $E = aN_tC_tV^2$, where C_t is the capacitance of a single transistor and V is the supply voltage. For 2μ CMOS, we use $C_t = 40\text{fF}$. This allows 10fF each for the source, drain, gate, and interconnect. Capacitance, and therefore energy per device, scales as $1/S$. For a 16×16-bit multiplier we get $E = 0.44 \times 16^2 \times 30\times40\text{e-}15\times5^2 = 3.4\text{nJ}$. Power is then Ef, where f is the operating frequency: a 4nJ multiplier dissipates 100mW at 25MHz.

6.4   Technology scaling

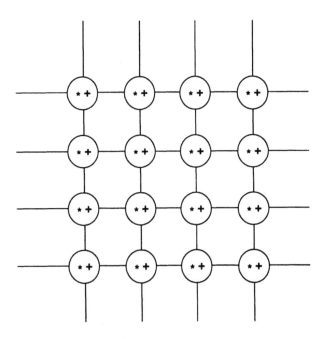

Figure 55: SP: one processor per synapse

We have applied the area and performance estimation techniques to three different inner product processor architectures. The first (SP; see Figure 55) has one processor per synapse. The second (NP; see Figure 56) has one processor per neuron, with the synaptic weights stored in memory local to each processor, similar to the Adaptive Solutions X1 architecture. The third (FP; see Figure 57) has a fixed number of processors on chip, and off-chip weights.

SP has the lowest synaptic storage density but the highest computational throughput. FP has the lowest throughput, but the highest synaptic storage density and unlimited capacity.

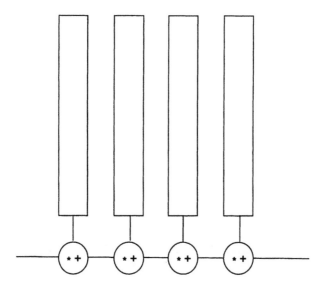

Figure 56: NP: one processor per neuron

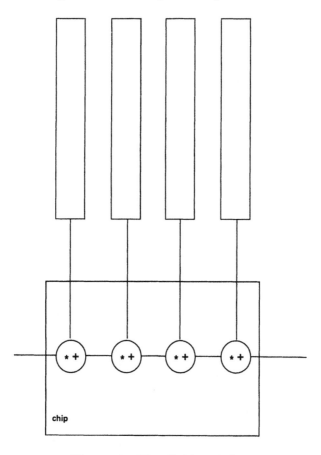

Figure 57: FP: off-chip weights

The number of processors which can be placed on a single FP is I/O limited. Assuming 4 bit weights, 128 pins would be required to support 32 processors. The I/O constraint can be substantially alleviated with multichip module (MCM) packaging [Joh90]; this allows far more flexibility in choosing die size and system partitioning.

6.5   Technology and performance scaling of inner product processors

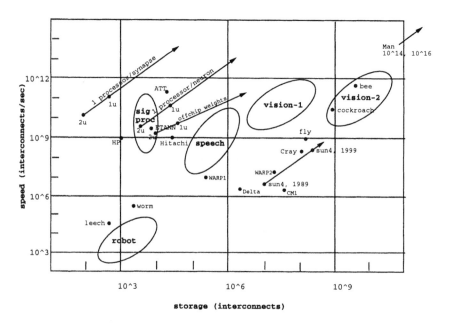

Figure 58: technology and performance scaling

In Figure 58 we plotted the technology scaling trendlines of the SP, NP, and FP architectures on the DARPA-style capacity-vs-performance graph, along with the DARPA application requirements.

In Figure 59, we plotted the waferscale integration trendlines of the three architectures. The FP architecture has two trendlines. Since its weights are offchip, a waferscale implementation has the option of replicating memory alone. In this case performance remains constant, but capacity improves. Interestingly, of the options shown, this one most closely matches the requirements of the DARPA applications.

In Figure 60, we derive the architecture from the application and the technology. In a very rough sense, an application can be characterized by its capacity and performance requirements. An architecture can be loosely defined in terms of the connections serviced by a processor. For example, the SP architecture has one connection per processor. The technology defines the available performance. According to Figure 60, the product of the available performance and the ratio of the desired capacity and performance determines the architecture.

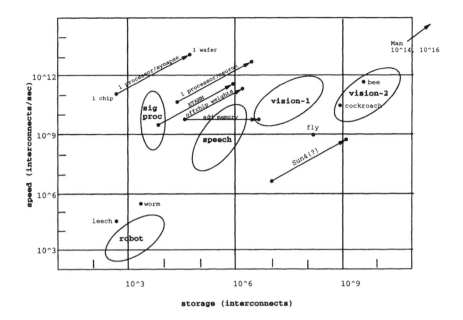

Figure 59: waferscale integration at 1.0μ

Figure 60 suggests that SP is most appropriate for the signal processing application in 2.0 micron CMOS, but that NP is more appropriate in 1.0 and 0.6 micron. By comparing chip DRAM capacities in the table (5 bit weights), we see that the FP architecture is best in all other cases except speech and vision-1 in 0.6 micron, where NP is best.

6.6   Technology scaling rules and examples

To scale area, performance, and power to a desired technology, we build into our spreadsheet the parameters of a base technology (2.0 micron CMOS in our case; $\lambda_0 = 1.0\mu$), compute $S = \lambda_0/\lambda$, and apply the equations in Table 11. We leave V, the supply voltage, explicit so we can compare the impact of "constant voltage" scaling and full scaling. These equations follow the development in Hodges and Jackson [HJ83].

Table 12 shows how actual 2μ numbers would look scaled for a range of technologies. In the table, R(12:2) is the effective resistance of a $12\lambda \times 2\lambda$ transistor. Cg, Cd, and Ci are nominal gate, diffusion, and wiring capacitances. Tgate is the propagation delay of a single gate. Tbrb is the propagation delay of a BRB adder.

Table 13 shows typical spice parameters for different technologies. The "exp" row in the table shows the exponent which would be supplied to spice. For example, for 2.0μ, tox=403e-10. cpa, $m1a$, and $m2a$ are area capacitance of polysilicon, metal1, and metal2 in farads/meter^2.

The information for 2.0, 1.6, and 1.2 μ technologies was obtained from MOSIS. 16 runs were averaged in each technology, and the deck closest to the mean was selected as nominal. The 0.8 micron information was

Applications and architectures

$$\text{connections/processor} \;=\; \frac{\text{connections/system}}{\text{connections/sec/system}} \;\times\; \text{connections/sec/proc}$$

$$\text{(architecture)} \qquad\qquad \text{(application)} \qquad\qquad \text{(technology)}$$

Application	connections	cps	connections / cps	cpp 2.0u	1.0u	0.6u
robot arm	3e3	3e4	1e-1	8e6	2e7	4e7
signal proc	3e4	1e10	3e-6	2.4	600	1500
speech	1e6	1e9	1e-3	8e4	2e5	5e5
vision-1	1e8	1e11	1e-3	8e4	2e5	5e5
vision-2	1e10	1e11	1e-1	8e6	2e7	5e7

Technology capacity and performance

Tech	Clock(S^1.5)	chip 3TDRAM	1TDRAM	wafer 3TDRAM	1TDRAM
2.0	80 MHz	1.5e3	4.5e3	1.5e5	4.5e5
1.0	200 MHz	6.0e4	1.8e5	6.0e6	1.8e7
0.6	500 MHz	2.0e5	6.0e5	2.0e7	6.0e7

Application	Nprocessors	Nchips (wafers)
robot arm	e-4	e-2 chips
signal proc	50	1 chip
speech	5	10 chips
vision-1	500	10 wafers
vision-2	500	1000 wafers

Figure 60: algorithms and architectures

param	scaling	description
tech	S	λ_0/λ
tox	$1/S$	gate oxide
cox	S	gate capacitance
uo	$1/S^{0.5}$	mobility
k	$S^{0.5}$	transconductance
r	$1/S^{0.5}/V$	resistance
xj	$V^{0.5}$	junction depth
cj	$S/V^{0.5}$	diffusion capacitance
cjsw	S	diffusion sidewall capacitance
hi	$1/S$	metal elevation
ti	$1/S$	metal thickness
ci	S	interconnect area capacitance
cisw	1	interconnect sidewall capacitance
R	$1/S^{0.5}/V$	device resistance
C	$1/S$	device capacitance
Q	V/S	device charge
I	$S^{0.5}(V-V_t)^2$	device current
tp	$V/(S^{1.5}(V-V_t)^2)$	propagation delay
i	$S^{2.5}(V-V_t)^2$	current density
p	$S^{0.5}(V-V_t)^2 V$	device power
P	$S^{2.5}(V-V_t)^2 V$	power density

Table 11: Technology scaling. V_t is the threshold voltage. Current through short channel devices is reduced by velocity saturation.

tech		2.0	1.6	1.2	1.0	0.8	0.6	
S		1.0	1.3	1.7	2.0	2.5	3.3	
S^2		1.0	1.7	2.9	4.0	6.3	11.0	
clk	S^2	20	34	58	80	125	220	MHz
clk	S^2	80	136	232	320	500	880	MHz
R(12:2)	$1/S$	6.0	4.6	3.5	3.0	2.4	1.8	kohm
Cg	$1/S$	20.0	15.4	11.8	10.0	8.0	6.1	ff
Cd	$1/S^2$	40.0	23.5	13.8	10.0	6.3	3.6	ff
Ci	$1/S$	40.0	30.8	23.6	20.0	16.0	12.2	ff
Tgate	$1/S^2$	1.38	0.81	0.48	0.35	0.23	0.13	nsec
Tbrb	$1/S^2$	12.5	7.35	4.31	3.13	1.98	1.14	nsec

Table 12: scaling example

tech	u0n	u0p	cgdon	cgdop	cjswn	cjswp	xjn	xjp	nsubn	nsubp
exp	0	0	-12	-12	-12	-12	-9	-9	+15	+15
2.0	631	237	298	285	548	334	250	50	5.76	6.24
1.6	583	186	573	494	588	184				
1.2	574	181	628	324	423	159				
0.8	447	101	229	271	200	200	157	138	85.58	79.65

tech	tox	cpa	cjn	cjp	m1a	m2a		
exp	-10	-6	-6	-6	-6	-6	tox	$1/S$
							cpa	S
2.0	403	388	130	262	26	19	cjn	$S^{3/2}$
1.6	250	573	140	432	35	23	cjp	S
1.2	209	644	293	481	36	24	m1a	S
0.8	170	794	547	570	79	31	m2a	$S^{1/2}$

Table 13: nominal process parameters for different technologies

obtained from a much smaller dataset.

These technologies all run at 5 volts, so S^2 performance is predicted by the formulas in Table 11. However, Table 13 shows that mobility, uo, is not constant. Comparing 2.0μ and 0.8μ, uo is scaling as $1/S^{1/2}$. Now R scales as $1/S/V/uo = 1/S^{1/2}/V$ and C as $1/S$ so propagation delay tp should then scale as $1/S^{3/2}/V$.

According to Hodges and Jackson, mobility is a function of substrate doping, N_A. N_A increases as λ decreases. As N_A increases, collisions become more likely so mobility decreases. If the increase in N_A is necessary to prevent breakdown in the presence of higher E fields in constant voltage scaling, then presumably NA could be kept constant if V were scaled.

tech	tpd	MHz constV	Vdd	MHz scaleV
2.00	10.83n	92	5	
1.60	7.58n	132	5	
1.40	6.12n	163	5	
1.20	4.78n	209	5	
1.00	3.57n	280	5	
0.80	2.50n	400	5	
0.60	1.58n	634	3.3	418
0.50	1.18n	849	3.3	560
0.40	825p	1213	3.3	801
0.30	520p	1921	3	1153
0.25	389p	2572	3	1543
0.10	90p	11143	2	4457

Table 14: BRB adder-based clock circuit performance

Table 14 shows predicted performance of self-timed BRB adder based clock driver [San89] in various technologies based on both constant V ($S^{3/2}$) and scaled V (S). The numbers are in good agreement with observed performance at 2.0, 1.6, and 0.8μ CMOS. Extrapolations below 0.5μ are highly speculative. They serve as an upper bound on expected performance.

Process variations can impact performance significantly. We have found the ratio of "fast-fast" models to "slow-slow" models to be a factor of two in performance, with "typical" models right in the middle.

$$ff/typ = typ/ss = \sqrt{2}$$

$$ff/ss = 2$$

Performance degrades by a factor of 1.8 from 25degC, 5V to 130degC, 4.5V. Worst case performance can therefore be as much as four times slower than best case when process and temperature variations are combined.

The technology scaling technique described in this section is sufficient to give a rough idea of performance in a target technology given measured performance in a known technology. However, actual performance can vary widely, especially at smaller feature sizes, depending on the optimization constraints.

7   Miscellaneous topics

7.1   Multichip modules

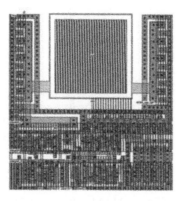

Figure 61: MCM pad vs MOSIS TinyChip pad. The MCM pad capacitance is 50fF. The TinyChip pad capacitance is 2pF (40×).

Multichip modules (MCMs), and especially 3D stacked MCMs, offer exceptional opportunities for implementing high performance, high capacity signal processing, especially since the high degree of parallelism, locality of communication, and regularity of signal processing structure can all be exploited in the MCM environment.

Power dissipation is a big concern due to the availability of massive parallelism in signal processing computation. MCMs reduce communication energy by reducing wire length, and increase available interchip communication bandwidth through area bonded I/Os. Silicon based approaches provide 25nF/cm intrinsic bypass capacitance, and very low inductance (0.1nH) solder bump pad connections. Power and ground connections, as well as system-wide clock distribution, can be delivered within the chip to the point of use. Pad driver size can be substantially reduced (see Figure 61).

MCMs can provide good noise isolation for low voltage operation, and naturally permit integration of level shifting interface components for communicating with a high voltage external world.

7.2 Energy and capacity limits in VLSI

technology	energy J/C	capacity C/cm³	performance CPS/cm³
optical	10^{-13}	10^{10}	10^{14}
0.8μ digital	10^{-9}	10^{8}	10^{12}
analog	10^{-12}	10^{7}	10^{13}
0.1μ digital	10^{-12}	10^{10}	10^{15}
biology	10^{-15}	10^{12}	10^{13}

Table 15: Energy, capacity, and performance density for various technologies. Assumptions: 100 cm² silicon per cm³; 8-bit synapses; 0.8μ technology: 5.0V; 0.1μ technology: 0.5V.

Table 15 was compiled at the 1992 Banff workshop on neural network hardware. It suggests that VLSI implementations are much closer to matching biological performance density than either energy or storage density.

The energy gap can be reduced in digital VLSI by lowering the supply voltage even further; this results in slower processors but achieves a better balance between the silicon area required for processing and the area required for weight storage. If arithmetic precision is reduced from 8 to 5 bits, then computation energy can be reduced by a factor of 2.5. If the supply voltage can be reduced from 500mV to 120mV within the neural processors, energy can be reduced another factor of 17. Together these two improvements reduce energy by a factor of 40, placing 0.1μ digital energy per connection within a factor of 20 of biology.

7.3 Low voltage digital logic

In the preceding section, we showed that there is a significant challenge in matching biological energy efficiency in neural network computation. This translates directly to a challenge in reducing power dissipation: at 1nJ per connection, 10^{16} connections per second is 10 megawatts. Energy scales as V^2/S. What are the limits on V? We are building demonstration CMOS circuits with supply and threshold voltages in the 100-500mV range with which we plan to study performance, yield, and reliability issues.

We have found that the performance does not degrade nearly as much as expected at low voltage if the process is optimized to operate at that voltage, especially for channel lengths under 0.5μ. The biggest reasons for this are: 1) velocity saturation is less severe at low voltage, 2) reduced substrate doping increases mobility, and 3) channel lengths can be reduced without having to worry about punchthrough.

We have also found that device performance can be substantially improved by biasing the substrate to induce thresholds. We have shown [BP91b] that minimum energy operation occurs when $V_{dd} = V_t$, and that the minimum energy V_t occurs at $V_t = nV_T\ln(Id/a)$, where $n = (C_{ox} + C_{dep})/C_{ox}$ is the gate coupling coefficient that determines what fraction of the gate voltage appears at the silicon surface due to the capacitive voltage divider between the gate capacitance and depletion region capacitance; $V_T = KT/q$ is the thermal voltage (26mV at 300degK); ld is the logic depth; and a is the activity. This V_t minimizes total energy by equating switching energy, which decreases as $V_d d$ is reduced, and leakage energy, which increases as V_t is reduced.

In section 5.1, we asserted that minimum energy occurs when $ld = 10$. In section 6.3, we said typical circuit activity is around 0.1. n is close to 1.4 for typical CMOS processes, giving a subthreshold slope m_s of 84mV/decade. However, for a process optimized for low voltage, and with back bias, n is close to 1.0, so $m_s = 60$mV/decade at 300degK. Then $V_t = 0.026 \times \ln(10/0.1) = 120$mV. A big question is how much variation in V_t to expect, and how much margin must be designed in at low voltage. Threshold variations will be reduced at low substrate doping concentrations.

For minimum energy, we set $V_{dd} = V_t = 120$mV. For minimum energy $\times$ time, we set $V_{dd} = 3V_t = 360$mV. If we can optimize the process, we choose the substrate doping so we get a 300mV threshold when the substrate bias is 3 volts. Using this approach, we expect submicron logic depth 10 circuits to operate in the 100MHz range and achieve on the order of 10^{-13}J per connection.

Bibliography

[AA90] Robert B. Allen and Joshua Alspector. Learning of stable states in stochastic asymmetric networks. Bellcore, 1990.

[AGA89] Joshua Alspector, Bhusan Gupta, and Robert B. Allen. Performance of a stochastic learning microchip. In *Advances in Neural Information Processing Systems*, pages 748–760, 1989.

[AH85] David H. Ackley and Geoffrey E. Hinton. A learning algorithm for Boltzmann Machines. *Cognitive Science*, 9:147–169, 1985.

[Als90] Joshua Alspector. CLC - A cascadeable learning chip. NIPS90 VLSI Workshop, December 1990.

[AMO+90] Yutaka Arima, Koichiro Mashiko, Keisuke Okada, Tsuyoshi Yamada, Atushi Maeda, Harufusa Kondoh, and Shinpei Kayano. A self-learning neural network chip with 125 neurons and 10K self-organization synapses. In *Symposium on VLSI Circuits*, pages 63–64, 1990.

[ANH+87] Masakazu Aoki, Yoshinobu Nakagome, Masahi Horiguchi, Shin'ichi Ikenaga, and Katsuhiro Shimohigashi. A 16-level/cell dynamic memory. *IEEE Journal of Solid-State Circuits*, pages 297–299, April 1987.

[ANY90] Aharon J. Agranat, Charles F. Neugebauer, and Amnon Yariv. A CCD based neural network integrated circuit with 64K analog programmable synapses. In *IJCNN International Joint Conference on Neural Networks*, pages II:551–555, 1990.

[AOA+90] Shingo Aizaki, Masayoshi Ohkawa, Akane Aizaki, Yasushi Okuyama, Isao Sasaki, Toshiyuki Shimizu, Kazuhiko Abe, Manabu Ando, and Osamu Kudoh. A 15ns 4Mb CMOS SRAM. In *IEEE International Solid-State Circuits Conference*, pages 126–127, 1990.

[AP89] M.O. Ahmad and D.V. Poornalah. Design of an efficient VLSI inner product processor for real-time DSP applications. *IEEE Transactions on Circuits and Systems*, 36:324–329, February 1989.

[AS90] Morteza Afghahi and Christer Svensson. A Unified Single-Phase Clocking Scheme for VLSI systems. *IEEE JSSC*, 25(1):225–233, February 1990.

[ASS+88] Syed B. Ali, Barmak Sani, Alex S. Shubat, Keyhan Sinai, Reza Kazerounian, Ching-Jen Hu, Yueh Yale Ma, and Boaz Eitan. A 50-ns 256K CMOS split-gate EPROM. *IEEE Journal of Solid-State Circuits*, pages 79–85, February 1988.

[Avi61] Algirdas Avizienis. Signed-digit number representations for fast parallel arithmetic. *IRE Transactions on Electronic Computers*, pages 389–400, September 1961.

[B+87] James B. Burr et al. A 20 MHz Prime Factor DFT Processor. Technical report, Stanford University, September 1987.

[Baa91] Bevan Baas. A pipelined memory system for an interleaved processor. Technical report, Stanford University, September 1991.

[Bak86] H. B. Bakoglu. *Circuit and system performance limits on ULSI: Interconnections and packaging*. PhD thesis, Stanford University, October 1986.

[BBP91] James B. Burr, James R. Burnham, and Allen M. Peterson. System-wide energy optimization in the MCM environment. In *IEEE Multichip Module Workshop*, pages 66–83, 1991.

[BJ87] A. G. Barto and M. I. Jordan. Gradient following without back propagation. In *IEEE First International Conference on Neural Networks*, 1987.

[BL84] Adir Bar-Lev. *Semiconductors and Electronic Devices*. Prentice Hall, 1984.

[BP91a] James B. Burr and Allen M. Peterson. Energy considerations in multichip-module based multiprocessors. In *IEEE International Conference on Computer Design*, pages 593–600, 1991.

[BP91b] James B. Burr and Allen M. Peterson. Ultra low power CMOS technology. In *NASA VLSI Design Symposium*, pages 4.2.1–4.2.13, 1991.

[Bru89] Erik Bruun. Reverse-voltage protection methods for CMOS circuits. *IEEE Journal of Solid-State Circuits*, pages 100–103, February 1989.

[BTOK91] Wolfgang Blazer, Masanobu Takahashi, Jun Ohta, and Kazuo Kyuma. Weight quantization in Boltzmann Machines. *Neural Networks*, 4:405–409, 1991.

[Bur70] A. W. Burks, editor. *Essays on Cellular Automata*. Univ. Illinois Press, 1970.

[Bur88] James B. Burr. Advanced simulation and development techniques. In *IREE 7th Australian Microelectronics Conference*, pages 231–238, 1988.

[Bur91a] James B. Burr. Digital Neural Network Implementations. In P. Antognetti and V. Milutinovic, editors, *Neural Networks: Concepts, Applications, and Implementations, Volume 2*, pages 237–285. Prentice Hall, 1991.

[Bur91b] James B. Burr. Energy, capacity, and technology scaling in digital VLSI neural networks. NIPS91 VLSI Workshop, May 1991.

[Bur92] James B. Burr. Digital Neurochip Design. In K. Wojtek Przytula and Viktor K. Prasanna, editors, *Digital Parallel Implementations of Neural Networks*. Prentice Hall, 1992.

[Bur93] James B. Burr. Stanford Ultra Low Power CMOS. to be presented at Hot Chips 93, April 1993.

[BWP91] James B. Burr, P. Roger Williamson, and Allen M. Peterson. Low power signal processing research at Stanford. In *NASA VLSI Design Symposium*, pages 11.1.1–11.1.12, 1991.

[CFWB88] D. Chengson, C. Frazao, H. W. Wang, and B. Billett. Signal delay in distributed RC tree networks. In *IEEE International Symposium on Circuits and Systems*, pages 2835–2837, 1988.

[CG89] Lawrence T. Clark and Robert O. Grondin. A pipelined associative memory implemented in VLSI. *IEEE Journal of Solid-State Circuits*, pages 28–34, 1989.

[CGSS71] L. Cohen, R. Green, K. Smith, and J. L. Seely. Single transistor cell makes room for more memory on a MOS chip. *Electronics*, page 69, Aug 2 1971.

[CMR+90] Alice Chiang, Robert Mountain, James Reinold, Jeoffrey LaFranchise, James Gregory, and George Lincoln. A programmable CCD signal processor. In *IEEE International Solid-State Circuits Conference*, pages 146–147, 1990.

[CP87] K. M. Chu and D. L. Pulfrey. A comparison of CMOS circuit techniques: Differential cascode voltage switch logic versus conventional logic. *IEEE Journal of Solid-State Circuits*, SC-22:528–532, 1987.

[CRS88] H. Jonathan Chao, Thomas J. Robe, and Lanny S. Smoot. A 140 Mbit/s CMOS LSI framer chip for a broad-band ISDN local access system. *IEEE Journal of Solid-State Circuits*, 23(1):133–141, February 1988.

[CSB92] Anantha P. Chandrakasan, Samuel Sheng, and Robert W. Broderson. Low power CMOS digital design. *IEEE Journal of Solid-State Circuits*, pages 473–484, September 1992.

[CTK+89] Shizuo Chou, Tsuneo Takano, Akio Kita, Fumio Ichikawa, and Masaru Uesugi. A 60-ns 16-Mbit DRAM with a minimized sensing delay caused by bit-line stray capacitance. *IEEE Journal of Solid-State Circuits*, pages 1176–1183, October 1989.

[CTTF79] Pallab K. Chatterjee, Geoffrey W. Taylor, Al F. Tasch, and Horng-Sen Fu. Leakage studies in high-density dynamic MOS memory devices. *IEEE Transactions on Electron Devices*, pages 564–575, April 1979.

[Dad89] Luigi Dadda. On serial-input multipliers for two's complement numbers. *IEEE Transactions on Computers*, pages 1341–1345, September 1989.

[Den] R. H. Dennard. Field-effect transistor memory. U. S. Patent 3 387 286, June 4, 1968.

[DLHP88] Sang H. Dhong, Nicky Chau-Chau Lu, Wei Hwang, and Stephen A. Parke. High-speed sensing scheme for CMOS DRAM's. *IEEE Journal of Solid-State Circuits*, pages 34–40, February 1988.

[DMS+86] B. J. Donlan, J. F. McDonald, R. H. Steinvorth, M. K. Dhodhi, G. F. Taylor, and A. S. Bergendahl. The wafer transmission module. *VLSI Systems Design*, pages 54–90, January 1986.

[DR85] P. Denyer and D. Renshaw. *VLSI Signal Processing: a Bit-Serial Approach.* Addison-Wesley, 1985.

[Ebe81] Carl Ebeling. *The Gemini user's manual.* Carnegie-Mellon University, 1981.

[Eib88] Alfred J. Eiblmeier. A reduced coefficient FFT butterfly processor. Technical report, Stanford University, October 1988.

[Enz89] C.C. Enz. *High precision CMOS micropower amplifiers.* PhD thesis, Ecole Polytechnique Federale de Lausanne, 1989.

[FD85] John P. Fishburn and Alfred E. Dunlop. TILOS: A posynomial programming approach to transistor sizing. In *IEEE International Conference on Computer Aided Design*, pages 326–328, 1985.

[Fis87] John P. Fishburn. Clock skew optimization. Technical report, ATT Bell Laboratories, 1987.

[FPH+90] Stephen Flannagan, Perry Pelley, Norman Herr, Bruce Engles, Tai sheng Feng, Scott Nogle, John Eagan, Robert Dunnigan, Lawrence Day, and Roger Kung. 8ns CMOS 64kx4 and 256kx1 SRAMs. In *IEEE International Solid-State Circuits Conference*, pages 134–135, 1990.

[Gai67] Brian R. Gaines. Stochastic computing. In *Spring Joint Computer Conference*, pages 149–156, 1967.

[GH86] T. A. Grotjohn and B. Hoefflinger. Sample-Set Differential Logic (SSDL) for complex high-speed VLSI. *IEEE Journal of Solid-State Circuits*, pages 367–369, April 1986.

[GH89] Conrad C. Galland and Geoffrey E. Hinton. Deterministic Boltzmann learning in networks with asymmetric connectivity. Technical Report CRG-TR-89-6, University of Toronto, December 1989.

[GH90a] James A. Gasbarro and Mark A. Horowitz. A single-chip, functional tester for VLSI circuits. In *IEEE International Solid-State Circuits Conference*, pages 84–85, 1990.

[GM83] N. Goncalves and H. J. De Man. NORA: A racefree dynamic CMOS technique for pipelined logic structures. *IEEE Journal of Solid-State Circuits*, SC-18:261–266, 1983.

[Gro76] Steven Grossberg. Adaptive pattern classification and universal recoding: Part I. Parallel development and coding of neural feature detectors. *Biological Cybernetics*, 23:121–134, 1976.

[Gro78] Steven Grossberg. A theory of visual coding, memory, and development. In *Formal Theories of Visual Perception*. Wiley, 1978.

[Ham90] Dan Hammerstrom. A VLSI architecture for high-performance, low-cost, on-chip learning. In *IJCNN International Joint Conference on Neural Networks*, pages II:537–544, 1990.

[HAN+88] Masahi Horiguchi, Masakazu Aoki, Yoshinobu Nakagome, Shin'ichi Ikenaga, and Katsuhiro Shimohigashi. An experimental large-capacity semiconductor file memory using 16-levels/cell storage. *IEEE Journal of Solid-State Circuits*, pages 27–33, February 1988.

[Heb49] Donald O. Hebb. *The Organization of Behavior*. Wiley, 1949.

[HFM+89] Hideto Hidaka, Kazuyasu Fujishima, Yoshio Matsuda, Mikio Asakura, and Tsutomu Yoshihara. Twisted bit-line architectures for multi-megabit DRAM's. *IEEE Journal of Solid-State Circuits*, pages 21–27, February 1989.

[HGDT84] L. G. Heller, W. R. Griffin, J. W. Davis, and N. G. Thoma. Cascode voltage switch logic: A differential CMOS logic family. In *IEEE International Solid-State Circuits Conference*, pages 16–17, 1984.

[HH76] Raymond A. Heald and David A. Hodges. Multilevel random-access memory using one transistor per cell. *IEEE Journal of Solid-State Circuits*, pages 519–528, August 1976.

[HH91] Jordan L. Holt and Jenq-Neng Hwang. Finite precisoin error analysis of neural network hardware. University of Washington, 1991.

[Hin89] Geoffrey E. Hinton. Deterministic Boltzmann learning performs steepest descent in weight-space. *Neural Computation*, 1:143–150, 1989.

[HJ83] David A. Hodges and Horace G. Jackson. *Analysis and Design of Digital Integrated Circuits*. McGraw-Hill, 1983.

[HJ87] Nils Hedenstierna and Kjell O. Jeppson. CMOS circuit speed and buffer optimization. *IEEE Transactions on Computer Aided Design*, CAD-6(2):270–281, March 1987.

[HJ91] John L. Hennessy and Norman F. Jouppi. Computer technology and architecture: An evolving interaction. *IEEE Computer Magazine*, pages 18–29, September 1991.

[HKM+90] Toshihiko Hirose, Hirotada Kuriyama, Shuji Murakami, Kojiro Yuzuriha, Takao Mukai, Kazuhito Tsutsumi, Yasumasa Nishimura, Yoshio Kohno, and Kenji Anami. A 20ns 4Mb CMOS SRAM with hierarchical word decoding architecture. In *IEEE International Solid-State Circuits Conference*, pages 132–133, 1990.

[HKT+90] Shigeyuki Hayakawa, Masakazu Kakumu, Hideki Takeuchi, Katsuhiko Sato, Takayuki Ohtani, Takeshi Yoshida, Takeo Nakayama, Shigeru Morita, Masaaki Kinugawa, Kenji Maeguchi, Kiyofumi Ochii, Jun'ichi Matsunaga, Akira Aono, Kazuhiro Noguchi, and Tetsuya Asami. A 1uA retention 4Mb SRAM with a thin-film-transistor load cell. In *IEEE International Solid-State Circuits Conference*, pages 128–129, 1990.

[HL89] Dajen Huang and Weiping Li. On CMOS exclusive OR design. In *IEEE 32nd Midwest Symposium on Circuits and Systems*, 1989.

[HNN+87] Yoshihisa Harata, Yoshio Nakamura, Hiroshi Nagase, Mitsuharu Takigawa, and Naofumi Takagi. A high-speed multiplier using a redundant binary adder tree. *IEEE Journal of Solid-State Circuits*, pages 28–34, February 1987.

[HP89] John L. Hennessy and David A. Patterson. *Computer Architecture A Quantitative Approach*. Morgan Kaufmann Publishers, 1989.

[HSA84] G.E. Hinton, T.J. Sejnowski, and D.H. Ackley. Boltzmann Machines: Constraint satisfaction networks that learn. Technical Report CMU-CS-84-119, Carnegie-Mellon University, May 1984.

[IO89] Mary Jane Irwin and Robert Michael Owens. Design issues in digit serial signal processors. In *IEEE International Symposium on Circuits and Systems*, pages 441–444, 1989.

[JKS87] Y. Jiren, I. Karlsson, and C. Svensson. A true single phase clock dynamic CMOS circuit technique. *IEEE Journal of Solid-State Circuits*, SC-22:899–901, 1987.

[jLS89] Thu ji Lin and Henry Samueli. A CMOS bit-level pipelined implementation of an FIR $x/sin(x)$ predistortion digital filter. In *IEEE International Symposium on Circuits and Systems*, pages 351–354, 1989.

[Joh90] Robert R. Johnson. Multichip modules: Next-generation packages. *IEEE Spectrum*, pages 34–48, March 1990.

[JP87] Najmi T. Jarwala and D. E Pradhan. An easily testable architecture for multi-megabit RAMs. In *IEEE International Test Conference*, pages 750–758, 1987.

[KA92] Jeffrey G. Koller and William C. Athas. Adiabatic switching, low energy computing, and the physics of storing and erasing information. Technical report, USC Information Sciences Institute, August 1992.

[Kan83] Akira Kanuma. CMOS circuit optimization. *Solid-State Electronics*, pages 47–58, 1983.

[Kar88] Ingemar Karlsson. True single phase clock dynamic CMOS circuit technique. In *IEEE International Symposium on Circuits and Systems*, pages 475–478, 1988.

[KBD+90] Howard Kalter, John Barth, John Dilorenzo, Charles Drake, John Fifield, William Hovis, Gordon Kelley, Scott Lewis, John Nickel, Charles Stapper, and James Yankosky. A 50ns 16Mb DRAM with a 10ns data rate. In *IEEE International Solid-State Circuits Conference*, pages 232–233, 1990.

[KK88] Thomas F. Knight and Alexander Krymm. A self-terminating low-voltage swing CMOS output driver. *IEEE Journal of Solid-State Circuits*, pages 457–464, April 1988.

[KKHY88] Shoji Kawahito, Michitaka Kameyama, Tatsuo Higuchi, and Haruyasu Yamada. A 32x32-bit multiplier using multiple-valued MOS current-mode circuits. *IEEE Journal of Solid-State Circuits*, pages 124–132, February 1988.

[KLL82] R. H. Krambeck, C. M. Lee, and H. S. Law. High-speed compact circuits with CMOS. *IEEE Journal of Solid-State Circuits*, pages 614–619, June 1982.

[Koh77] Teuvo Kohonen. *Associative Memory*. Springer-Verlag, 1977.

[Koh84] Teuvo Kohonen. *Self-Organization and Associative Memory*. Springer-Verlag, 1984.

[KS88] Steven D. Kugelmass and Kenneth T. Steiglitz. A probabilistic model for clock skew. In *International Conference on Systolic Arrays*, pages 545–554, 1988.

[Kun85] H. T. Kung. Why systolic architectures. *IEEE Transactions on Computers*, 15(1):37–46, 1985.

[Kun88] S. Y. Kung. *VLSI Array Processors*. Prentice Hall, 1988.

[L+89] Ivan Linscott et al. The MCSA II - A broadband, high resolution 60 Mchannel spectrometer. In *23rd Asilomar Conference on Signals and Systems*, October 1989.

[Lag88] University of California, Berkeley. *LagerIV distribution 1.0 silicon assembly system manual*, June 1988.

[LB] Weiping Li and James B. Burr. Parallel multiplier accumulator using 4-2 adders. US patent pending, Application number 088,096, filing date August 21, 1987.

[LB86] L.R.Rabiner and B.H.Juang. An introduction to Hidden Markov Models. pages 4–16, January 1986.

[LB87] Weiping Li and James B. Burr. An 80 MHz Multiply Accumulator. Technical report, Stanford University, September 1987.

[LBP88] Weiping Li, James B. Burr, and Allen M. Peterson. A fully parallel VLSI implementation of distributed arithmetic. In *IEEE International Symposium on Circuits and Systems*, pages 1511–1515, June 1988.

[LBP92] Yen-Wen Lu, James B. Burr, and Allen M. Peterson. Permutation on the mesh with reconfigurable bus. submitted to IPPS93, August 1992.

[LBP93] Yen-Wen Lu, James B. Burr, and Allen M. Peterson. Permutation on the mesh with reconfigurable bus: Algorithms and practical considerations. 1993.

[LC91] P. Peggy Li and David W. Curkendall. Parallel three dimensional perspective rendering. Jet Propulsion Laboratory, Pasadena CA, 1991.

[LCH+88] Nicky C.C. Lu, Hu Chao, Wei Hwang, Walter Henkels, T. Rajeevakumar, Hussein Hanafi, Lewis Terman, and Robert Franch. A 20ns 512Kb DRAM with 83MHz page operation. In *IEEE International Solid-State Circuits Conference*, pages 240–241, 1988.

[Li88] Weiping Li. *The Block Z transform and applications to digital signal processing using distributed arithmetic and the Modified Fermat Number transform*. PhD thesis, Stanford University, January 1988.

[LM87] Edward A. Lee and David G. Messerschmitt. Pipeline interleaved programmable DSP's: Architecture. *IEEE Transactions on Acoustics, Speech, and Signal Processing*, pages 1320–1333, September 1987.

[LR91] P. Landman and J.M. Rabaey. Low-power system design techniques for speech coding. In *International Conference on Microelectronics*, 1991.

[LS93] Dake Liu and Christer Svensson. Trading speed for low power by choice of supply and threshold voltages. *IEEE Journal of Solid-State Circuits*, pages 10–17, January 1993.

[LTC90] Weiping Li, D. W. Tufts, and H. Chen. VLSI implementation of a high resolution spectral estimation algorithm. In *VLSI Signal Processing. IV*, November 1990.

[Lu88a] Shih-Lien Lu. A safe single-phase clocking scheme for CMOS circuits. *IEEE Journal of Solid-State Circuits*, 23(1):280–283, February 1988.

[Lu88b] Shih-Lien Lu. Implementation of iterative networks with CMOS differential logic. *IEEE Journal of Solid-State Circuits*, pages 1013–1017, August 1988.

[LV83] W. K. Luk and J. E. Vuillemin. Recursive implementations of optimal time VLSI integer multipliers. In F. Anceau and E. J. Aas, editors, *VLSI*, pages 155–168. Elsevier Science Publishers B. V. (North Holland), 1983.

[LW88] Kevin J. Lang and Michael J. Witbrock. Learning to tell two spirals apart. In *Connectionist Models Summer School*, pages 52–59, 1988.

[MBS+92] Michael Murray, James B. Burr, David G. Stork, Ming-Tak Leung, Kan Boonyanit, Gregory J. Wolff, and Allen M. Peterson. Scalable deterministic Boltzmann machine VLSI can be scaled using multichip modules. In *Application Specific Array Processors*, pages 206–217, 1992.

[MC80] Carver Mead and Lynn Conway. *Introduction to VLSI Systems*. Addison-Wesley, 1980.

[MG86] David P. Marple and Abbas El Gamal. Area-delay optimization of programmable logic arrays. In *Conference on Advance Research in VLSI*, pages 171–194, 1986.

[Mil88] Veljko Milutinovic. GaAs realization of a RISC processor. In *IREE 7th Australian Microelectronics Conference*, pages 115–116, 1988.

[MK86] Richard S. Muller and Theodore I. Kamins. *Device Electronics for Integrated Circuits*. Wiley, 1986.

[MKO+90] Yashuhiko Maki, Shinnosuke Kamata, Yoshinori Okajima, Tsunenori Yamauchi, and Hiroyuki Fukuma. A 6.5ns 1 Mb BiCMOS ECL SRAM. In *IEEE International Solid-State Circuits Conference*, pages 136–137, 1990.

[Mor90] Nelson Morgan, editor. *Artificial Neural Networks: Electronic Implementations.* Computer Society Press, 1990.

[MP69] M. Minsky and S. Papert. *Perceptrons.* MIT Press, 1969.

[MP81] P. Mars and W. J. Poppelbaum. *Stochastic and deterministic averaging processors.* IEE, 1981.

[MPKRQ88] R. Miller, V.K. Prasanna-Kumar, D. Reisis, and Q.F.Stout. Meshes with reconfigurable buses. In *Advanced Research in VLSI. Proceedings of the Fifth MIT Conference*, pages 163–178, March 1988.

[MR79] Carver A. Mead and Martin Rem. Cost and performance of VLSI computing structures. *IEEE Transactions on Electron Devices*, pages 533–540, April 1979.

[MW79] Timothy C. May and Murray H. Woods. Alpha-particle-induced soft errors in dynamic memories. *IEEE Transactions on Electron Devices*, pages 2–9, January 1979.

[MW85] C. Mead and J. Wawrzynek. A new discipline for CMOS design: An architecture for sound synthesis. In *Chapel Hill Conference in VLSI*, pages 87–104, 1985.

[Nag75] L. W. Nagel. SPICE2: A computer program to simulate semiconductor circuits. Technical report, University of California, Berkeley, May 9 1975. Memo ERL-M520.

[ND89] Yang Ni and F. Devos. A 4-transistor static memory cell design with a standard CMOS process. In *IEEE International Symposium on Circuits and Systems*, pages 162–166, 1989.

[O⁺] J. K. Ousterhout et al. *1985 VLSI tools: More works by the original artists.* University of California, Berkeley. VLSI Tools Distribution.

[OP92] A.O. Ogunfunmi and A.M. Peterson. On the implementation of the frequency-domain lms adaptive filter. *IEEE Transactions on Circuits and Systems*, 39(5):318–322, May 1992.

[PA87] C. Peterson and J. R. Anderson. A mean field theory learning algorithm for neural networks. *Complex Systems*, 1:995–1019, 1987.

[Par89] Keshab K. Parhi. Nibble-serial arithmetic processor designs via unfolding. In *IEEE International Symposium on Circuits and Systems*, pages 635–640, 1989.

[PAW91] Aleksandra Pavasovic, Andreas G. Andreou, and Charles R. Westgate. Characterization of CMOS process variations by measuring subthreshold current. In R.E. Green and C.O. Rudd, editors, *Nondestructive Characterization of Materials IV*. Plenum Press, 1991.

[PC90] Scott R. Powell and Paul M. Chau. Estimating power dissipation in VLSI signal processing chips: the PFA technique. In *VLSI Signal Processing IV*, pages 251–259, 1990.

[PF88] M. D. Plumbley and F. Fallside. An information-theoretic approach to unsupervised connectionist models. In *Connectionist Models Summer School*, pages 239–245, 1988.

[Pfi84] James R. Pfiester. *Performance limits of CMOS very large scale integration*. PhD thesis, Stanford University, June 1984.

[PH89] Carsten Peterson and Eric Hartman. Explorations of the mean field theory learning algorithm. In *Neural Networks*, volume 2, pages 475–494. Pergamon Press, 1989.

[PK88] Dhiraj K. Pradhan and Nirmala R. Kamath. RTRAM: Reconfigurable and testable multi-bit RAM design. In *IEEE International Test Conference*, pages 263–278, 1988.

[PRB+88] Raymond Pinkham, Donald Russell, Anthony Balisteri, Thanh Nguyen, Troy Herndon, Dan Anderson, Aswin Mahta, H. Sakurai, Seishi Hatakoshi, and Andre Guillemaud. A 128K x 8 70 MHz Video RAM with auto register reload. In *IEEE International Solid-State Circuits Conference*, pages 236–237, 1988.

[PSS87] John H. Pasternak, Alex S. Shubat, and C. Andre T. Salama. CMOS Differential Pass-Transistor Logic design. *IEEE Journal of Solid-State Circuits*, pages 216–222, April 1987.

[QWHHGC88] John P. Quine, Harold F. Webster, II Homer H. Glascock, and Richard O. Carlson. Characterization of via connections in silicon circuit boards. *IEEE Transactions on Microwave Theory and Techniques*, pages 21–27, January 1988.

[Rab89] Lawrence R. Rabiner. A tutorial on Hidden Markov Models and selected applications in speech recognition. pages 200–257, February 1989.

[Ram92] Rambus. Rambus architectural overview. Technical Report DL0001-00, Rambus Inc., 1992.

[Rid79] V. Leo Rideout. One-device cells for dynamic random-access memories: A tutorial. *IEEE Transactions on Electron Devices*, pages 839–852, June 1979.

[Rit86] Lee W. Ritchey. Does surface mount make sense for commercial VLSI design. *VLSI Systems Design*, pages 70–71, February 1986.

[RMG86] David E. Rumelhart, James L. McClelland, and PDP Research Group, editors. *Parallel Distributed Processing, Volume 1: Foundations*. MIT Press, 1986.

[San89] Mark R. Santoro. *Design and Clocking of VLSI Multipliers*. PhD thesis, Stanford University, October 1989.

[SB81] R. S. Sutton and A. G. Barto. Toward a modern theory of adaptive networks: Expectation and prediction. *Psychological Review*, 88:135–170, 1981.

[SC91] Harold S. Stone and John Cocke. Computer architecture in the 1990s. *IEEE Computer Magazine*, pages 30–38, September 1991.

[Sej90] Terrance J. Sejnowski. private communication. December 1990.

[SGI89] Kentaro Shimizu, Eiichi Goto, and Shuichi Ichikawa. CPC (Cyclic Pipeline Computer) - an architecture suited for Josephson and Pipelined-Memory machines. *IEEE Transactions on Computers*, 38(6):825–832, June 1989.

[SH88] Mark Santoro and Mark Horowitz. A pipelined 64X64b iterative array multiplier. In *IEEE International Solid-State Circuits Conference*, pages 35–36, February 1988.

[SH89] Mark Santoro and Mark Horowitz. SPIM: A pipelined 64X64-bit iterative multiplier. *IEEE Journal of Solid-State Circuits*, pages 487–493, April 1989.

[Sho86] Masakazu Shoji. Elimination of process-dependent clock skew in CMOS VLSI. *IEEE Journal of Solid-State Circuits*, SC-21(5):875–880, October 1986.

[SHU+88] Katsuro Sasaki, Shoji Hanamura, Kiyotsugu Ueda, Takao Oono, Osamu Minato, Kotaro Nishimura, Yoshio Sakai, Satoshi Meguro, Masayoshi Tsunematsu, Toshiaki Masuhara, Masaaki Kubotera, and Hiroshi Toyoshima. A 15ns 1Mb CMOS SRAM. In *IEEE International Solid-State Circuits Conference*, pages 174–175, 1988.

[SIY+90] Katsuro Sasaki, Koichiro Ishibashi, Toshiaki Yamanaka, Katsuhiro Shimohigashi, Nobuyuki Moriwaki, Shigeru Honjo, Shuji Ikeda, Atsuyoshi Koike, Satoshi Meguro, and Osamu Minato. A 23ns 4Mb CMOS SRAM with 0.5uA standby current. In *IEEE International Solid-State Circuits Conference*, pages 130–131, 1990.

[SM88] Roy E. Scheuerlein and James D. Meindl. Offset word-line architecture for scaling DRAM's to the gigabit level. *IEEE Journal of Solid-State Circuits*, pages 41–47, February 1988.

[Smi81] Burton J. Smith. Architecture and applications of the HEP multiprocessor computer system. In *SPIE, Real-Time Signal Processing IV*, pages 241–248, 1981.

[Smi88] Burton J. Smith. The Horizon Supercomputer. *Supercomputing*, October 1988.

[SMM87] M. Sivilotti, M. Mahowald, and C. Mead. Realtime visual computations using analog CMOS processing arrays. In *Proceedings of the 1987 Stanford Conference*, 1987.

[SOA73] Y. Suzuki, K. Odagawa, and T. Abe. Clocked CMOS calculator circuitry. *IEEE Journal of Solid-State Circuits*, SC-8:462–469, 1973.

[Spe67] Donald F. Specht. Generation of polynomial discriminant functions for pattern recognition. *IEEE Transactions on Electronic Computers*, pages 308–319, June 1967.

[SR86] Terrance J. Sejnowski and Charles R. Rosenberg. NETtalk: A parallel network that learns to read aloud. Technical Report JHU/EECS-86/01, Johns Hopkins University, 1986.

[ST83] T. Sakurai and K. Tamaru. Simple formulas for two- and three-dimensional capacitances. *IEEE Transactions on Electron Devices*, pages 183–185, February 1983.

[Sve86] C. Svensson. Signal resynchronization in VLSI systems. *Integration*, 4:75–80, 1986.

[SW78] D. T. Shen and A. Weinberger. 4-2 carry-save adder implementation using send circuits. In *IBM Technical Disclosure Bulletin*, pages 3594–3597, February 1978.

[SW88] William K. Stewart and Stephen A. Ward. A solution to a special case of the Synchronization Problem. *IEEE Transactions on Computers*, 37(1):123–125, January 1988.

[Swa74] Richard M. Swanson. *Complementary MOS transistors in micropower circuits*. PhD thesis, Stanford University, December 1974.

[Swa89] Earl E. Swartzlander. *Waferscale Integration*. Kluwer, 1989.

[Sze83] S. M. Sze, editor. *VLSI Technology*. McGraw-Hill, 1983.

[Ter82] Chris J. Terman. User's guide to NET, PRESIM, and RNL/NL. Technical Report VLSI 82-112, Massachusetts Institute of Technology, 1982.

[TH88] Stuart K. Tewksbury and L. A. Hornak. Communication network issues and high-density interconnects in large-scale distributed computing systems. *IEEE Journal on Selected Areas in Communications*, pages 587–609, apr 1988.

[TMO+88] Nobuo Tamba, Shuuichi Miyaoka, Masanori Odaka, Katsumi Ogiue, Kouichirou Tamada, Takahide Ikeda, Mitsuru Hirao, Hisayuki Higuchi, and Hideaki Uchida. An 8ns 256K BiCMOS RAM. In *IEEE International Solid-State Circuits Conference*, pages 184–185, 1988.

[TNT+90] Masahide Takada, Kazuyuki Nakamura, Toshio Takeshima, Kouichirou Furuta, Tohru Yamazaki, Kiyotaka Imai, Susumu Ohi, Yumi Fukuda, Yukio Minato, and Hisamitsu Kimoto. A 5ns 1Mb ECL BiCMOS SRAM. In *IEEE International Solid-State Circuits Conference*, pages 138–139, 1990.

[TOO+90] Kenji Tsuchida, Yukihito Oowaki, Masako Ohta, Daisaburo Takashima, Shigeyoshi Watanabe, Kazunori Ohuchi, and Fujio Masuoka. The stabilized reference-line (SRL) technique for scaled DRAM's. *IEEE Journal of Solid-State Circuits*, pages 24–29, February 1990.

[TTS+88] Toshio Takeshima, Masahide Takada, Toshiyuki Shimizu, Takuya Katoh, and Mitsuru Sakamoto. Voltage limiters for DRAM's with substrate-plate-electrode memory cells. *IEEE Journal of Solid-State Circuits*, pages 48–52, February 1988.

[TYW+90] Leilani Tamura, Tsen-Shau Yang, Drew Wingard, Mark Horowitz, and Bruce Wooley. A 4-ns BiCMOS translation-lookaside buffer. In *IEEE International Solid-State Circuits Conference*, pages 66–67, 1990.

[vdM73] C. von der Malsberg. Self-organizing of orientation sensitive cells in the striate cortex. *Kybernetik*, 14:85–100, 1973.

[Vit85] Eric A. Vittoz. Micropower techniques. In Y. Tsividis and P. Antognetti, editors, *Design of MOS VLSI Circuits for Telecommunications*. Prentice-Hall, 1985.

[Vog88] T. P. Vogl. Accelerating the convergence of the back-propagation method. In *Biological Cybernetics*, volume 59, pages 257–263, 1988.

[Vui83] Jean Vuillemin. A very fast multiplication algorithm for VLSI implementation. *Integration, the VLSI Journal*, pages 39–52, 1983.

[Wag88] Kenneth D. Wagner. Clock system design. *IEEE Design and Test of Computers*, pages 9–27, October 1988.

[Wal86] L. Waller. How MOSIS will slash the cost of IC prototyping. *Electronics*, 59(9), March 3 1986.

[Waw92] John Wawrzynek. A VLIW/SIMD microprocessor for artificial neural network computations. Banff workshop on neural networks hardware, March 1992.

[WE85] Neil Weste and Kamran Eshraghian. *Principles of CMOS VLSI Design - A Systems Perspective*. Addison-Wesley, 1985.

[Wei88] Carol Weiszmann, editor. *DARPA Neural Network Study, October 1987 - February 1988*. AFCEA International Press, 1988.

[Wer74] P. Werbos. *Beyond Regression: New T*. PhD thesis, Harvard, August 1974.

[WF82] Shlomo Waser and Michael J. Flynn. *Introduction to Arithmetic for Digital Systems Designers*. CBS College Publishing, 1982.

[WH60] B. Widrow and M. E. Hoff. Adaptive switching circuits. In *Wescon*, pages IV:96–104, 1960.

[Wol86] S. Wolfram, editor. *Theory and Applications of Cellular Automata.* World Scientific, 1986.

[Wor88] George A. Works. The creation of Delta: A new concept in ANS processing. In *IEEE International Conference on Neural Networks,* pages II:159–164, 1988.

[WS87] John P. Wade and Charles G. Sodini. Dynamic cross-coupled bit-line content addressable memory cell for high-density arrays. *IEEE Journal of Solid-State Circuits,* pages 119–121, February 1987.

[WSKK89] Takumi Watanabe, Yoshi Sugiyama, Toshio Kondo, and Yoshi-hiro Kitamura. Neural network simulations on a massively parallel cellular array processor: AAP-2. In *IJCNN International Joint Conference on Neural Networks,* pages II:155–161, 1989.

[WWT87] Chung-Yu Wu, Jinn-Shyan Wang, and Ming-Kai Tsai. The analysis and design of CMOS Multidrain Logic and Stacked Multidrain Logic. *IEEE Journal of Solid-State Circuits,* pages 47–56, February 1987.

[YKHT88] T. K. Yu, S. M. Kang, I. N. Hajj, and T. N. Trick. iEDISON: An interactive statistical design tool for MOS VLSI circuits. In *IEEE International Conference on Computer Aided Design,* pages 20–23, 1988.

[YMA+89] Moritoshi Yasunaga, Noboru Masuda, Mitsuo Asai, Minoru Yamada, Akira Masaki, and Yuzo Hirai. A wafer scale integration neural network utilizing completely digital circuits. In *IJCNN International Joint Conference on Neural Networks,* pages II:213–217, 1989.

[YMY+90] Moritoshi Yasunaga, Noboru Masuda, Masayoshi Yagyu, Mitsuo Asai, Minoru Yamada, and Akira Masaki. Design, fabrication and evaluation of a 5-inch wafer scale neural network LSI composed of 576 digital neurons. In *IJCNN International Joint Conference on Neural Networks,* pages II:527–535, 1990.

[YS89] Jiren Yuan and Christer Svensson. High-speed CMOS circuit technique. *IEEE Journal of Solid-State Circuits,* 24(1):62–70, February 1989.

[YTYY89] Alex Yuen, Peter Tsao, Patrick Yin, and Albert T. Yuen. A 32K ASIC synchronous RAM using a two-transistor basic cell. *IEEE Journal of Solid-State Circuits,* pages 57–61, February 1989.

[YYN+90] Kazuo Yano, Toshiaki Yamanaka, Takashi Nishida, Masayoshi Saito, Katsuhiro Shimohigashi, and Akihiro Shimizu. A 3.8-ns CMOS 16x16-b multiplier using Complementary Pass-Transistor Logic. *IEEE Journal of Solid-State Circuits,* pages 388–394, April 1990.

Recurrent Neural Networks for Adaptive Filtering

Simon Haykin

Communications Research Laboratory
McMaster University
Hamilton, Ontario, Canada L8S 4K1

I. INTRODUCTION

The term *filter* is often used to describe a device in the form of a piece of physical hardware or computer software that is applied to a set of noisy data for the purpose of extracting information about a prescribed quantity of interest. The noise may arise from a variety of factors. For example, the sensor used to gather the data may be noisy, or the data available for processing may represent the output of a noisy channel. In any event, the filter is used to perform three basic information-processing tasks:

- *Filtering*, which refers to the extraction of useful information at some time t by employing data measured up to and including time t.
- *Smoothing*, which differs from filtering in that information about the quantity of interest need not be available at time t, and data measured later than time t can in fact be used to obtain this information. Smoothing is expected to be more accurate than filtering, since it involves the use of more data than that available for filtering.
- *Prediction*, which is the forecasting side of information processing. The aim here is to derive information about the quantity of interest at time $t+\tau$ for some $\tau>0$ by using data up to and including time t.

Much has been written on the classical approach to the *linear optimum filtering problem*, assuming the availability of second-order statistical parameters (i.e., mean and covariance function) of the useful signal and unwanted additive noise. In this statistical approach, an *error signal* is usually defined between some desired response and the actual filter output, and the filter is designed to

Copyright © 1995 by Academic Press, Inc.
All rights of reproduction in any form reserved.

minimize a *cost function* defined as the mean-squared value of the error signal. For stationary inputs, the resulting optimum filter is commonly known as the *Wiener filter* in recognition of the pioneering work done by Norbert Wiener in the 1940s (Wiener, 1949). A more general solution to the linear optimum filtering problem is provided by the *Kalman filter*, so named after its originator (Kalman, 1960). Compared to the Wiener filter, the Kalman filter is a more powerful device, with a wide variety of engineering applications.

The design of Wiener filters and Kalman filters requires *a priori* knowledge of the second-order statistics of the data to be processed. These filters are therefore optimum in their individual ways only when the assumed statistical model of the data exactly matches the actual statistical characteristics of the input data. In most practical situations, however, we simply do not have the statistical parameters needed to design a Wiener filter or Kalman filter, in which case we have to resort to the use of an *adaptive filter*. By such a device we mean one that is self-designing in the sense that the adaptive filter relies for its operation on a *recursive algorithm*, which makes it possible for the filter to operate satisfactorily in an environment where knowledge of the relevant signal characteristics is not available. The algorithm starts from some prescribed *initial conditions* that represent complete ignorance about the environment; thereafter it proceeds to adjust the free parameters of the filter in a step-by-step fashion, such that after each step the filter becomes more knowledgeable about its environment (Haykin, 1991; Widrow and Stearns, 1985). Basically, the process involved in the parameter adjustments follows some form of *error-correction learning*, the purpose of which is to minimize the error signal in some statistical sense.

II. CLASSIFICATION OF ADAPTIVE FILTERS

An adaptive filter is formally defined as a *self-designing device with time-varying parameters that are adjusted recursively in accordance with the input data*. Consequently, an adaptive filter is in reality nonlinear in the sense that it does *not* obey the principle of superposition.

Adaptive filters may be classified in three different ways, depending on the feature of interest, as described next.

A. Linear Versus Nonlinear Adaptive Filters

Notwithstanding the inherently nonlinear behavior of adaptive filters, they are commonly classified into linear and nonlinear adaptive filters, depending on whether the basic computational units used in their construction are linear or not. Specifically, in a *linear adaptive filter* the estimate of a quantity of interest is computed at the filter output as a linear combination of the available set of observations applied to the filter input; because of its very nature, a linear adaptive filter involves a single computational unit for each output. The ubiquitous *least-mean-square (LMS) algorithm*, originated by Widrow and Hoff (1960), is an important example of linear adaptive filtering. Indeed, the LMS algorithm is the work horse of traditional forms of the ever-expanding field of adaptive signal processing. Another important example of linear adaptive filtering is the *recursive least-squares (RLS) algorithm*, which may be viewed as a special case of Kalman filtering. The LMS algorithm is "stochastic", providing an approximation to Wiener filtering formulated in accordance with the method of steepest descent. On the other hand, the RLS algorithm is "exact", providing a recursive solution to the linear filtering problem formulated in accordance with the method of least squares that goes back to Gauss. The LMS algorithm offers simplicity of implementation and a robust performance in tracking statistical variations of a nonstationary environment. The RLS algorithm, on the other hand, offers a faster rate of convergence (defined below) than the LMS algorithm at the expense of increased computational complexity; however, its tracking behavior is usually worse than that of the LMS algorithm. Various schemes, known collectively as *fast algorithms* (Haykin, 1991), have been devised to improve the computational efficiency of recursive least-squares estimation. In any event, a major limitation of all linear adaptive filtering algorithms is the inability to exploit higher-order statistics of the input data, which, in turn, restricts the scope of their practical applications.

In contrast, *nonlinear adaptive filters* involve the use of nonlinear computational elements, which make it possible to exploit the full information content of the input data. Examples of nonlinear adaptive filters include Volterra filters and neural networks. Naturally, the presence of nonlinearity makes it difficult to analyse mathematically the behavior of nonlinear adaptive filters in a way comparable to their linear counterpart.

B. Nonrecursive Versus Recursive Adaptive Filters

Another way of classifying adaptive filters is in terms of whether or not their physical construction involves any form *of feedback*. We thus speak of nonrecursive and recursive adaptive filters. A nonrecursive adaptive filter has finite memory, whereas a recursive adaptive filter has infinite memory that fades with time. A tapped-delay-line filter (i.e., a discrete-time filter with finite-duration impulse response) operating in accordance with the LMS algorithm is an example of a nonrecursive adaptive filter. On the other hand, an adaptive scheme using an infinite-duration impulse response (IIR) filter (Shynk, 1989) is an example of a recursive adaptive filter; in this latter example, the adaptive filter uses a single computational unit with feedback built into its design. Note also that, according to this classification, the term "recursive" does not refer to the algorithmic adjustment of free parameters of the filter, but rather, as already stated, to the presence of some form of feedback in the filter's physical construction.

C. Supervised Versus Unsupervised Adaptive Filters

Yet another way in which adaptive filters may be classified is in terms of how the desired response (needed for executing the error-correction learning process) is actually provided.

Specifically, we may speak of a *supervised adaptive filter*, the operation of which requires a "teacher" for supplying the desired response. Examples of supervised adaptive filtering tasks include the following:

- *Identification*, where the adaptive filter is used to provide a *model* that represents the best fit (in some statistical sense) to an unknown plant (e.g., control system). In this class of applications, the unknown plant and the adaptive filter share a common input, and the output of the plant provides the desired response for the adaptive filter; both the input and output represent histories over time.

- *Inverse modeling*, where the function of the adaptive filter is to provide an *inverse model* connected in cascade with the unknown plant; in this case, some practical mechanism has to be found to supply the desired response. *Adaptive equalization* of an unknown communication channel is an important example of inverse modeling. The objective here is to accommodate the highest possible rate of digital data transmission through the channel, subject to a specified reliability that is usually measured in terms of the *error rate* (i.e., average probability of symbol error). Typically, data transmission through the channel is limited by *intersymbol interference* (ISI) caused by dispersion in the communication system. Adaptive equalization provides a powerful method to control ISI and also to combat the effects of channel noise. Indeed, every modem (modulator-demodulator), designed to facilitate the transmission of computer data over a voice-grade telephone channel, employs some form of adaptive equalization (Qureshi, 1985).

- *Interference cancellation*, where the adaptive filter is used to suppress an unknown interference contained (alongside an information-bearing signal of interest) in a *primary signal* that serves as the desired response. The input to the adaptive filter is represented by a *reference signal* that is dominated essentially by a correlated version of the unknown interference. To achieve these functional requirements, an interference cancellation system uses two separate sensors: one sensor supplies the primary signal, and the other sensor is designed specifically to supply the reference signal. *Adaptive beamforming* (Compton, 1988) is an important example of interference cancellation; this operation is a form of spatial filtering, the purpose of which is to distinguish between the spatial properties of a target signal of interest and an interfering signal. Another example is that of *echo cancellation* (Sondhi and Berkley, 1980) used in long-distance telephone communications over a satellite channel; in this latter application, adaptive (temporal) filtering is used to synthesize a replica of the echo experienced on the satellite channel due to unavoidable impedance mismatch at the receiving point and then to subtract it from the received signal.

On the other hand, we may speak of an *unsupervised* or *self-organized* *adaptive filter*, in which the error-correction learning process proceeds without the need for a separate input (i.e., teacher) supplying the desired response.

Important examples of self-organized adaptive filtering tasks include the following:

- *Prediction*, where the function of the adaptive filter is to provide the optimum prediction (in some statistical sense) of the present value of an input signal, given a past record of the input. In this application, the present value of the input signal serves as the desired response.

- *Blind deconvolution*, the need for which arises in hands-free telephone, blind equalization, seismic deconvolution and image restoration, just to mention a few examples. The use of a *hands-free telephone* is severely limited by the barrel effect due to acoustic reverberation produced in the surrounding environment of the near-end talker; the adaptive cancellation of this reverberation is a blind deconvolution problem. In *blind equalization* the requirement is to develop the inverse model of an unknown channel, given only the signal measured at the channel output. The *seismic deconvolution* problem is complicated by the fact that the exact waveform of the actual excitation responsible for the generation of the received signal is usually unknown. In *image restoration* we have an unknown system that represents blurring effects caused by photographic or electronic imperfections or both. In all these situations, the system of interest is unknown and its input is inaccessible; hence, a precise knowledge of the actual signal applied to the input of the system is not available for processing. To perform blind deconvolution, we are however permitted to invoke some reasonable assumptions about the statistics of the input signal. Given this prior knowledge, we proceed to solve the blind deconvolution problem by designing an adaptive filter that operates on the (available) received signal to produce an output signal whose statistics match those of the (unobservable) original signal as closely as possible (Haykin, 1994c).

III. CONTINUOUS LEARNING

As mentioned previously, an adaptive filter undergoes a form of error-correction learning in the course of adjustments applied to its free parameters. More specifically, this learning process is of a *continuous* nature, which means that the adaptive filter learns continuously about its environment while the input signal is being processed; in other words, the learning process never stops. This form of learning should be carefully distinguished from the learning process that takes place in a certain class of neural networks, known as multilayer perceptrons trained with the back-propagation algorithm (Rumelhart and McClelland, 1986; Haykin, 1994a). In the latter case the neural network is first trained with a set of input-output examples representative of the environment in which it operates; when the training is completed the free parameters of the network are all *fixed*, and the network is then ready to undergo testing with input data not seen before. On the other hand, in continuous learning the free parameters of the filter undergo adjustments all the time.

There are two important issues involved in the study of continuous learning, namely, convergence and tracking; these two issues are considered in the sequel in turn.

(a) *Convergence.* Typically, the design of a learning algorithm includes a *learning-rate parameter*, which controls the adjustments applied to the free parameters of the adaptive filter from one iteration to the next. For the learning algorithm to be convergent, certain constraints are imposed on the value assigned to the learning-rate parameter. In particular, if the learning-rate parameter exceeds a critical value, the learning algorithm diverges (i.e., the adaptive filter becomes unstable). The most stringent criterion for convergence is usually *convergence in the mean square.* According to this criterion, the mean squared error, defined as the mean-squared value of the difference between the desired response and the actual output of the adaptive filter, should approach a constant value as the number of iterations approaches infinity.

Of course, for such a criterion to be valid, the stochastic process from which the input data are picked would have to be *stationary.* When operating in such an environment, the adaptive filter starts from an arbitrary point on the error performance surface (i.e., a multidimensional plot of the mean-squared error versus the free parameters of the filter), and then moves in a step-by-step manner towards a minimum point of the surface. In the case of a linear adaptive filter, the minimum point is usually the global minimum of the error performance surface. On the other hand, in a nonlinear adaptive filter the minimum point may be a local minimum or a global minimum.

The difference between the actual mean-squared error realized by the filter and the absolute minimum mean-squared error, expressed as a percentage of the latter quantity, is referred to as *percentage misadjustment.* Clearly, it is highly desirable to make the percentage misadjustment as small as possible.

The number of iterations required by the learning algorithm to reach a steady state is referred to as the *rate of convergence.* Ordinarily, the smaller we make the learning-rate parameter, the smaller the misadjustment becomes, and the better is the filtering accuracy. Unfortunately, this is usually attained at the expense of a relatively slow rate of convergence.

(b) *Tracking.* When the input signal is a sample function of a nonstationary process, as is frequently the case in practice, we have a more complicated situation on our hands. The error performance surface now executes a random motion of its own about some minimum point. Consequently, the adaptive filter has the task of not only finding a minimum point of the error performance surface but also tracking its motion with time. The important point to note is that the underlying mechanisms responsible for the convergence and tracking modes of operation of an adaptive filter are quite different. More specifically, a fast convergence performance does not necessarily guarantee a good tracking behavior.

IV. RECURRENT NEURAL NETWORKS

Our interest in this chapter is in a particular class of nonlinear adaptive filters using *recurrent neural networks*. A recurrent neural network is a network made up of "artificial" neurons with one or more feedback loops. The artificial neuron model of interest to us here consists of the following elements, as described in Fig. 1(a):

- A set of synapses characterized by *weights* that are fed by respective input signals.
- A summer that produces a linear combination of the input signals, as shown by

$$v_j = \sum_{i=0}^{p} w_{ji} x_i \qquad (1)$$

where w_{ji} denotes the weight of synapse i, and x_i is the corresponding input signal. The index j refers to the neuron in question. The particular weight w_{j0}, corresponding to a fixed input $x_0 = +1$, defines the bias applied to neuron j.

- A sigmoidal activation function that squashes the output signal v_j at the summer's output in accordance with the logistic function

$$y_j = \frac{1}{1 + \exp(-v_j)} \qquad (2)$$

or the hyperbolic tangent function

$$y_j = \tanh(v_j) \qquad (3)$$

The neuron model described in Fig. 1(a) is static (memoryless) in nature. A dynamic version of this model includes, in addition to the elements described in Fig. 1(a), the parallel combination of a resistor and a capacitor connected across the input terminals of the nonlinear element as shown in Fig. 1(b). This latter model, called an *additive model*, is basic to the operation of recurrent neural networks such as the continuous Hopfield network (Hopfield, 1984) and recurrent back-propagation learning (Pineda, 1989). We will confine our attention largely to the model described in Fig. 1(a); some notes pertaining to the additive model of Fig. 1(b) are presented in the Discussion in Section VIII.

Specifically, we are interested in the *real-time recurrent learning* (RTRL) *algorithm* (Williams and Zipser, 1989; McBride and Narendra, 1965). The network configuration for the algorithm is depicted in Fig. 2 for the example of a single output neuron and two hidden neurons; the input layer of the network shown in this example consists of three feedback inputs and four external inputs;

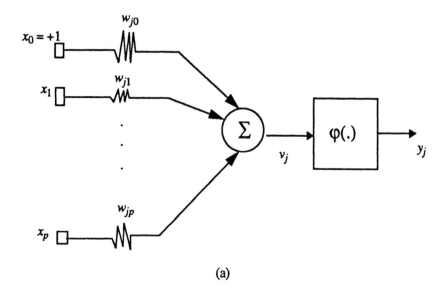

(a)

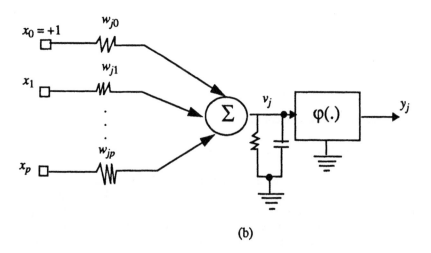

(b)

Figure 1: (a) Static, and (b) Dynamic models of artificial neurons

the blocks labeled z^{-1}, connected to the feedback paths, represent one-unit delay elements. The network is fully recurrent in that the output of each neuron in the network is fed back to the input. The abundant presence of feedback in the network gives it a distinctive dynamic behavior that naturally accounts for *time*, an essential dimension of learning. This dynamic behavior is completely different from that attained by the use of finite-duration impulse response (FIR) filters for the synaptic connections of a multilayer perceptron as described in Wan (1994).

Li (1992) has shown that a neural network trained with the RTRL algorithm can be a universal approximator of a differential trajectory on a compact time interval. Moreover, the RTRL algorithm does not require *a priori* knowledge of time dependences between the input data samples. However, a major limitation of the RTRL algorithm is that it is computationally intensive. In particular, the computational complexity in training grows as N^4, where N is the total number of neurons in the network. Accordingly, the computational requirement of the algorithm can be prohibitive when solving a difficult learning task (e.g., one-step prediction of a speech signal) that needs the use of a large N.

Catfolis (1993) describes an improved implementation of the RTRL algorithm that makes it possible to increase the performance of the algorithm during the training phase by exploiting some *a priori* knowledge about the time necessities of the task at hand; the resulting reduction in computational complexity of the training phase enables the algorithm to handle more complex tasks.

Wu and Niranjan (1994) describe another simplification of the RTRL algorithm, as briefly summarized here. Let $y_j(n)$ denote the output of neuron j in the fully recurrent neural network at time (iteration) n. Define the gradient

$$\pi_{kl}^j(n) = \frac{\partial y_j(n)}{\partial w_{kl}(n)} \tag{4}$$

where $w_{kl}(n)$ is any synaptic weight in the network. In the RTRL algorithm as originally postulated by Williams and Zipser (1989), the calculation of gradient $\pi_{kl}^j(n)$ follows the recursive evolution process

$$\pi_{kl}^j(n+1) = \varphi'(v_j(n)) \left[\sum_{i \in B} w_{ji}(n) \, \pi_{kl}^j(n) + \delta_{kl} u_l(n) \right] \tag{5}$$

where $\varphi'(.)$ is the derivative of the activation function $\varphi(.)$ with respect to its argument, and δ_{kl} is a Kronecker delta; $u_l(n)$ is defined by

$$u_l = \begin{cases} x_l(n) & \text{if } l \in A \\ y_l(n) & \text{if } l \in B \end{cases}, \tag{6}$$

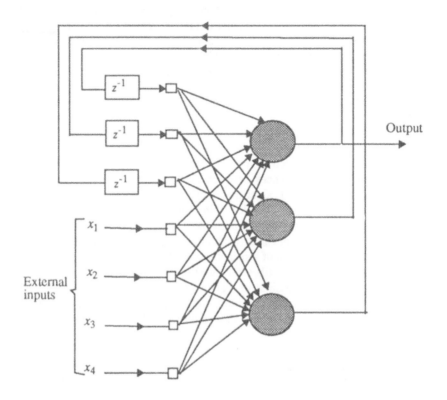

Figure 2: Real-time recurrent leaving network

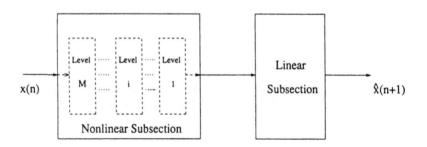

Figure 3: Pipelined nonlinear adaptive predictor

where A refers to the set of input nodes and B refers to the set of output neurons. Wu and Niranjan (1994) have proposed a simplification of the RTRL algorithm by replacing Eq. (5) with the following

$$\pi_{kl}^{j}(n+1) = \alpha_{jk}(n)\, z_{kl}(n-1) + \delta_{jk} z_{kl}(n) \tag{7}$$

where

$$\alpha_{jk}(n) = \varphi'(v_j(n))\, w_{jk}(n) \tag{8}$$

$$z_{kl}(n) = \varphi'(v_k(n))\, u_l(n) \tag{9}$$

Thus, the summation over all recurrent neurons in Eq. (5) vanishes, thereby saving a factor N in the order of computation requirement. Wu and Niranjan (1994) present simulation results using speech signals that demonstrate that their simplified RTRL algorithm can learn essentially the same tasks as the conventional RTRL algorithm.

In the next section we present a pipelined modification of the RTRL algorithm that is motivated by neurobiological considerations, the use of which can make a dramatic reduction in computational complexity when learning difficult tasks; the reduction in complexity is significantly greater than that achieved by any other technique known to this author. The first version of this modification was described by Haykin and Li (1993). The new structure, called a *pipelined recurrent neural network (PRNN)*, is particularly well suited for the one-step nonlinear prediction of nonstationary signals, nonlinear adaptive equalization, the identification of unknown nonlinear dynamic systems, and adaptive noise cancellation. In the next section, we consider the case of one-step prediction.

V. THE PIPELINED ADAPTIVE PREDICTOR

Construction

Figure 3 shows a block diagram of a pipelined nonlinear adaptive predictor, consisting of a nonlinear subsection with many levels of recurrent processing, followed by a linear subsection. The nonlinear subsection performs a "global" mapping by virtue of feedback built into its design, and the linear subsection fine-tunes the final result by performing a "local" mapping.

A detailed structure of the nonlinear subsection is shown in Fig. 4(a), involving a total of M levels of processing. Each level has a module and a comparator of its own. Every module consists of a recurrent neural network, which has N neurons. Figure 4(b) shows the detailed structure of module i. In addition to the p external inputs, there are N feedback inputs. To accommodate a bias for each neuron, besides the $p + N$ inputs, we include one input whose value is always +1. For each module, a total of N - 1 outputs are fed back to its input,

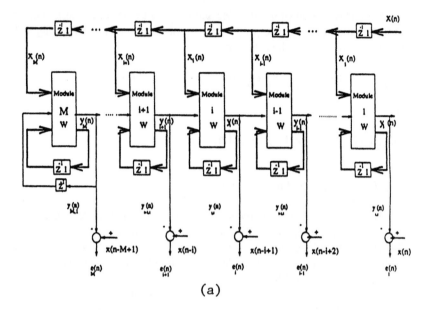

(a)

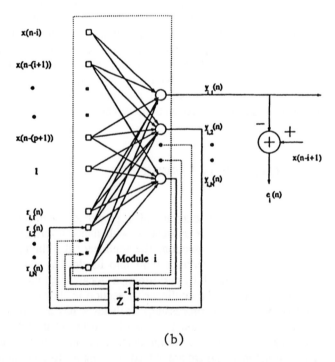

(b)

Figure 4: (a) Nonlinear subsection: a pipelined neural network
(b) The construction of Level i

and the remaining output is applied directly to the next module. In the case of module M, a one-unit delayed version of the module's output is also fed back to the input. Thus, all the modules operate similarly in that they all have the same number of external inputs and feedback nodes. Note that the timing of the external input vector

$$\mathbf{x}_i(n) = \mathbf{x}(n-i) \tag{10}$$

$$= [x(n-i), x(n-i-1), ..., x(i-p+1)]^T$$

applied to module i coincides with the timing of the output $y_{i,1}(n\text{-}i)$ computed by the next module $i+1$ in the chain. Moreover, note that all the modules are designed to have exactly the *same* synaptic weight matrix.

An overall cost function for the pipelined recurrent network of Fig. 4(a) is defined by

$$\varepsilon(n) = \sum_{i=1}^{M} \lambda^{i-1} e_i^2(n) \tag{11}$$

where the prediction error $e_i(n)$ is defined by

$$e_i(n) = x(n-i+1) - \hat{y}_{i,1}(n)$$

and λ is an exponential forgetting factor that lies in the range $(0 < \lambda \le 1)$. The term λ^{i-1} is, roughly speaking, a measure of the memory of the pipelined current neural network. By minimizing the cost function $\varepsilon(n)$, the real-time supervised learning algorithm is used to calculate the change $\Delta \mathbf{W}$ to the weight matrix $\mathbf{W}$ along the negative of the gradient of $\varepsilon(n)$ with respect to $\mathbf{W}$.

Turning next to the linear subsection, it consists of a tapped-delay-line filter whose tap inputs are defined by the present output $y_{1,1}(n)$ of the nonlinear subsection and q-1 past value $y_{1,1}(n\text{-}1)$, $y_{1,1}(n\text{-}2)$,...,$y_{1,1}(n\text{-}q+1)$. With the desired response defined as the input $x(n + 1)$ one step into the future, the tap weights of the linear subsection are adjusted in accordance with the LMS algorithm.

For details of the algorithm used to design the nonlinear and linear subsections of the pipelined adaptive predictor, the reader is referred to the Appendix at the end of the chapter.

Neurobiological Considerations

The design of the pipelined neural network structure follows an important engineering principle, namely, the *principle of divide and conquer*:

To solve a complex problem, break it into a set of simpler problems.

According to Van Essen et al. [1992], this same principle is also reflected in the design of the brain, as follows:

- Separate modules are created for different subtasks, permitting the neural architecture to be optimized for particular types of computation.
- The same module is replicated several times over.
- A coordinated and efficient flow of information is maintained between the modules.

The importance of modularity as an important principle of learning is also stressed by Houk (1992).

In a rather loose sense, all three elements of the principle of divide and conquer, viewed in a biological context also feature in the pipelined recurrent neural network (PRNN) of Fig. 4, as described here:

- The PRNN is composed of M separate modules, each of which is designed to perform a one-step nonlinear prediction on an appropriately delayed version of the input signal vector.
- The modules are identical, each designed as a recurrent neural network with a single output neuron.
- Information flow into and out of the modules proceeds in a synchronized fashion.

How to deal with multiple time series

In some applications, we have to deal with the analysis of multiple time series that are correlated with each other. For example, in stock market data analysis we have different time series representing the highest, lowest, and closing daily values of a particular stock. Naturally, these different time series are correlated with each other, and the issue of interest is how to use them all to make a prediction of the closing value of the stock in question for the following day. Such a problem in one-step prediction is one of "data fusion".

We may readily deal with this problem by expanding the input layer and therefore the set of synaptic weights corresponding to each module in the pipelined recurrent neural network in the manner shown in Fig. 5 pertaining to module i. Specifically, the external input applied to module i consists of the p-by-1 vectors $x_{1,i}(n)$, $x_{2,i}(n)$,...,$x_{g,i}(n)$, where g is the total number of time series to be considered. Assuming that the vector $x_{1,i}(n)$ (whose elements are denoted by $x_1(n-i)$, $x_1(n-i-1)$,...,$x_1(n-i-p)$) refers to the principal time series of interest, module i is designed to make a prediction of $x_1(n - i + 1)$. The cost function for the pipelined recurrent structure is essentially unchanged, retaining the mathematical form described in Eq. (11) except for a minor change in notation. Basically, the synaptic weights of module i are thus adjusted in the same way as before.

Virtues of the Pipelined Recurrent Neural Network

The pipelined recurrent neural network (PRNN) described herein offers the following features, with positive consequences of their own:

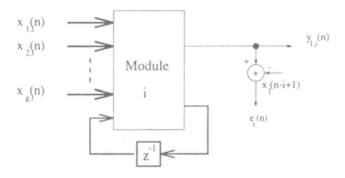

Figure 5: A Module i with g multiple input signal vectors

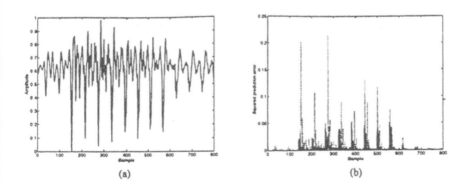

Figure 6: (a) A male speech signal (800 samples)
(b) Resulting squared prediction errors using the linear
and nonlinear adaptive predictors

1. *Improved Stability (Convergence).* According to Atiya (1988), the necessary
    condition for a recurrent neural network of any kind to converge to a unique
    fixed-point attractor is to satisfy the condition

    $$\| \mathbf{w} \|^2 < \frac{1}{(\max|\varphi'|)^2} \qquad (12)$$

    where φ' is the derivative of the nonlinear activation function with respect
    to its argument, and $\| \mathbf{w} \|$ is the Euclidean norm of the full synaptic weight
    matrix of the network. The latter quantity is defined by

    $$\| \mathbf{w} \| = \left(\sum_j \sum_i w_{ji}^2 \right)^{1/2} \qquad (13)$$

    where w_{ji} denotes the synaptic weight of neuron j connected to neuron i. In
    Eq. (12) it is assumed that all the neurons use the same kind of activation
    function $\varphi(.)$ which is the most common case encountered in practice.
    Typically, the PRNN uses two or three neurons per module, which is much
    smaller than the corresponding number of neurons in a single recurrent
    network for solving a difficult signal processing task. This means that the
    Euclidean weight matrix $\mathbf{W}$ of each module in the PRNN is usually smaller
    than the corresponding value for a single recurrent network of comparable
    performance. Accordingly, for a given activation function $\varphi(.)$, the
    PRNN is more likely to satisfy the stability criterion of Eq. (12) than a single
    recurrent neural network of comparable performance in the same number of
    neurons.

2. *Weight Sharing.* Another important virtue of the PRNN is that all of its
    modules share exactly the same synaptic weight matrix $\mathbf{W}$. Thus, with M
    modules and N neurons per module, the computational complexity of the
    PRNN is $O(MN^4)$. On the other hand, a conventional recurrent structure of
    training the same size has MN neurons, and its computational complexity is
    therefore $O(M^4N^4)$. Clearly, for $M > 1$ learning algorithm of the PRNN has a
    significantly reduced computational complexity compared to RTRL
    algorithm of a single recurrent neural network of the same size; the
    reduction in computational complexity becomes more pronounced with
    increasing M.

3. *Nested Nonlinearity.* The overall input-output relation of the PRNN exhibits
    a form of nested nonlinearity, similar to that found in a multilayer
    perceptron. Specifically, in the case of the PRNN depicted in Fig. 4, we may
    write

$$y_{1,1}\,(n) = \varphi(\mathbf{W},\mathbf{x}\,(n\text{-}1)\,,...,\varphi(\mathbf{W},\mathbf{x}\,(n\text{-}M)\,,y_{M,1}\,(n\text{-}1)\,)...) \tag{14}$$

This characteristic has the beneficial effect of enhancing the computing power of the PRNN.

4. *Overlapping Data Windows.* From Eq. (14) it is readily apparent that the individual modules of the PRNN operate on data windows that overlap with each other by one sampling period. Specifically, at each iteration the PRNN processes a total of $p + M$-1 input samples, where p is the order of each vector $\mathbf{x}$ and M is the number of modules. On the other hand, a single recurrent network processes only p input samples at each iteration. It follows therefore that the amount of information processed by the PRNN at each time instant may exceed the corresponding value pertaining to a single recurrent structure, depending on the correlation properties of the input data.

5. *Smoothed Cost Function.* The cost function of the PRNN is defined as the *exponentially weighted sum of squared estimation errors* computed by the individual modules, as shown in Eq. (11). Hence, it may be argued that, in comparison with the conventional RTRL algorithm, the cost function $\varepsilon\,(n)$ so defined is "closer" in form to that used in deriving the more elaborate steepest descent version of the learning algorithm originally formulated for recurrent neural networks by Williams and Zipser (1989).

6. *Data Fusion.* A multiplicity of time series pertaining to a particular phenomenon of interest can be processed simultaneously simply by expanding the size of the input layer as described in the previous subsection, thereby exploiting the correlation that may exist among the various time series. Of course, this same remark also applies to a conventional recurrent network.

VI. ONE-STEP PREDICTION OF A SPEECH SIGNAL

In this section, we illustrate the application of the real-time nonlinear adaptive prediction described herein to the important case of a male speech signal: **when recording audio data....** The recorded time series corresponding to this speech signal, sampled at 8 kHz, is made up of 10,000 points.

The pipelined nonlinear predictor used for this speech signal has the following parameters.

- Number of modules, $M = 5$.
- Number of neurons per module, $N = 2$.
- Number of external inputs, $p = 4$.
- Length of tapped-delay-line filter, $q = 12$.
- Exponential forgetting factor, $\lambda = 0.9$.

Figure 6(a) displays the waveform of the actual speech signal (800 samples). The solid curve in Fig. 6(b) shows the squared prediction error using the pipelined nonlinear predictor, and the dotted curve shows the corresponding results obtained using a 12-tap linear predictor operating in accordance with the LMS algorithm. The results presented in Fig. 6(b) clearly shows that the prediction of a speech signal using a nonlinear adaptive predictor provides a much better approximation to the actual speech signal than the linear predictor.

For a quantitative evaluation of prediction performance, we may use the following gain measure expressed in decibels:

$$R_p = 10 \, \log_{10} (\delta_s^2 / \delta_p^2) \tag{15}$$

where δ_s^2 is the average-mean-squared value of the speech signal at the transmitter input, and δ_p^2 is the corresponding value of the prediction error at the output of the predictor. For 10,000 speech samples, by using the PRNN-based nonlinear predictor R_p is about 25.06 dB; by using the linear predictor, R_p is about 22.01 dB. This provides a quantitative demonstration of the superior performance of the nonlinear predictor over the linear predictor.

Figure 7(a) shows the power spectrum of the resulting prediction errors. The power spectral density is fairly constant across a band from 200 Hz to 3000 Hz. The solid line presents the average power spectrum on each frequency point. The two dashed lines correspond to the 95 percent confidence range. Moreover, Figure 7(b) shows the histogram of the prediction error, which is found to have a β distribution. The conclusion to be drawn from these observations is that the (nonlinear) prediction error may be closely modeled as a white and approximately Gaussian noise process. This is testimony to the efficiency of the neural network-based predictor in extracting the information content of the speech signal almost fully.

Finally, to demonstrate the computational efficiency of the PRNN as a one-step predictor, we computed the squared prediction error for the same speech signal of Fig. 6(a), this time however using the conventional RTRL algorithm for the following network parameters:

- Number of neurons, $N = 10$
- Number of external inputs, $p = 4$
- Length of tapped-delay line filter, $q = 12$
- Number of modules, $M = 1$

The result of this latter computation is shown as the dotted curve in Fig. 8. This result is slightly worse than that obtained by using the PRNN, which is represented by the solid curve in Fig. 8. Most importantly, the computational complexity of the RTRL algorithm is, for this example, on the order of $N^4 = 10^4$. This is at least two orders of magnitude greater than the corresponding computational complexity of training the PRNN, namely, $M(N^4) = 5 \times 2^4 = 80$.

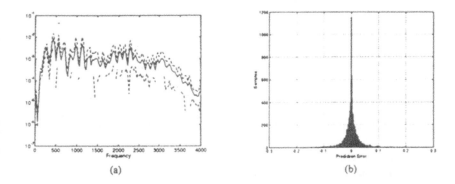

(a)   (b)

Figure 7: (a) Power spectrum of overall nonlinear prediction errors.
(b) Histogram of overall nonlinear prediction errors

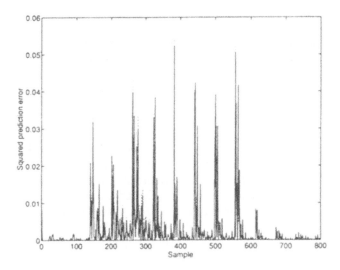

Figure 8: Comparison of prediction errors with a larger size recurrent
network (---) and a pipelined network (—)

VII. APPLICATIONS

A fully recurrent neural network is endowed with a dynamic behavior, which makes it a powerful adaptive signal processing device for a variety of applications that involve time as an essential dimension of learning. Applications of recurrent neural networks reported in the literature include neurobiological modeling (Anastasio, 1991), linguistic tasks such as grammatical inference (Giles et al., 1992; Zeng et al., 1994), phoneme probability estimation in large vocabulary speech recognition (Robinson, 1994), neurocontrol of nonlinear dynamical systems (Puskorius and Feldkamp, 1994), adaptive differential pulse-code modulation of speech signals (Haykin and Li, 1993), code-excited nonlinear predictive coding of speech signals (Wu et al., 1993), and adaptive equalization of communication channels (Kechriotis et al., 1994). In what follows we present highlights of the latter three applications.

Coding Speech at Low-bit Rates

A straightforward method of digitizing speech signals for transmission over a communication channel is through the use of pulse-code modulation (PCM), which operates at the standard rate of 64 kb/s. This high data rate demands a higher channel bandwidth for its implementation. However, in certain applications (e.g., secure communication over radio channels that are of low capacity), channel bandwidth is at a premium. In applications of this kind, there is a definite need for speech coding at low bit rates, while maintaining an acceptable fidelity or quality of reproduction (Jayant and Noll, 1984). In this context, two coding methods for speech that come to mind are adaptive differential pulse-code modulation and coded-excited prediction.

An *adaptive differential pulse-code modulation* (ADPCM) system consists of an encoder and a decoder separated by a communication channel. The encoder, located in the transmitter, uses a feedback scheme that includes an adaptive quantizer at the forward path and an adaptive predictor in the feedback path, as depicted in Fig. 9(a). The predictor, acting on a quantized version of the incoming speech signal, produces a one-step prediction of this signal. The prediction error, defined as the difference between the actual speech signal and the one-step prediction so produced, is in turn quantized, thereby completing the feedback loop (Cutler, 1952; Haykin, 1994b). The encoded version of the quantized prediction error constitutes the transmitted signal. This transmitted signal represents a compressed version of the original speech signal by virtue of removing the redundant (i.e., predictable) portion of the speech signal. The decoder, located in the receiver, employs an exact replica of the adaptive predictor used in the transmitter, as depicted in Fig. 9(b). In the absence of noise, the reconstructed speech signal at the receiver output is exactly the same as the original speech signal, except for noise due to the quantization process in the transmitter. The challenge in designing an ADPCM system is to perform signal reconstruction without transmitting any side-information, that is, to ensure that the receiver merely requires the (quantized) prediction error for its operation; the configuration described in Fig.9 makes it possible to realize this challenge.

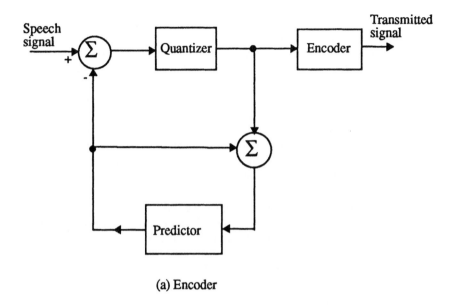

(a) Encoder

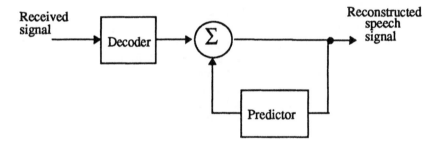

(b) Decoder

Figure 9: Adaptive differential pulse-code modulation system

The traditional form of ADPCM uses a linear adaptive predictor (Jayant and Noll, 1985). More specifically, in a 32 kb/s ADPCM system, accepted internationally as a standard coding technique for speech signals, the linear predictor consists of an infinite-duration impulse response filter whose transfer function has 6 zeros and 2 poles, and the free parameters which are adapted in accordance with a novel coefficient update algorithm that minimizes mistracking (Cointor, 1982). Yet it is widely recognized that a speech signal is the result of a dynamic process that is both nonlinear and nonstationary. It should not therefore be surprising to find that the use of a nonlinear predictor in the construction of an ADPCM system would provide a significant improvement in its performance. Indeed, this was first demonstrated by Haykin and Li (1993), using the PRNN-based prediction for the design of the nonlinear predictor. A more detailed study of the superior performance of an ADPCM using a PRNN-based predictor, compared to the AT&T version of the system using a linear predictor, is presented in a doctoral thesis by Li (1994). This latter study is supported by extensive quantitative evaluations and subjective tests.

Turning next to the *code-excited predictive coding* of speech signals, the preferred multipath search coding procedure is that of code-book coding. In general, code-book coding is impractical because of the large size of the code book needed. However, exhaustive search of the code book to find the optimum innovation (prediction-error) sequence for encoding short segments of the speech signal becomes possible at very low bit rates (Schroeder and Atal, 1985). The encoding part of the system uses a speech synthesizer that consists of two time-varying filters, each with a predictor in the feedback loop, as shown in Fig. 10. The first feedback loop includes a long-delay (pitch) predictor that generates the pitch period of the voiced speech, whereas the second feedback loop includes a short-delay predictor to restore the spectral envelope (Schroeder and Atal, 1985). In the case of *code-excited linear prediction* (CELP) for high-quality speech at very low bit rates, both of these predictors are linear. Wu and Niranjan (1994) have investigated the use of nonlinear predictors for code-excited predictive speech coding. Specifically, they used a recurrent neural network trained with a simplified version of the RTRL algorithm described earlier. Here also it is reported that the use of a nonlinear predictor results in improved speech coding performance.

Adaptive equalization of communication channels

As mentioned previously, adaptive equalization may be viewed as a form of inverse modeling, such that the adaptive equalizer connected in cascade with a communication channel of interest approximates an ideal distortionless transmission system (i.e., one with a constant amplitude response and a linear phase response). In the case of a linear communication channel (e.g., a telephone channel used for digital data transmission), the traditional form of an adaptive equalizer consists of a relatively long tapped-delay line filter whose tap-weights are adjusted in accordance with the LMS algorithm (Qureshi, 1985). During training, the desired response for the adaptive equalizer is provided by means of a pseudo-noise (PN) sequence generator, which is located in the receiver and

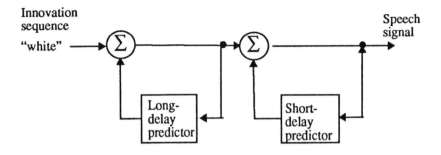

Figure 10: Code-excited predictive coding encoder

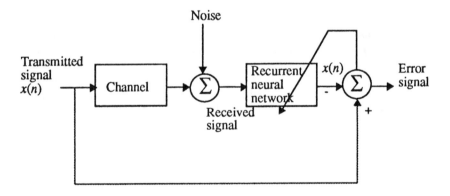

Figure 11: Recurrent neural network-based adaptive equalizer

which operates in synchronism with another PN sequence generator that supplies
the input data applied to the transmitter input, as depicted in Fig. 11.

In a recent paper, Kechriotis et al. (1994) have proposed recurrent neural
network (RNN)-based adaptive equalizers for both trained adaptation and blind
(self-organized) adaptation. The results presented therein may be summarized as
follows (Kechriotis et al., 1994):

• *Linear nonminimum-phase channels*. For a simple linear minimum-phase
  channel with a discrete transfer function $H(z)$ having a single zero at $z=-0.7$,
  where z^{-1} denotes the unit-delay element, the RNN-based nonlinear
  equalizer achieves a comparable or even smaller bit error rate (BER) than its
  linear counterpart (using the RLS algorithm). The slight improvement in
  performance is due to the fact that the impulse response of the RNN-based
  equalizer is infinitely long and therefore capable of inverse modeling the
  channel more accurately than is possible with the linear equalizer having a
  finite-duration impulse response. Moreover, the RNN-based equalizer can
  construct nonlinear decision regions, which is what is actually called for in
  the presence of channel noise (Chan et al., 1990).

• *Partial-response channels*. In a partial-response channel, an example of
  which is encountered in magnetic recording, the discrete transfer function
  $H(z)$ has zeros on the unit circle in the z-plane. In situations of this kind, it is
  not feasible to build a linear adaptive equalizer with a satisfactory
  performance. Kechriotis et al. (1994) considers the example of a partial-
  response channel whose discrete transfer function $H(z)$ has double zeros at
  $z=+1$. They show that a nonlinear equalizer, using a fully recurrent neural
  network with a single output neuron and a single hidden neuron, provides a
  bit error rate of 10^{-3} for a signal-to-noise ratio of 20 dB at the receiver input;
  such a performance is well beyond the capability of a linear adaptive
  equalizer.

• *Nonlinear channels*. When the communication channel is both dispersive
  and nonlinear, we naturally expect that a nonlinear adaptive equalizer would
  outperform a linear one. Kechriotis et al. (1994) present examples
  demonstrating this property.

The points made thus far have been in the context of a RNN-based
equalizer supplied with a desired response for trained adaptation. Kechriotis et al.
(1994) also considers the use of such an equalizer structure for the blind
equalization of a communication channel. Specifically, if perfect equalization is
attained, then the equalizer's output would have exactly the same statistical
moments as the data sequence applied to the transmitter input. It is assumed that
the latter sequence is drawn from an independent and identically distributed (iid)
zero-mean distribution; that is, the higher-order moments of the transmitter input
are known. Accordingly, the objective function is now defined by

$$\varepsilon(n+1) = \sum_{k=1}^{4} \alpha_k (E[x^k] - E_{n+1}[\hat{x}^k])^2 \qquad (16)$$

where $E[x^k]$, is the kth moment of the sequence applied to the transmitter input, and $E_{n+1}[\hat{x}^k]$ is the estimated kth moment of the equalizer output using its present and past sample values. The α_k are fixed positive constants. The RTRL algorithm is now formulated so as to minimize the objective function $\varepsilon(n+1)$. Kechriotis et al. (1994) present results that demonstrate the superiority of the performance of the RNN-based adaptive equalizer over a linear blind equalizer using the Godard criterion (Godard, 1980). The computer experiments described therein pertain to (1) a linear mixed-phase channel whose discrete transfer function $H(z)$ has a zero inside the unit circle and a zero outside the unit circle, and (2) a nonlinear channel.

The conclusions to be drawn from the series of computer-oriented experiments described by Kechriotis et al. (1994) are: a nonlinear adaptive equalizer using a recurrent neural network exhibits a performance comparable to that of a traditionally linear equalizer when operating over a linear channel with a relatively mild interference, and it outperforms the linear equalizer by orders of magnitude when the channel has spectral nulls or severe nonlinear distortion.

VIII. DISCUSSION

In this chapter we have identified a fully recurrent neural network as a nonlinear dynamical system that is ideally suited to adaptive filtering applications such as identification, equalization (inverse modeling), prediction, and noise cancellation. For the adjustment of the free parameters (i.e., synaptic weights and bias values) of the network, we may use the real-time recurrent learning algorithm. To improve computational efficiency of the algorithm in a significant way, we may use a pipelined structure with multiple recurrent levels of processing. The need for this latter approach arises when tackling computationally difficult tasks that may require the use of a large number of neurons.

The traditional form of the real-time recurrent learning algorithm, originally formulated by Williams and Zipser (1989), applies to real-valued data. In certain adaptive filtering applications (e.g., the adaptive equalization of a communication channel involving the use of M-ary phase-shift keying or M-ary quadrature-amplitude modulation), the baseband signal is complex valued. The free parameters of the recurrent neural network would then assume complex values too. To deal with situations of this kind, we may double the size of the input layer and that of the output layer to accommodate the fact that the input and output signals of the recurrent neural network consist of real and imaginary components. A more efficient procedure, however, is to adopt complex-valued synaptic weights for the characterization of the individual neurons in the network, and to use the complex version of the real-time recurrent learning algorithm for their computation. Such an algorithm is described by Kechriotis et al. (1994).

Another point of interest is that the conventional form of the RTRL algorithm is based on the memoryless model of a neuron depicted in Fig. 1(a). For a more elaborate version of the RTRL algorithm, we may use the additive (RC) model of a neuron shown in Fig. 1(b). It may be shown that this latter move has the effect of replacing the first-order recursion of Eq. (5), involving the gradient π^j_{kl}, with a second-order one. Such an increase in computational complexity may improve the tracking behavior of the RTRL algorithm. However, in some experiments carried out on the one-step prediction of speech signals, we have found that the improvement in tracking performance is not large enough to justify the additional computational complexity.

Finally, mention should be made of a paper by Bengio et al. (1994), in which three basic requirements are postulated for a parametric dynamical system to learn and store relevant state information:

1. That the system be able to store information for an arbitrary time duration.
2. That the system be resistant to noise.
3. That the system be trainable (i.e., its free parameters be computable) in a reasonable time duration.

Bengio et al. (1994) show that gradient-based learning algorithms (e.g., the RTRL algorithm) fail to meet the first two requirements and, consequently, face an increasingly difficult problem as the duration of dependences to be captured is increased. To design recurrent networks with improved memory retention, they propose alternative methods to standard gradient descent; such methods include simulated annealing, multi-grid random search, and time-weighted Pseudo-Newton optimization methods; the improvement is however attained at the cost of a much longer time to train the network. It may prove fruitful to examine the nonlinear adaptive prediction of speech signals and the nonlinear adaptive equalization of communication channels in light of the points made by Bengio et al. (1994).

ACKNOWLEDGEMENTS

The author is grateful to the Natural Sciences and Engineering Research Council of Canada for financial support, and to his Ph.D. student Liang Li for many stimulating discussions and for providing the experimental results reported in Figs. 6 to 8.

APPENDIX: Algorithm for on-line training of the PRNN

1. Real-time Adaptive Calculation of Nonlinear Subsection

Prior prediction processing

At the instant nth time point, using the input data, obtain the input vectors $x_1(n),...,x_M(n)$. The prior prediction is taken from Level M to Level 1. The output vector of Module i and the error signal of Level i are defined by, respectively,

$$y_i(n) = \varphi(\mathbf{W},\ \mathbf{x}_i(n), y_{i+1,1}(n-1))$$

$$e_i(n) = x(n-i+1) - y_{i,1}(n)$$

where $y_i(n)$ denotes the prior prediction output of the ith module during the training process, and $e_i(n)$ is the corresponding error signal. After every module of PRNN finishes its prior prediction calculation, a series of error signals are obtained, namely, $e_1(n), e_2(n),..., e_M(n)$.

Updating the weight vector of modules

An overall cost function for the pipelined recurrent neural network is defined by

$$\varepsilon(n) = \sum_{i=1}^{M} \lambda^{i-1} e_i^2(n) \tag{17}$$

By minimizing the cost function $\varepsilon(n)$, the gradient estimation algorithm is used to calculate the change $\Delta\mathbf{W}$ to the weight matrix $\mathbf{W}$ along the negative of the gradient of $\varepsilon(n)$ with respect to $\mathbf{W}$.

The change applied to the klth element in the weight matrix can be written as

$$\Delta w_{k,l} = -\eta \frac{\partial}{\partial w_{k,l}} \left(\sum_{i=1}^{M} \lambda^{i-1} e_i^2(n) \right) \tag{18}$$

where η is a fixed learning-rate parameter, $1 \le k \le N$ and $1 \le l \le (p+N+1)$. Finally, the weight matrix is updated as

$$\tag{19}$$

$$\mathbf{W} \leftarrow \mathbf{W} + \Delta\mathbf{W}$$

Filtering Process

Using the same input signal and updated weight matrix, the filtering calculation of a module is the same as the prior prediction. For example, the filtering process of module i is defined by

$$n) = \varphi((\mathbf{W}, \ x_i)(n), y_{i+1, 1}(n-1$$

(20)

Sequentially, the filtering operation is executed from module M to module 1. The output $y_{1,1}(n)$ of Module 1 is sent to the tapped-delay-line filter as an overall output $y(n)$.

2. Adaptive Operation of Linear Subsection

Outputting the optimal estimate

The standard least-mean-square algorithm is used so as to produce the optimum estimate of $\hat{x}(n+1)$:

$$\hat{y}(n+1) = \mathbf{w}_l^T \mathbf{y}(n)$$

(21)

where

$$\mathbf{y}(n) = [y_{1,1}(n), y_{1,1}(n\text{-}1), ..., y_{1,1}(n-q+1)]^T$$

(22)

Updating the weight of the linear filter

Input the new sample of signal $x(n + 1)$ as the desired signal. We define the estimation error or residual $e(n + 1)$ as the difference between the desired response $x(n + 1)$ and the estimated output, as shown by

$$e(n+1) = x(n+1) - \mathbf{w}_l^T \mathbf{y}(n)$$

(23)

The updated weight vector of the tapped-delay-line filter is written as:

$$\mathbf{w}_l \leftarrow \mathbf{w}_l + \mu \mathbf{y}(n) e(n+1)$$

(24)

where μ is the step-size parameter.

3. Recursive Calculation

Let $n = n + 1$ and return to step 1. Repeat the real-time adaptive prediction until the input data stops.

REFERENCES

Anastasio, T.J., 1991. "A recurrent neural network model of velocity storage in the vestibulo-ocular reflex". In *Advances in Neural Information Processing Systems 3* (R.P. Lippmann, J.E. Moody, and D.S. Touretzky, eds.), pp. 32-38. San Mateo, CA: Morgan Kaufmann.

Atiya, A.F., 1988. "Learning on a general network", In *Neural Information Processing Systems* (D.Z. Anderson, editor), pp. 22-30, American Institute of Physics.

Bengio, Y., P. Simard, and P. Frasconi, 1994. "Learning long-term dependences with gradient descent is difficult", *IEEE Transactions on Neural Networks* 5, pp. 157-166.

Catfolis, T., 1993. "A method for improving the real-time recurrent learning algorithms", *Neural Networks* 6, pp. 807-822.

Compton, R.T., Jr., 1988. "Adaptive Antennas: Concepts and Performance", Prentice-Hall.

Cutler, C.C., 1952. "Differential quantization of communication signals", United States Patent No. 2-505-361.

Giles, C.L., et al., 1992. "Learning and extracting finite state automate with second-order recurrent neural networks", *Neural Computation* 4, pp. 393-405.

Godard, D.M., 1980. "Self-recovering equalization and carrier tracking in a two-dimensional data communication system", *IEEE Transactions on Communications*, COM-28, pp. 1867-1875.

Haykin, S., 1991, *Adaptive Filter Theory*, Second Edition, Englewood Cliffs, NJ: Prentice-Hall.

Haykin, S., 1994a. *Neural Networks: A Comprehensive Foundation*, Macmillan.

Haykin, S., 1994b. *Communication Systems, Third Edition*, New York: Wiley.

Haykin, S., (editor) 1994c. *Blind Deconvolution*, Englewood Cliffs, NJ: Prentice-Hall.

Haykin, S., and L. Li (1993). "16 kb/s adaptive differential pulse code modulation of speech", In *Applications of Neural Networks to Telecommunications* (J. Alspector, R. Goodman, and T.X. Brown, eds.), pp. 132-137, Lawrence Erlbaum

Hopfield, J.J. (1984). "Neurons with graded response have collective computational properties like those of two-state neurons", *Proc. National Academy of Sciences, USA*, vol. 81, pp. 3088-3092.

Houk, J.C. (1992). "Learning in modular networks", In *Proceedings of the Seventh Yale Workshop on Adaptive and Learning Systems*, pp. 80-84. New Haven, CT: Yale University.

Jayant, N.S., and P. Noll, 1984. *Digital Coding of Waveforms*. Englewood Cliffs, NJ: Prentice Hall.

Kalman, R.E., 1960. "A new approach to linear filtering and prediction problems", Journal of Basic Engineering **82**, pp. 35-45.

Kechriotis, G., E. Zerves, and E.S. Manalakos, 1994. "Using recurrent neural networks for adaptive communication channel equalization", *IEEE Transactions on Neural Networks* **5**, pp. 267-278.

Li, L. (1994). "Nonlinear adaptive prediction of nonstationary signals and its application to speech communications", Ph.D. dissertation, McMaster University, Canada.

Li, L.K., 1992. "Approximation theory and recurrent networks", *International Joint Conference on Neural Networks*, vol, 2, pp. 266-271, Baltimore, MD.

McBride, L.E., Jr., and K.S. Narendra, 1965. "Optimization of time-varying systems", *IEEE Transactions on Automatic Control* **AC-10**, pp. 289-294.

Pineda, F.J. (1989). "Recurrent backpropagation and the dynamical approach to adaptive neural computation", *Neural Computation* **1**, pp. 161-172.

Puskorius, G.V., and L.A. Feldkamp, 1994. "Neurocontrol and nonlinear dynamical systems with Kalman-filter trained recurrent networks", *IEEE Transactions on Neural Networks* **5**, pp. 279-297.

Qureshi, S., 1985. "Adaptive equalization", *Proceedings of the IEEE* **73**, pp. 1349-1387.

Robinson, T., 1994. "An application of recurrent nets to phone probability estimation", *IEEE Transactions on Neural Networks* **5**, pp. 298-305.

Rumelhart, D.E., and J.L. McClelland, eds., 1986. *Parallel Distributed Processing: Explorations in the Microstructure of Cognition*, Vol. 1, Cambridge, MA: MIT Press.

Schroeder, M.R., and B.S. Atal, 1985. "Code-excited linear prediction (CELP): High-quality speech and very low bit rates", *International Conference on Acoustics, Speech, and Signal Processing*, pp. 1672-1676.

Schynk, J.J., 1989. "Adaptive IIR filtering", *IEEE ASSP Magazine*, **6**, pp. 4-21.

Sondhi, M.M., and D.A. Berkley, 1980. "Silencing echoes in the telephone network", *Proceedings of the IEEE* **68**, pp. 948-963.

Van Essen, D.C., et al. (1992). "Information processing in the primate visual system: An integrated systems perspective", *Science* **255**, pp. 419-423.

Wan, E.A., 1994. "Time series prediction by using a connectionist network with internal delay lines". In *Time Series Prediction: Forecasting the Future and Understanding the Past* (A.S. Wingard and N.A. Gershenfeld, eds.), pp. 195-217, Addison-Wesley.

Widrow, B., and S.D. Stearns, 1985. *Adaptive Signal Processing*. Englewood Cliffs, NJ: Prentice Hall.

Widrow, B., and M.E. Hoff, Jr., 1960. "Adaptive switching circuits", *IRE WESCON Convention Record*, pp. 96-104.

Wiener, N., 1949. *Extrapolation, Interpolation, and Smoothing of Stationary Time Series with Engineering Applications*. Cambridge, MA: MIT Press. (This was originally issued as a classified National Defense Research Report, February 1942.)

Williams, R.J., and D. Zipser, 1989. "A learning algorithm for continually running fully recurrent neural networks", *Neural Computation* **1**, pp. 270-280.

Wu, L., and M. Niranjan, 1994. "On the design of nonlinear speech predictors with recurrent nets", *International Conference on Acoustics, Speech, and Signal Processing*, vol. II, pp. 529-532, Adelaide, Australia.

Zeng, Z., R.M. Goodman, and P. Smyth, 1994. "Discrete recurrent neural networks for grammatical inferences", *IEEE Transactions on Neural Networks* **5**, pp. 320-330.

Multiscale Signal Processing: From QMF to Wavelets

Albert Benveniste*

Abstract

This is a tutorial[1] intended to relate to each other the following topics: multirate filtering and maximally decimated QMF filter banks, multiresolution analysis and multiscale signal analysis and associated wavelet transforms.

I. Introduction

Multirate filtering [11] is now recognized as an area of increasing importance in digital signal processing. The key issue in this area is how to handle properly the aliasing due to sampling below the Nyquist rate. Maximally decimated filter banks have been introduced [19, 29, 30] that allow to design filter banks with an exact saving of the global sampling rate (e.g., a 2–filter bank must involve downsampling by a factor of 2 in each subband).

Multiscale signal analysis [14, 13] appears as an emerging alternative technique to Fourier analysis. Wavelet and related transforms are becoming increasingly popular in this area. *Multiscale signal recognition*, i.e., performing multiscale pattern recognition on signals, is certainly a desirable objective, although no

---

*A. Benveniste is with IRISA-INRIA, Campus de Beaulieu, 35042 RENNES CEDEX, FRANCE, benvenist@yin.irisa.fr, (33) 99 36 20 00
[1]EDICS: 4.8, 6.1.3

CONTROL AND DYNAMIC SYSTEMS, VOL. 68
Copyright © 1995 by Academic Press, Inc.
All rights of reproduction in any form reserved.

well established approach is available today for this purpose (the interested reader may however be referred to [1, 2] for an attempt toward this direction).

Finally, *multiresolution analyses* of L^2-spaces of functions or kernels, and related wavelets [27], recently proved an extremely powerful toolbox for highly demanding mathematical problems in harmonic and functional analysis. They provide a very effective approach to derive new approximations and expansions of functions, or operators, that are classically difficult to handle (integral equations, singular integral, pseudodifferential operators). These theories also provide a new theoretical support to the general area of *multigrid methods* [22, 25].

It has been recently recognized that these apparently different topics are closely related to each other, see [32], an interesting expository paper. This fact is precisely the subject of this short tutorial, mostly intended to readers with signal processing and otherwise general mathematical background. For the sake of simplicity, we decided to concentrate on the essential features, leaving aside unnecessary technicalities: $1D$–domains, 2–filter banks, and infinite signals are only considered here. *Openings* are briefly presented which concern more advanced results or generalizations. To emphasize the essentials of the topic, we also tried to separate as much as possible in our presentation the algebraic aspects (mostly related to the theory of *Quadrature Mirror Filters –QMF–* in signal processing) from those relevant to harmonic and functional analysis (issues of convergence and approximation). This is the originality of the paper. Finally, to avoid overlength and tedious calculations, we decided to have a relatively compact presentation of the mathematics, so that this paper is indeed worth of a pencil-and-paper aided reading.

The paper is organized as follows. Section II. is devoted to the presentation of the polyphase approach to QMF banks, this section has been much inspired by [30]. QMF syntheses are discussed in section III., both unitary and non unitary. Section IV. is an important one : orthonormal decompositions of l^2-spaces of signals are provided that result from QMF unitary banks. Up to this point, purely algebraic techniques are used (involving algebraic manipulations of filters and down/upsampling operators. Techniques from (functional and harmonic) analysis are first used in section V. where various kinds of wavelet bases are introduced ; this section mostly relies on [12, 13]. A simple example of an application to issues of approximations in functional analysis is illustrated in section VI., where some of the basic principles of the so-called "BCR" (Beylkin-Coifman-Rokhlin) method of approximating integral operators are sketched. Finally an interesting use of QMF banks associated with wavelets for multiresolution processing of random signals is presented in section VII., based on [8].

II. QMF Banks and the Polyphase approach

In this section we present the polyphase approach promoted by Crochiere and Rabiner [11] to multirate processing of signals. This approach has also been extensively used by Vaidyanathan [30] for the design of maximally decimated filter banks and we largely follow this very nice paper in this section. It turns out that it is also extremely powerful to introduce orthonormal wavelets and their numerous clones as we shall see later.

A. Down– and up-sampling

The above diagram will represent down–sampling by a factor of 2 throughout this paper, i.e. $y_n = x_{2n}$. Accordingly

$$X(z) = \sum_n x_n z^{-n}$$

$$Y(z^2) = \sum_n y_n z^{-2n} = \sum_n x_{2n} z^{-2n}$$

whence

$$Y(z^2) = \frac{1}{2}\left[X(z) + X(-z)\right] \tag{1}$$

Notice that the downsampling operator is *not* a filter for it is not time invariant (it relies on a particular choice of the 0 instant). Similarly

$$Y \longrightarrow \boxed{\uparrow 2} \longrightarrow X$$

represents up–sampling by a factor of 2, so that

$$X(z) = \sum_n x_n z^{-n} = \sum_n y_n z^{-2n} = Y(z^2) \tag{2}$$

holds.

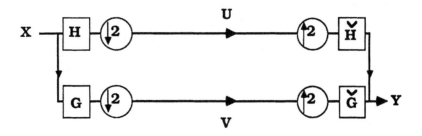

Fig. 1. Maximally decimated filter bank

B.   Maximally decimated filter banks

The diagram of the figure 1 depicts such a filter bank. Using formulae (1,2), the input–output map is written as follows:

$$
\begin{aligned}
Y(z) \;=\;& \check{H}(z)U(z^2) + \check{G}(z)V(z^2) \\
=\;& \frac{1}{2}\left[\check{H}(z)H(z) + \check{G}(z)G(z)\right] X(z) \qquad (3) \\
+\;& \frac{1}{2}\left[\check{H}(z)H(-z) + \check{G}(z)G(-z)\right] X(-z) \qquad (4)
\end{aligned}
$$

In this formula, the expression (4) represents the *aliasing* component, whereas (3) represents the linear transfer component. Hence, for this linear map to be an (*aliasfree*) time–invariant filter, we must have

$$
\check{H}(z)H(-z) + \check{G}(z)G(-z) = 0 \qquad (5)
$$

and, for *perfect reconstruction*

$$
\check{H}(z)H(z) + \check{G}(z)G(z) = 1 \qquad (6)
$$

up to a delay.

A sketch of history.

The first paper to consider this problem is [19], where the following solution is proposed to satisfy the condition (5) for antialising:

$$
\check{H}(z) = G(-z) , \; \check{G}(z) = -H(-z)
$$

Then, the following choice for H

$$G(z) = H(-z)$$

results in the following condition for perfect reconstruction:

$$H^2(z) - H^2(-z) = 1$$

for which no pleasant exact solution does exist. The first satisfactory solution was due to Smith & Barnwell in 1983 [29] and is explained next. Select H such that

$$H(z)H(z^{-1}) + H(-z)H(-z^{-1}) = 1$$

(i.e. white noise $\rightarrow$ $\boxed{H}$ $\rightarrow$ $\boxed{\downarrow 2}$ $\rightarrow$ white noise). Then the following choice

$$
\begin{array}{ccccc}
H(z) & H(z^{-1}) & + & H(-z) & H(-z^{-1}) & = & 1 \\
\updownarrow & & & \updownarrow & \updownarrow \\
\check{H}(z) & & & z^{-1}\check{G}(z) & zG(z)
\end{array}
$$

satisfies both antialiasing (6) and perfect reconstruction (5) conditions. The reader is referred to [30] for further historical information.

C. The polyphase approach

Consider the diagram (a) of the figure 2: It certainly satisfies both antialiasing and perfect reconstruction conditions. On the other hand, this diagram is clearly equivalent to the next one (b) of the same figure provided that

$$E(z)\check{E}(z) = I \tag{7}$$

holds. This last diagram is now redrawn as in diagram (c) of the same figure, or, equivalently, as in the diagram (d) by setting

$$E(z) = \begin{bmatrix} H_0(z) & H_1(z) \\ G_0(z) & G_1(z) \end{bmatrix}, \quad \check{E}(z) = \begin{bmatrix} \check{H}_1(z) & \check{G}_1(z) \\ \check{H}_0(z) & \check{G}_0(z) \end{bmatrix} \tag{8}$$

and

$$
\begin{aligned}
H(z) &= H_0(z^2) + z^{-1}H_1(z^2) \\
G(z) &= G_0(z^2) + z^{-1}G_1(z^2)
\end{aligned}
\tag{9}
$$

and similarly for $\check{H}, \check{G}$. Checking for solutions to (9) and then applying formulae (8,7) is known as the *polyphase approach to the QMF problem*. In the sequel, we shall mainly concentrate on matrix *polynomial* solutions to (9) to get FIR maximally decimated filter banks.

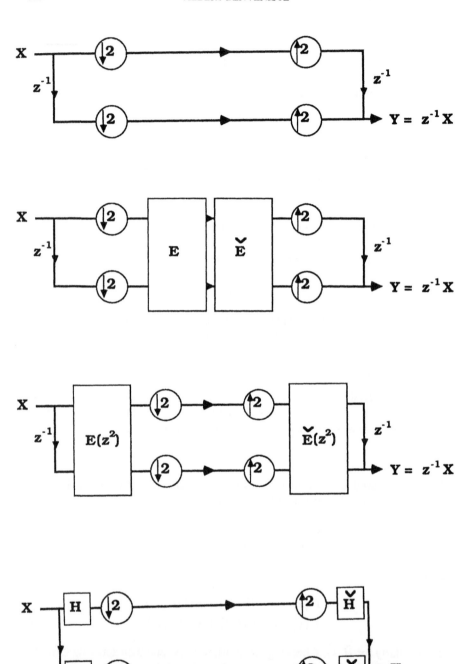

Fig. 2. The polyphase approach: (a) stage 1, (b) stage 2, (c) stage 3, (d) stage 4

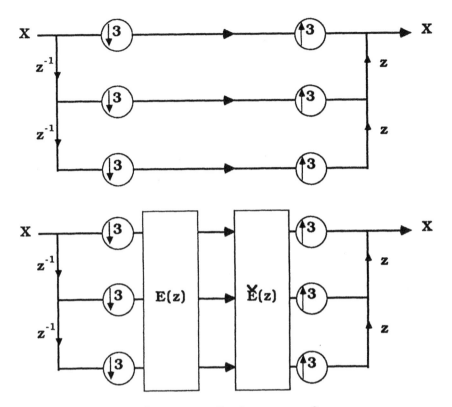

Fig. 3. Downsampling by a rate $p = 3$

D. Openings

The polyphase approach may be generalized in various ways. Rather than presenting a general framework for this, we shall scan some of these generalizations and discuss them. Some of these extensions are depicted in the figures 3 to 6. In these figures, the delays at the input of the polyphase filter bank are compensated by *inverse* delays which are noncausal, but corresponding causal delays such as provided in the figure 2 may be used as well whenever needed. The figure 3 corresponds to multirate signal processing with a sampling rate $p > 2$: thus the filter bank to be synthesized is composed of p filters and the matrix E is now $p \times p$. The figure 4 corresponds to multirate image processing with a sampling rate of $2 \times 2 = 4$: the downsampling operator denoted by $\downarrow 4$ consists of selecting the
• pixels of the grid ; the matrix $E(z_1, z_2)$ is 4×4 with entries that are transfer functions in the two variables z_1 (horizontal shift) and z_2 (vertical shift). Separable filters $E(z_1, z_2) = E_1(z_1)E_2(z_2)$ may be synthesized in this case. The case of figure 5 is more interesting, it has been proposed by Fauveau [20]. The downsam-

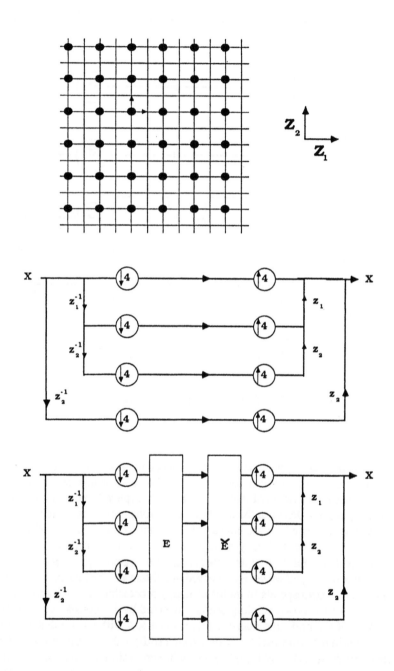

Fig. 4. Separable downsampling of Z^2

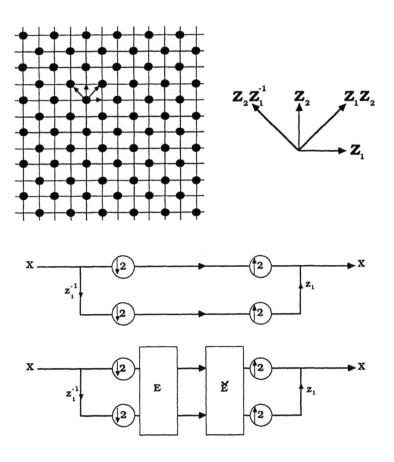

Fig. 5. Interleaved downsampling of Z^2

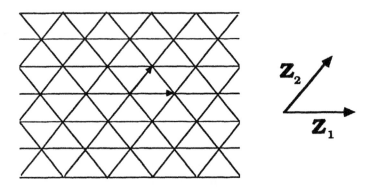

Fig. 6. Hexagonal sampling of the plane.

pling operator denoted by $\downarrow 2$ consists of selecting the $\bullet$ pixels of the grid, i.e. half of the pixels. One can see that this downsampling operator *rotates* the grid by $\pi/2$. Accordingly, when the 2×2 filter E moves across the downsampling operator (as done in figure 2-(b,c)), *a change of variable* $(z_1 , z_2) \longmapsto (z_1 z_2 , z_1 z_2^{-1})$ has to be performed ; performing this downsampling twice produces thus the change of variable $(z_1 , z_2) \longmapsto (z_1 z_2 , z_1 z_2^{-1}) \longmapsto (z_1^2 , z_2^2)$, i.e. is equivalent to performing downsampling as in the figure 4. Thus synthesizing separable filters E will result in filter banks that are separable in the variables $[z_1 z_2, z_1 z_2^{-1}]$, this remark has been used by Fauveau in his implementation. Finally the figure 6 is just an "oblique" redrawing of figure 4. Other index sets could be considered as well, provided that they be associated with some kind of "regular grid" and appropriate pairs $(E, \check{E})$ of operators can be synthesized.

III. QMF synthesis

In this section, we check for 2–port transfer functions E and $\check{E}$ satisfying

$$E\check{E} = I \qquad (10)$$

Several solutions exist. Of particular interest are *polynomial* solutions we shall discuss throughtout this tutorial.

A. Non-unitary QMF synthesis

This word refers to any solution $(E, \check{E})$ of equation (10). Such a solution does exist if and only if the 2–port polynomial matrix E is a unit in the ring of polynomial matrices, i.e. if it is *unimodular*, say[2]

$$1 \equiv \det E(z) = H_0(z)G_1(z) - H_1(z)G_0(z)$$

---

[2]up to a delay and a constant gain we don't care here

Then we take

$$\check{E} = E^{-1} = \begin{bmatrix} G_1 & -H_1 \\ -G_0 & H_0 \end{bmatrix}$$

which yields, up to a delay,

$$\begin{aligned} \check{H}(z) &= -G_0(z^2) - z^{-1}G_1(z^2) \\ \check{G}(z) &= H_0(z^2) - z^{-1}H_1(z^2) \end{aligned} \tag{11}$$

Finally, given H, the problem reduces to a *Bezout or Diophantine equation* to find G. A solution does exist if and only if the pair (H_0, H_1) is coprime. Cascade parametrizations of unimodular E matrices are provided in [28], which structurally guaranty *linear phase* properties for the analysis and synthesis filter banks. Further results along these lines may be found in [32], together with a very nice introduction to wavelets for signal processing people.

B.   Unitary QMF synthesis

The question is now the following: find E *lossless*, i.e. such that

$$E(z)E^T(z^{-1}) = I$$

and take $\check{E}(z) = E^T(z^{-1})$ up to a delay. Now, the following equivalences are easy to check[3]:

$$E = \begin{bmatrix} H_0 & H_1 \\ G_0 & G_1 \end{bmatrix} \text{ lossless}$$

$$\Updownarrow \text{ for } z = e^{i\omega}$$

$$\begin{aligned} 1 &= |H_0|^2 + |G_0|^2 = |H_1|^2 + |G_1|^2 \\ &= |H_0|^2 + |H_1|^2 = |G_0|^2 + |G_1|^2 \end{aligned} \tag{12}$$

$$0 = \overline{H}_0 H_1 + \overline{G}_0 G_1 = H_0\overline{G}_0 + H_1\overline{G}_1 \tag{13}$$

$$\Updownarrow$$

$$\frac{1}{\sqrt{2}} \begin{bmatrix} H(z) & G(z) \\ H(-z) & G(-z) \end{bmatrix} \overset{\Delta}{=} J \text{ lossless} \tag{14}$$

Now, from (13) we deduce the following equalities

$$\frac{G_1(e^{i\omega})}{H_0(e^{-i\omega})} = -\frac{H_1(e^{i\omega})}{G_0(e^{-i\omega})} = -\frac{G_0(e^{i\omega})}{H_1(e^{-i\omega})} = e^{i\phi(\omega)}$$

---

[3]$\overline{z}$ denotes the complex conjugate of z

where the last one is obtained by noticing that $\bar{z} = 1/z \Rightarrow |z| = 1$. As a consequence there must exist an all-pass filter $I(z)(|I(e^{i\omega})|^2 = 1)$ such that

$$
\begin{aligned}
G_1(z) &= I(z)H_0(z^{-1}) \\
G_0(z) &= -I(z)H_1(z^{-1})
\end{aligned}
$$

whence

$$
\begin{aligned}
G(z) &= G_0(z^2) + z^{-1}G_1(z^2) \\
&= z^{-1}I(z^2)\left(H_0(z^{-2}) - zH_1(z^{-2})\right) \\
&= z^{-1}I(z^2)H(-z^{-1})
\end{aligned}
$$

If G polynomial is wanted, we must take $I(z) =$ pure delay. Combining this with J lossless yields

$$
H(z)H(z^{-1}) + H(-z)H(-z^{-1}) = 2 \tag{15}
$$

or, equivalently

$$
|H\left(e^{i\omega}\right)|^2 + |H\left(e^{i(\omega+\pi)}\right)|^2 = 2 \tag{16}
$$

These equivalent conditions will be referred to in the sequel as the *Quadrature Mirror Filter (QMF) condition.*

Summary for unitary QMF synthesis

$$E \text{ lossless}$$

$$\updownarrow$$

$$J \text{ lossless}$$

$$\updownarrow$$

$$\boxed{\begin{array}{l} \text{QMF: } H(z)H(z^{-1}) + H(-z)H(-z^{-1}) = 2 \\ \qquad G(z) = z^{-1}H(-z^{-1}) \end{array}}$$

Finally, take, up to a delay,

$$\breve{H}(z) = H(z^{-1}), \quad \breve{G}(z) = G(z^{-1}) \tag{17}$$

C. Synthesis of 2-port lossless FIR polynomial transfer functions.

Direct cascade parametrization of lossless 2–port transfer functions.

The following elementary block

$$E_\alpha(z) \triangleq \begin{bmatrix} \cos \alpha & \sin \alpha \\ -\sin \alpha & \cos \alpha \end{bmatrix} \begin{bmatrix} 1 & 0 \\ 0 & z^{-1} \end{bmatrix}$$

is lossless. Hence, so is

$$E(z) = \prod_{m=n}^{1} E_{\alpha_m}(z) \qquad (18)$$

Conversely, we are given

$$E = \begin{bmatrix} H_0 & H_1 \\ G_0 & G_1 \end{bmatrix}$$

lossless. We can assume $\det E \equiv 1$, which implies

$$H_{0,n} G_{1,n} = H_{1,n} G_{0,n} \ (n = \deg E) \qquad (19)$$

($H_{i,m}$ coefficient of z^{-m} in H_i). Consider

$$\begin{bmatrix} \tilde{H}_0 & \tilde{H}_1 \\ \tilde{G}_0 & \tilde{G}_1 \end{bmatrix} \triangleq \begin{bmatrix} \cos \alpha & -\sin \alpha \\ z \sin \alpha & z \cos \alpha \end{bmatrix} \begin{bmatrix} H_0 & H_1 \\ G_0 & G_1 \end{bmatrix}$$

From (19) we deduce

$$\exists_1 \alpha = \alpha_n : d^\circ \tilde{H}_i \leq n - 1, i = 0, 1 \qquad (20)$$

On the other hand, with $\alpha = \alpha_n$, and using (20), we get

$$\left. \begin{array}{rcl} G_1(z) & = & z^{-n} H_0(z^{-1}) \\ G_0(z) & = & -z^{-n} H_1(z^{-1}) \end{array} \right\} \Rightarrow \tilde{G}_i = z^{-1} \tilde{\tilde{G}}_i, i = 0, 1$$

for some $\tilde{\tilde{G}}_i$. Hence, setting $E = E_n$, we get the decomposition

$$E_n = \begin{bmatrix} \cos \alpha_n & \sin \alpha_n \\ -\sin \alpha_n & \cos \alpha_n \end{bmatrix} \begin{bmatrix} 1 & 0 \\ 0 & z^{-1} \end{bmatrix} E_{n-1},$$

$$\deg E_{n-1} \leq n - 1$$

showing that the cascade decomposition (18) may be used as a general parametrization of 2–port lossless polynomial transfer functions; such a factorization is originally due to Potapov in the fifties. See also [21] for an overview of wave digital
filter synthesis, where most of the classical lossless digital filter design techniques

are presented following a circuit theoretic point of view. An alternative related parametrization is

$$E(z) = \prod_{m=n}^{1} \begin{bmatrix} 1 & z^{-1}k_m \\ -k_m & z^{-1} \end{bmatrix}$$

which is equivalent to (18) up to a constant normalization gain. This parametrization is used by Vaidyanathan [31] to synthetize a {low–pass, high–pass} QMF pair using optimization techniques.

I. Daubechies' QMF unitary synthesis.

{Low–pass, high–pass} QMF pairs can be synthetized in an alternative way. Take $H(z)$ of the form

$$H(z) = (1 + z^{-1})^N \tilde{H}(z) \quad \text{(typically low-pass)}$$

Then the condition $2 = H(z)H(z^{-1}) + H(-z)H(-z^{-1})$ is rewritten as

$$\begin{aligned} 1 &= \left(1 + \frac{z + z^{-1}}{2}\right)^N \tilde{H}(z)\tilde{H}(z^{-1}) \\ &+ \left(1 - \frac{z + z^{-1}}{2}\right)^N \tilde{H}(-z)\tilde{H}(-z^{-1}) \end{aligned}$$

Consider the equation

$$1 = \left(1 + \frac{z + z^{-1}}{2}\right)^N Q(z) + \left(1 - \frac{z + z^{-1}}{2}\right)^N Q(-z) \qquad (21)$$

where $Q(z^{-1}) = Q(z)$. I. Daubechies [12, 13] proved the following important result: *there exists a unique solution to (21) of degree $2N$, and this solution satisfies*

$$Q\left(e^{i\omega}\right) \geq 0$$

Then QMF synthesis reduces to a polynomial spectral factorization problem.

Openings.

We have presented here the classical synthesis techniques for 2-port lossless polynomial transfer functions. It may be of interest however to recall the work done in the area of scattering theory, see [16, 17, 18]. This approach is based on the following fact. A 4×4 transfer matrix Θ is said to be J-lossless if its poles are strictly inside the unit disc and if it satisfies[4]

$$\Theta(z)J\Theta(z)^* \begin{cases} \leq J & \text{for } |z| > 1 \\ = J & \text{for } |z| = 1 \end{cases} \quad \text{where } J = \begin{bmatrix} I & 0 \\ 0 & -I \end{bmatrix} \qquad (22)$$

---

[4] A^* denotes the hermitian transpose of the matrix A.

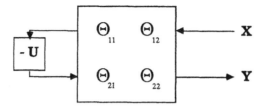

Fig. 7. Synthesis of E lossless via the inverse scattering approach

and I is the 2×2 identity matrix. Partitioning Θ as

$$\Theta = \left[\begin{array}{cc} \Theta_{11} & \Theta_{12} \\ \Theta_{21} & \Theta_{22} \end{array} \right]$$

we derive from (22) the following inequalities where $U(z)$ is an arbitrary lossless 2×2 matrix transfer function :

$$\left[\begin{array}{cc} U(z) & I \end{array} \right] \Theta(z) J \Theta(z)^* \left[\begin{array}{c} U(z)^* \\ I \end{array} \right] \leq \left[\begin{array}{cc} U(z) & I \end{array} \right] J \left[\begin{array}{c} U(z)^* \\ I \end{array} \right] \leq 0 \tag{23}$$

for $|z| \geq 1$ with equalities on the unit circle. But it turns out that (23) is equivalent to

$$E(z)E(z)^* \quad \left\{ \begin{array}{ll} \leq I & \text{for } |z| > 1 \\ = I & \text{for } |z| = 1 \end{array} \right. \quad \text{where} \tag{24}$$

$$E = -(U\Theta_{12} + \Theta_{22})^{-1}(U\Theta_{11} + \Theta_{21}) \tag{25}$$

but (24) exactly means that E is lossless. Such a synthesis of the filter E is depicted in the figure 7. The interest of this approach is that pairs (Θ, U) can be found such that the McMillan degree of U be strictly smaller than that of E, so that applying recursively this procedure yields a cascade synthesis of E. A complete solution to this problem may be found in [16, 17, 18] in terms of cascade transfer function matrices. Furthermore interpolation conditions of the Nevanlinna-Pick type [16, 18] may be satisfied as well, which might be useful as we shall see later in sections V. and VI. when discussing "vanishing moment conditions" for wavelets.

IV. Hilbert space structures of *orthonormal* QMF syntheses.

In this section, we investigate how orthonormal QMF banks give raise to some orthonormal decompositions of l^2–spaces of signals. Such decompositions turn out to be useful (for instance) for coding purposes.

A. Basic decompositions

In what follows, we select a given discrete time index set that will serve as a reference for the various sampling rates we shall consider in the sequel. We denote by $\mathbf{Z}$ this discrete time reference. We denote by $l^2(n\mathbf{Z})$ (n integer) the space of square integrable signals x_k such that $x_k \neq 0$ only if k is a multiple of n. Similarly, we denote by $l^2(\frac{1}{n}\mathbf{Z})$ the space of square integrable signals that are *upsampled* at a rate n. Using these notations, the following diagram

induces an operator

$$\mathcal{H} \; : \; l^2(\mathbf{Z}) \longrightarrow l^2(2\mathbf{Z})$$

and similarly for $\mathcal{G}$. Vice-versa

induces an operator

$$\check{\mathcal{H}} \; : \; l^2(2\mathbf{Z}) \longrightarrow l^2(\mathbf{Z})$$

Now, if we take the case $E = I$ in the diagram of the figure 2-b we get immediately the identities

$$\begin{aligned}
I &= \check{\mathcal{H}}\mathcal{H} + \check{\mathcal{G}}\mathcal{G} \\
I &= \mathcal{H}\check{\mathcal{H}} = \mathcal{G}\check{\mathcal{G}} \\
0 &= \mathcal{H}\check{\mathcal{G}} = \mathcal{G}\check{\mathcal{H}} \\
\check{\mathcal{H}} &= \mathcal{H}^* , \; \check{\mathcal{G}} = \mathcal{G}^*
\end{aligned} \tag{26}$$

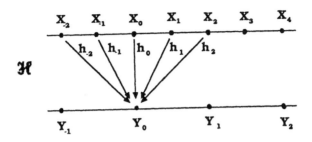

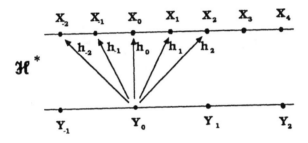

Fig. 8. $\mathcal{H}$ and $\mathcal{H}^*$

where $\cdot^*$ denotes the adjoint. Introducing E lossless yields an orthonormal change of basis in $l^2(2\mathbb{Z}) \times l^2(2\mathbb{Z})$, hence the above identities remain valid. Consequently, using the Mason rule, we may draw the $\mathcal{H}$ and $\mathcal{H}^*$ operators as in the figure 8. The formulae (26) yield the decomposition

$$\mathcal{V}_0 \stackrel{\Delta}{=} l^2(\mathbb{Z}) = \mathrm{Im}(\mathcal{H}^*) \oplus \mathrm{Im}(\mathcal{G}^*)$$
$$\stackrel{\Delta}{=} \mathcal{V}_{-1} \oplus \mathcal{W}_{-1} \tag{27}$$

we shall use in different ways in the sequel.

B. Some coarse–scale decompositions

1. A "wave-packet" coarse–scale decomposition

Using recursively (27), the diagram of the figure 9 yields on its leaves a decomposition into signals at coarser scales, namely

$$\mathcal{V}_0 = \bigoplus_{w \in \{0,1\}^n} \mathrm{Im}\left(\mathcal{T}_{w_n}^* \dots \mathcal{T}_{w_1}^*\right) \tag{28}$$

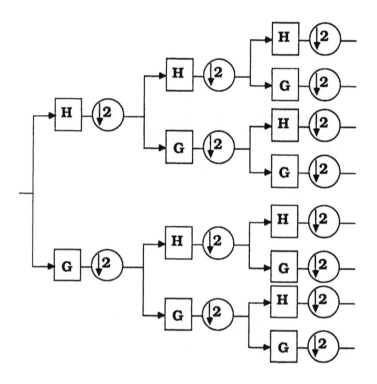

Fig. 9. The wave-packet tree

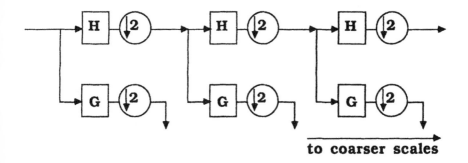

Fig. 10. The wavelet tree

where $w = w_n...w_1$, $w_i \in \{0, 1\}$ ranges over all words of length n over the alphabet $\{0, 1\}$, and

$$T_0^* = \mathcal{H}^*, T_1^* = \mathcal{G}^* \tag{29}$$

The name of the decomposition will be justified later on.

2. A wavelet coarse–scale decomposition

Using again (27), the alternative diagram of the figure 10 yields the decomposition

$$\begin{aligned} V_0 &= W_{-1} \oplus W_{-2} \oplus ... \oplus W_{-n} \oplus V_{-n} \\ W_{-m} &= \text{Im}\left((\mathcal{H}^*)^{m-1}\mathcal{G}^*\right) \end{aligned} \tag{30}$$

In this case it is usual to take for $\{H, G\}$ a $\{$low–pass,high–pass$\}$ pair.

3. The smoothness issue

In using analysis–and–synthesis QMF banks such as introduced above the following *robustness* issue emerges: suppose that the channel linking the analysis and the synthesis banks together is not perfect, i.e. (hopefully little) degradations may be caused on the signals. This is certainly the case in coding applications, since at least quantization noise is introduced (and channel errors as well but infrequently). Then this additional disturbance is processed by the synthesis bank. In the case of the wavelet filter bank corresponding to figures 10 (analysis) and 12 (synthesis), the additional disturbance is processed by the operator $(\mathcal{H}^*)^n$. It is thus important to synthesize QMF banks that have the desirable property of smoothing out the effect of the disturbances: this is precisely the issue of smoothness for the wavelet associated with this QMF pair as we shall see in the section to follow. Similarly, it is important that the same condition be satisfied by the operators T_w^* for the wave-packet filter bank corresponding to figure 9 (analysis) and the dual one (for synthesis).

C. Openings:

1. Draw the tree of the figure 9 until infinity. In this figure, it is cut according to vertical lines. In the figure 10 the cut was taken along a parallel to the upper branch. It should be clear from these comments that *many other cuts could be considered as well to yield different orthogonal decompositions.* In general we get in this way decompositions of the form

$$V_0 \;=\; \bigoplus_{w \in W} \mathrm{Im}\,(T_w^*) \qquad (31)$$

$$T_w^* \;=\; T_{w_n}^* \cdots T_{w_1}^*$$

where W is any *cut* of the tree, i.e. any subset of the set $\{0, 1\}^*$ of all finite words over $\{0, 1\}$ such that

- no $w \in W$ is a prefix of another element of W,
- any element of $\{0, 1\}^*$ is either a prefix of an element of W, or has some element of W as one of its prefixes.

An example of a cut is shown in the figure 11. This allows us to *decompose l^2-spaces of discrete time signals into orthonormal subspaces in various ways.* This remark has been used in [7, 33] to select decompositions that are best fitted to a given signal, according to some "entropy" criterion (in this setting, the best basis provides the shortest coding), with applications to speech coding reported successfull in [33]. This remark has also been used in [7] to generate various alternative bases of L^2 that are different from the wavelet bases (in particular the so-called "wave-packet" bases), see section V..

2. Assume we consider now a *non-unitary* QMF bank following section A.. Then all but the last formulae (26) remain valid. Hence (27) should be replaced by

$$V_0 \;\overset{\Delta}{=}\; l^2(\mathbf{Z}) = \mathrm{Im}(\widecheck{\mathcal{H}}) + \mathrm{Im}(\widecheck{\mathcal{G}})$$

$$\overset{\Delta}{=}\; V_{-1} + W_{-1}$$

where $+$ refers here to complementary (but not necessary orthogonal) subspaces of V_0. Hence both the wave-packet tree of the figure 9 and the wavelet tree of the figure 10 could be reinterpreted in this weaker sense, giving raise to non-orthogonal decompositions of $l^2(\mathbf{Z})$ into complementary subspaces of signals at coarser scales.

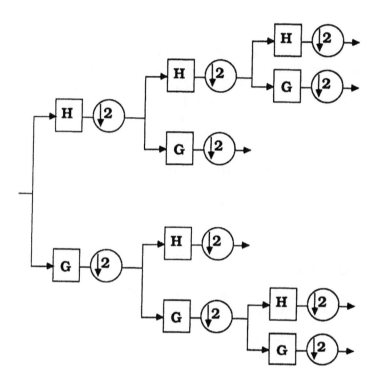

Fig. 11. A cut of the tree.

3. We have introduced the operator $\mathcal{H}$ as mapping $l^2(\mathbf{Z})$ into $l^2(2\mathbf{Z})$, and we have iterated this operator when considering the wavelet and wave-packet trees of operators. It would be more convenient to consider $\mathcal{H}$ as an operator mapping l^2-signals at any given level of resolution into downsampled l^2-signals, i.e. to consider $\mathcal{H}$ as an operator

$$\mathcal{H} : \bigoplus_{n=-\infty}^{+\infty} l^2(2^n\mathbf{Z}) \longmapsto \bigoplus_{n=-\infty}^{+\infty} l^2(2^n\mathbf{Z})$$

but this latter Hilbert space of signals is just $l^2(\mathcal{T})$ where $\mathcal{T}$ is the *dyadic tree* defined as follows: the nodes of $\mathcal{T}$ are the binary numbers, and $t \to s$ is a branch of $\mathcal{T}$ if and only if t is obtained via canceling the last bit in s. QMF pairs can thus be considered as operators defining orthonormal decompositions of $l^2(\mathcal{T})$. The algebra generated by such QMF pairs is studied in [3] from a system theoretic point of view.

V. Introducing orthonormal wavelets

In this section, we shall introduce orthonormal wavelets via the asymptotic analysis of QMF banks.

A. Fine scale asymptotic behaviour of the orthonormal QMF bank, and multiresolution analysis of $L^2(\mathbf{R})$

This subsection is mainly based on [12, 13]. Consider again the QMF analysis filter bank as shown in the figure 10. It yields a coarse–scale decomposition

$$\begin{aligned} l^2(\mathbf{Z}) &= \mathcal{V}_0 \\ &= \left(\bigoplus_{m=-1}^{-n} \mathcal{W}_m \right) \oplus \mathcal{V}_{-n} \end{aligned}$$

that can be also interpreted as a fine–scale decomposition

$$\begin{aligned} l^2(2^{-n}\mathbf{Z}) &= \mathcal{V}_n \\ &= \left(\bigoplus_{m=0}^{n-1} \mathcal{W}_m \right) \oplus \mathcal{V}_0 \end{aligned} \tag{32}$$

(just change the name of the input signal space). This suggests to consider the "infinite length" synthesis bank of the figure 12, i.e. to take the limit for fine scaling. The corresponding infinite block–diagram of the figure 12 exhibits a single infinite branch, namely the top one (all other branches cross this one at the

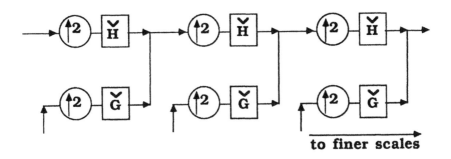

Fig. 12. Infinite length synthesis filter bank

first stage). Hence understanding this infinite behaviour amounts to study the top rightgoing branch, i.e. to ask for

$$\lim_{n \to \infty} \left(\mathcal{H}^* \right)^n . \delta_0 = ? \tag{33}$$

Recall that $\mathcal{H}$ is a (partial, i.e. non invertible) isometry. It should be clear that (33) is not the proper way to study the infinite behaviour of the synthesis filter bank: in this bank, the resolution is multiplied by 2 at each stage of the cascade, so that we expect to end up with an "infinite" resolution, something that is better represented with *continuous time index* rather than discrete one. Hence, what we shall do from the beginning is to imbed the discrete time index $l^2(2^{-n}\mathbf{Z})$–spaces into $L^2(\mathbf{R})$ by considering functions that are constant on the dyadic intervals of corresponding length. More specifically, denote by L_n^2 the Hilbert space of the L^2–functions that are constant on the dyadic intervals of length 2^{-n}, and associate $l^2(2^{-n}\mathbf{Z})$ with L_n^2. The upsampling operator $\uparrow 2 \ : \ L_n^2 \to L_{n+1}^2$ is carried on as shown in the figure 13. In this picture, each grey rectangle represent one element of the canonical orthonormal basis of L_n^2 (top) and L_{n+1}^2 (bottom) respectively. The scaling by $\sqrt{2}$ has been introduced to make $\uparrow 2$ a *unitary* injective operator. The translation of (33) in this new framework is presented now. Introduce

$$\chi_0 \stackrel{\Delta}{=} \chi_{[-\frac{1}{2},+\frac{1}{2}[} \quad \text{(characteristic function)}$$

and consider the following recursion:

$$\begin{aligned} \chi_1 &\stackrel{\Delta}{=} \overline{\mathcal{H}}^* \chi_0 \\ &= \sqrt{2} \sum_m h_m \chi_{[(m-\frac{1}{2})2^{-1},(m+\frac{1}{2})2^{-1}[} \end{aligned}$$

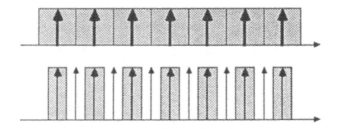

Fig. 13. Upsampling in continuous time (note the scaling by a factor of $\sqrt{2}$).

$$\chi_2 \triangleq \overline{\mathcal{H}}^* \chi_1$$

$$= 2 \sum_m h_m \left[\sum_p h_p \chi_{[(2m+p-\frac{1}{2})2^{-2},(2m+p+\frac{1}{2})2^{-2}[} \right]$$

and so on. These two recursion stages are depicted in the figure 14, where the merging arrows refer to additive superposition. The corresponding recursion involves the following operator $\overline{\mathcal{H}}^*$:

$$\overline{\mathcal{H}}^* f(x) \triangleq \sqrt{2} \sum_m h_m f(2x - m) \qquad (34)$$

Let us for the moment assume that we have proved the following :

$$\text{the limit } \phi \triangleq \lim_{n \to \infty} (\overline{\mathcal{H}}^*)^n \chi_0 \text{ exists in the } L^2\text{-sense,} \qquad (35)$$

$$\{\tau^m \phi\}_{m \in \mathbf{Z}} \text{ orthonormal system in } L^2. \qquad (36)$$

where $\tau f(x) = f(x + 1)$ denotes the translation by 1. Then the following algebraic properties can be derived based on the properties of the orthonormal QMF synthesis bank:

Algebraic properties of the limit ϕ

1. the function ϕ satisfies the following fixpoint equation:

$$\phi(x) = \sqrt{2} \sum h_m \phi(2x - m) \text{ , i.e. } \phi = \overline{\mathcal{H}}^* \phi \qquad (37)$$

2. introducing

$$\psi(x) = \sqrt{2} \sum g_m \phi(2x - m) \triangleq \overline{\mathcal{G}}^* \phi \qquad (38)$$

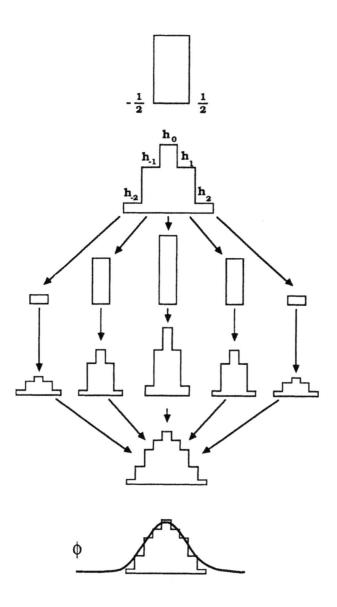

Fig. 14. $\chi_0, \chi_1, \chi_2, \cdots, \phi$

we have

$$\psi \perp \phi \text{, and} \tag{39}$$
$$\{\tau^m \psi\}_{m \in \mathbf{Z}} \text{ orthonormal system in } L^2$$

3. Introducing the following translation and dilation operators (both are isometries of L^2),

$$\tau f(x) = f(x+1) \text{ (translation)} \tag{40}$$
$$\sigma f(x) = \sqrt{2} f(2x) \text{ (dilation)}, \quad \sqrt{2}\tau\sigma = \sigma\tau^2$$

the following orthogonal decompositions of L^2 into "successive scales" hold [12, 13]:

$$L^2(\mathbf{R}) = \bigoplus_{n=-\infty}^{+\infty} \overline{W}_n \tag{41}$$

$$\overline{V}_n = \bigoplus_{m=-\infty}^{n-1} \overline{W}_m \tag{42}$$

where

$$\overline{V}_n = \text{span}\{\sigma^n \tau^m \phi, m \in \mathbf{Z}\}$$
$$\overline{W}_n = \text{span}\{\sigma^n \tau^m \psi, m \in \mathbf{Z}\}$$

Proof (sketch of).

Point 1. is immediate. For point 2., we use the formula

$$[\tau^k \phi](x) = \sqrt{2} \sum_m h_m \phi(2x + 2k - m)$$

$$= \sqrt{2} \sum_m h_m \left[\tau^{2k-m} \phi\right](2x)$$

$$= \sigma\left(\sum_m h_m \left[\tau^{2k-m} \phi\right]\right)(x)$$

and the corresponding one for ψ with g_m instead of h_m. This can be rewritten as

$$\sigma^{-1}\tau^k \phi = \sum_m h_m \tau^{2k-m} \phi \tag{43}$$

$$\sigma^{-1}\tau^l \psi = \sum_m g_m \tau^{2l-m} \phi \tag{44}$$

Hence, denoting by $< .,. >$ the inner product in L^2 and using the fact that the $\tau^m \phi$ form an orthonormal system in L^2, we may write

$$
\begin{aligned}
< \tau^k \phi, \tau^l \psi > &= < \sigma^{-1} \tau^k \phi, \sigma^{-1} \tau^l \psi > \\
\text{(using (43,44))} &= \mathcal{H}\mathcal{G}^* \delta_{k-l} = 0
\end{aligned}
\tag{45}
$$

due to the identities (26). Similarly, using again (26), we have

$$
\begin{aligned}
< \tau^k \psi, \tau^l \psi > &= < \sigma^{-1} \tau^k \psi, \sigma^{-1} \tau^l \psi > \\
\text{(using (44))} &= \mathcal{G}\mathcal{G}^* \delta_{k-l} = \delta_{k-l}
\end{aligned}
\tag{46}
$$

which finishes to prove the point 2, and furthermore proves that the systems that span the spaces $\overline{V}_n$ and $\overline{W}_n$ are orthonormal.

To prove (42) we remark that (43) implies $\overline{V}_{-1} \subseteq \overline{V}_0$. On the other hand, for $\overline{V}_0 \ni f = \sum_k c_k \sigma^{-1} \tau^k \phi$, we have also thanks to (43) $f = \sum_k c_k' \tau^k \phi$ where $c' = \mathcal{H}c$. A similar result holds for $\overline{W}_{-1}$ and $\mathcal{G}$ respectively. Thus the identity (27) carries on to yield

$$
\overline{V}_0 = \overline{V}_{-1} \oplus \overline{W}_{-1}
$$

and (42) follows by induction from the corresponding properties of the orthonormal QMF bank. To derive (41) — i.e. to prove that our decomposition actually spans the whole L^2-space — requires more technical work however, and we refer the reader to [12, 13, 10]. Note that this latter result states formally that adding finer scales is a valid approximation procedure.□

By the way, what we have got is a

multiresolution analysis of $L^2(\mathbf{R})$

in the sense of S. Mallat [23, 24] as reported in [12, 13] and the system $\{\sigma^n \tau^m \psi, m, n \in \mathbf{Z}\}$ we have so obtained is termed an *orthonormal wavelet basis*.

REMARK:

when a non–unitary QMF bank is used, a more careful use of similar techniques allows to derive the *bi–orthogonal* wavelets due to Albert Cohen and Ingrid Daubechies, cf. remark 2 of section 2..

B. Existence and properties of the limit ϕ: proof of (35,36)

Here we switch from *algebra* to *analysis*.

Existence of ϕ: proof of (35).

Introducing the Fourier transform

$$\hat{f}(\omega) = \frac{1}{\sqrt{2\pi}} \int_{-\infty}^{\infty} f(x)e^{-i\omega x}\,dx$$

the formula

$$\phi = \lim_{n \to \infty} (\overline{\mathcal{H}}^*)^n \chi_0 \qquad (47)$$

is rewritten as follows,

$$\hat{\phi}(\omega) = \frac{1}{\sqrt{2\pi}} \prod_{n=1}^{\infty} H(e^{i2^{-n}\omega})$$

and the limit is defined pointwise. This limit belongs to L^2 since, thanks to the QMF property, we know that the finite products $(\overline{\mathcal{H}}^*)^n \chi_0$ are uniformly bounded in L^2.

Proof of (36).

The following equivalences hold:

$$\{\tau^m \phi, m \in \mathbf{Z}\} \text{ orthonormal} \Leftrightarrow \sum_{m \in \mathbf{Z}} |\hat{\phi}(\omega + m2\pi)|^2 \equiv 1$$

Introduce

$$\Phi(\omega) \overset{\Delta}{=} \sum_{m \in \mathbf{Z}} |\hat{\phi}(\omega + m2\pi)|^2$$

$$u(\omega) \overset{\Delta}{=} |H(e^{i\omega})|^2$$

Then

$$\phi = \overline{\mathcal{H}}^* \phi \quad \Leftrightarrow \quad \hat{\phi}(2\omega) = \sqrt{2}H(e^{i\omega})\hat{\phi}(\omega)$$
$$\Rightarrow \quad P_u \Phi = \Phi$$

where the operator P_u is defined by

$$2P_u f(2\omega) = u(\omega)f(\omega) + u(\omega + \pi)f(\omega + \pi) \qquad (48)$$

But, since the filter H satisfies the orthonormal QMF condition, we have

$$u(\omega) + u(\omega + \pi) = 2 \qquad (49)$$

i.e P_u is the *transition probability of a Markov chain on the unit circle*. Then, knowing that Φ is P_u–invariant, we want to deduce that *it must be a constant*. Using the following equivalences

$$P_u\Phi = \Phi \ \Rightarrow\ \Phi \equiv 1$$

$$\Uparrow$$

continuous P_u–invariant functions must be constant

$$\Updownarrow$$

$$P_u \text{ ergodic}$$

J-P Conze, A. Raugi (IRMAR Rennes) gave in [8] necessary and sufficient conditions on u for this Markov chain to be ergodic. Recall that ergodicity is a generic situation for Markov chains, so that QMF banks generally yield orthonormal wavelets. Similar results have been obtained in [5] by A. Cohen (CEREMADE, Paris) with different techniques.

Smoothness conditions

are useful when the so obtained basis of L^2 is used for harmonic or functional analysis. They also play a fundamental role in the success of wavelets as applied to coding of signals or images, see section 3.. For such applications, it is extremely important that the n-th iterate of the operator $\overline{\mathcal{H}}^*$ introduced in formula (34) converge to a smooth function (see the figure 14 for a picture of this) : this condition ensures a proper smoothing of quantization noise that it introduced by any coding scheme, it also improves robustness against channel errors. The following theorem holds [12, 13] :

Theorem 1 (I. Daubechies) *Assume H is selected of the following form*

$$H(z) = \left[\frac{1}{2}(1 + z^{-1})\right]^N \tilde{H}(z) \tag{50}$$

where, in addition to be selected for H to satisfy the orthonormal QMF property, the filter $\tilde{H}$ satisfies the following conditions:

$$\sum |\tilde{H}_n||n|^\epsilon < \infty, \epsilon > 0$$

$$\sup |\tilde{H}(e^{i\omega})| = B < 2^{N-1}$$

then ϕ satisfies

$$|\hat{\phi}(\omega)| \leq C(1 + |\omega|)^{-N+(\log B)/(\log 2)}$$

and the convergence is pointwise.

This last inequality guaranties that ϕ has its M–th derivative in L^2 for some M, but this index M is much smaller than N (in practice, for the wavelets obtained by the method of [12, 13], one has approximately $M \simeq 0.2N$). Better results on the regularity of ϕ given the decay rate of H may be found in [15, 13, 6, 9, 10].

C. Openings.

1. (50) and the QMF condition together imply

$$
\begin{aligned}
G(e^{i\omega}) &= G'(e^{i\omega}) = ... = G^{(N)}(e^{i\omega}) \\
&= 0 \text{ for } \omega = 0
\end{aligned}
\tag{51}
$$

or, equivalently,

$$
\hat{\psi}(0) = ... = \hat{\psi}^{(N)}(0) = 0
$$

and, finally, via inverse Fourier transform,

$$
\int x^n \psi = 0, \quad n \le N
\tag{52}
$$

Such vanishing moment conditions appear to be extremely useful for approximating functions, or integral operators when a $2D$–theory is considered, see section VI..

Translate conditions (51) in terms of the lossless matrix transfer function E introduced in equation (8), we get

$$
0 = G(1) = G'(1) = ... = G^{(n)}(1)
\tag{53}
$$

$$
G(z) = \begin{bmatrix} 0 & 1 \end{bmatrix} E(z^2) \begin{bmatrix} 1 \\ z^{-1} \end{bmatrix}
$$

It turns out that synthesizing E's satisfying this condition is an instance of a Nevanlinna-Pick interpolation problem such as mentioned in the "opening" paragraph of section C.. It is a nonstandard one however since the constraints (53) involve multiple roots that are on the unit circle [18]. Further work has to be performed to check how this synthesis method may be used for systematically constructing wavelets with various kinds of vanishing moment conditions.

2. In subsection C., we have discussed how various cuts of the tree of figure 9 may be considered, to yield various orthonormal decompositions of l^2-spaces of signals that are associated with them. Since we have seen that wavelet bases of L^2 have been constructed that are associated with the wavelet tree, it may be guessed that another basis of L^2 could be associated with the wave-packet tree of figure 9, and maybe as well that a different basis

may be associated with each "cut" of this tree. This informal conjecture received a formal positive answer in the papers [7, 33]: the *wave-packet* basis is introduced in [7], which corresponds to our "wave-packet" tree (hence the name of it). And other bases are presented as well, that are the counterpart of all possible "cuts" of our tree. In fact, the wave-packet basis consists of the following family of functions [7]

$$\left\{ \overline{T}_w^* \phi(x - k) \; : \; w \in \{0, 1\}^*, k \in \mathbf{Z} \right\}$$

where $\overline{T}_0^* = \overline{\mathcal{H}}^*$, $\overline{T}_1^* = \overline{\mathcal{G}}^*$, and notations similar to those of formulae (31) have been used. Moreover, restricting this family to all words w of length n yields a basis of $\overline{\mathcal{W}}_n$.

VI. Efficient approximations of L^2–functions

I. Daubechies proved that, with further algebraic conditions on H, the following conditions may be satisfied:

$$\exists x_o \in \mathbf{R} : \int \phi(x + x_o) x^m dx = 0 : 1 \leq m < M \tag{54}$$

Recall that, on the other hand,

$$\int \phi dx = 1$$

Again could these vanishing moment conditions be translated in terms of the E lossless matrix transfer function introduced in equation (8) to yield particular Nevanlinna-Pick interpolation problems. We shall use these vanishing moment conditions to compute efficient expansions of functions of the M-th order Sobolev space. For this purpose, we consider the diagram of the figure 15. This diagram describes a two–step approximation procedure:

1. project f onto the space $\overline{V}_n$ of L^2–functions at scale 2^{-n},

2. use the QMF analysis bank to further decompose the projection into an orthogonal expansion.

Since the second stage has already been investigated, only the following problem remains to be investigated: compute

$$< f, \sigma^n \tau^m \phi >$$

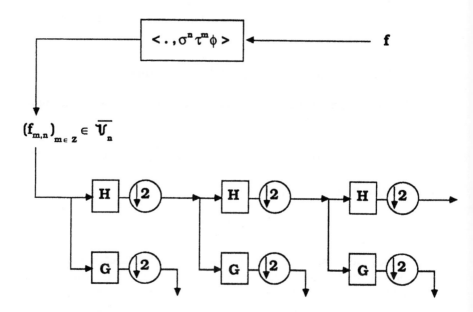

Fig. 15. Expanding f.

efficiently. This is explained next. Write the following equalities:

$$< f, \sigma^n \phi > = 2^{\frac{n}{2}} \int f(x) \phi(2^n x) dx$$

$$= 2^{-\frac{n}{2}} \int f(2^{-n} x) \phi(x) dx$$

$$= 2^{-\frac{n}{2}} \int f(2^{-n} (x + x_o)) \, \phi(x + x_o) dx$$

A Taylor expansion of f around $2^{-n} x_o$ yields:

$$f(2^{-n}(x + x_o)) = f(2^{-n} x_o)$$
$$+ 2^{-n} x f'(2^{-n} x_o) + \ldots$$
$$+ 2^{-(M-1)n} x^{M-1} f^{(M-1)}(2^{-n} x_o)$$
$$+ O\left(\frac{2^{-Mn}}{M!} |x|^M\right)$$

so that, thanks to the vanishing moment conditions (54), we get the following *single point approximation* !

$$< f, \sigma^n \phi > = 2^{-\frac{n}{2}} f(2^{-n} x_o) + O\left(\frac{2^{-n(M+1/2)}}{M!} |x|^M\right)$$

$2D$–generalizations of this are useful to approximate integral Kernels involved in some integral equations (BCR algorithm, due to Beylkin, Coifman, and Rokhlin, see [4] [26]).

VII.   Application of orthonormal QMF banks to random signals

Consider again the "wave-packet" tree of the figure 16. The following question has been considered by J-P Conze and A. Raugi [8]:

What are the statistics of the signals
on the leaves of the "wave-packet" tree ?

More precisely, we want to investigate this question for very long trees.

A.   Filtering and decimation of signals

Decimation of signals results in the following change of the spectral measure of the input signal:

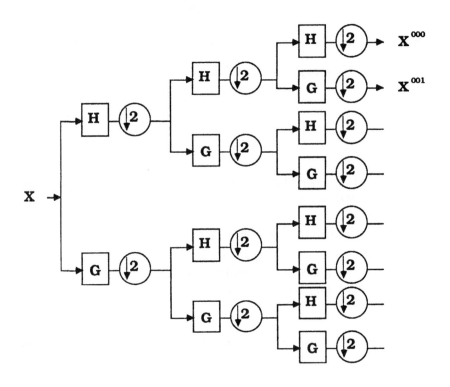

Fig. 16. Signals through the "wave-packet" tree

$$X \longrightarrow \boxed{\downarrow 2} \longrightarrow Y$$

Introduce the spectral measure of the process X:

$$\mathbf{E} X_0 X_m = \int_0^{2\pi} e^{i\omega m} \mathcal{R}_X(d\omega)$$

we have

$$\int f(\omega)\mathcal{R}_Y(d\omega) = \int f(2\omega)\mathcal{R}_X(d\omega) \tag{55}$$

If X has a spectral density, we have, denoting by $R_X(z)$ the spectrum of X:

$$R_Y(z) = \frac{1}{2}\left[R_X(z) + R_X(-z)\right] \tag{56}$$

Now, considering the case of filtering and decimation, we get

$$\longrightarrow \boxed{H} \boxed{\downarrow 2} \longrightarrow$$

$$\int f(\omega)\mathcal{R}_Y(d\omega) = \int f(2\omega)|H(e^{i\omega})|^2 \mathcal{R}_X(d\omega) \tag{57}$$

Again, if X has a spectrum we denote by R_X, we get:

$$R_Y(z) = \frac{1}{2}\left[R_{HX}(z) + R_{HX}(-z)\right] \text{ where}$$
$$R_{HX}(z) = H(z)R_X(z)H(z^{-1}) \tag{58}$$

Case of a pure frequency

$$\mathcal{R}_X = \delta_{\omega_o} + \delta_{-\omega_o}$$
$$\mathcal{R}_Y = u(2\omega_o)\left[\delta_{2\omega_o} + \delta_{-2\omega_o}\right] \tag{59}$$
$$u(\omega) \triangleq |H(e^{i\omega})|^2$$

Case of a spectral density

$$\mathcal{R}_X(d\omega) = \mathcal{R}_X(e^{i\omega})d\omega \triangleq F(\omega)d\omega$$
$$\mathcal{R}_Y(d\omega) = \mathcal{R}_Y(e^{i\omega})d\omega \triangleq P_u F(\omega)d\omega \tag{60}$$

where P_u is the operator we already introduced:

$$2P_u f(\omega) = u(\omega)f(\omega) + u(\omega + \pi)f(\omega + \pi)$$

B.   Use of the "wave-packet" tree

Consider again the figure 16, and introduce the following notations:

$$
\begin{aligned}
X &= (X_n), \text{ with spectral measure } \mathcal{R}_X(d\omega) \\
T_0 &= \mathcal{H},\ T_1 = \mathcal{G} \\
T_w &= T_{w_n}...T_{w_1}, \text{ where } w = w_n...w_1 \in \{0,1\}^n \\
X^w &= T_w X
\end{aligned}
$$

In other words, we encode the paths of the tree via a dyadic coding; when infinite paths are considered, the associated coding is the *dyadic expansion of the real number defined by this infinite path*. We are thus interested in the limit of the statistics of X^w when the length $|w|$ of the word w tends to infinity. Again, we consider two separate cases:

Case of a pure frequency.

The following formula holds:

$$\mathcal{R}_{X^w} = \left[\prod_{m=1}^{|w|} u_{w_m}(2^m \omega_o)\right] \left[\delta_{2^{|w|}\omega_o} + \delta_{-2^{|w|}\omega_o}\right] \tag{61}$$

where we used the notation

$$u_0(\omega) = |H(e^{i\omega})|^2, \quad u_1(\omega) = |G(e^{i\omega})|^2$$

Now, we make the following key remark:

$$2^m \omega_o \bmod 1 \ \text{ yields the dyadic expansion of } \omega_o \tag{62}$$

Now, assume $\{H, G\}$ is a {low-pass / high-pass} pair (cf figure 17): From the above condition and the formulae (62,61), we get the following asymptotic result, which holds for $|w|$ "infinitely" large:

$$\mathcal{R}_{X^w} \neq 0 \Leftrightarrow w_m = 2^m \omega_o \bmod 1 \tag{63}$$

From this result, we derive the following

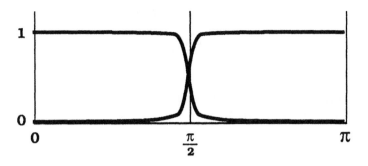

Fig. 17. Low-pass / high-pass QMF bank.

TEST:

the energy concentrates on the path encoding the dyadic expansion of ω_0. By the way, what we got is a

> **multiscale isolation of pure frequencies**

Case of a density $F(\omega)$ for X.

Introduce the following notations:

$$P_i \stackrel{\Delta}{=} P_{u_i}, \ i = 0, 1$$
$$P_w \stackrel{\Delta}{=} P_{w_n} ... P_{w_1}$$
$$F_w(\omega) \stackrel{\Delta}{=} P_w F(\omega) \text{ spectral density of } X^w$$

What we want to investigate here is

$$\lim_{|w| \to \infty} F_w = ???$$

A new difficulty arises here, namely P_w is a *non homogeneous* Markov Transition kernel (recall that, thanks to QMF, we do have $P_w 1 = 1$). Such non homogeneous operators have been studied by I. Daubechies and J. Lagarias in [15, 13] via tight bounds based on elementary linear operator algebra. On the other hand, Conze and Raugi [8] have imbedded this problem into that of the analysis of a single homogeneous Markov transition kernel: this is done by *randomizing the paths on the tree*, i.e. for each successive decimation, select H or G at random using a (fair) coin tossing. Hence w is randomized. Now consider the spectral density $F_w = P_w F$ of the process X^w: it is an L^2–function on the unit circle, and we have the following theorem:

Theorem 2 (Conze–Raugi) *For almost all infinite path* w,

$$P_w F \; \xrightarrow[n \to \infty]{L^2(\mathbf{T})} \; C_w$$

where C_w *is a constant spectral density with a power depending on the path* w, *and the convergence holds in the* L^2*-sense on the unit circle* $\mathbf{T}$*. In other words,* $X^w \to$ *white noise with power* C_w.

Case of several frequencies in coloured noise.

Since $C_w \ll$ Dirac, an iterative multiscale procedure to isolate frequencies in coloured noise can be derived from the theorem above.

VIII.   Conclusion

We have provided an account of those concepts of multiscale signal processing that are concerned with QMF techniques and orthonormal wavelet transforms. As opposed to most classical presentations of this subject, we started from discrete time signal processing and QMF banks and exlpoited their properties as far as possible to approach the construction of wavelets. Aside from providing a new insight on this topic, we think this presentation enlightened some questions, namely:

- How to exploit the suggestion of section 2. to generate other orthonormal decompositions of signal spaces and L^2-spaces? A proposal for this has been provided in [33] where an entropy criterion is presented to select a basis that is best fitted to a particular signal. Here, "best" refers to the mentioned entropy criterion. This method is applied to speech coding and the authors report successful results.

- Is it possible to derive parametrizations of loss-less 2–port transfer functions in the style of section C. to additionnally guaranty vanishing moment conditions such as provided by I. Daubechies' "explicit" examples? We have suggested to rely for this on the so-called "scattering theory" approach by translating such conditions into proper Nevanlinna-Pick interpolation problems. Further work has to be performed in this direction.

ACKNOWLEDGEMENT: this short tutorial is certainly an outcome of the "Analyse Multirésolution" working group at Rennes and surroundings[5], and it is worth to mention here Albert Cohen and Yves Meyer (both from CEREMADE, Paris), Patrick Flandrin (from Lyon), Loic Hervé, Albert Raugi, and also Ingrid Daubechies who was occasionnally invited to attend the group. Daniel Alpay is

---

[5] surroundings of Rennes include Rouen, Lyon, and Paris

acknowledged for helpful discussions on inverse scattering techniques to synthesize lossless filters. But, first and mostly, the author is indebted to Jean-Pierre Conze, who performed an outstanding work in rephrazing the work of the classics on wavelet transforms to make it accessible to nonprofessionnal mathematicians, and furthermore provided exciting new uses and points of views.

IX. REFERENCES

1. M. BASSEVILLE, A. BENVENISTE, A.S. WILLSKY, "Multiscale Autoregressive Processes, Part I : Schur-Levinson Parametrizations, and Part II : Lattice Structures for Whitening and Modelling", *IEEE Trans. on Signal Processing*, vol SP-40 n°8, 1915–1954, 1992.

2. M. BASSEVILLE, A. BENVENISTE, K.C. CHOU, S.A. GOLDEN, R. NIKOUKHAH, A.S. WILLSKY, " Modelling and Estimation of Multiresolution Stochastic Processes", *IEEE Trans. on Information Theory*, special issue on Wavelet Transforms and Multiresolution Signal Analysis, vol IT-38 n°2, 766–784, 1992.

3. A. BENVENISTE, R. NIKOUKHAH, A.S. WILLSKY, "Multiscale system theory", Proc. of the CDC'90, Honolulu, 1990 (extended version submitted for publication)

4. G. BEYLKIN, R. COIFMAN, V. ROKHLIN, "Fast wavelet transforms", preprint Yale University, 1989.

5. A. COHEN, "Ondelettes, analyses multirésolution, et filtres miroir en quadrature", *Ann. Inst. Poincaré*, vol 7, 439–459, 1990.

6. A. COHEN, "Construction de bases d'ondelettes α–Hölderiennes", to appear in *Revista Matematica Iberoamericana*, 1991.

7. R.R. COIFMAN, M.V. WICKERHAUSER, "Entropy-based algorithms for Best Basis Selection", *IEEE Trans. on Information Theory*, special issue on Wavelet Transforms and Multiresolution Signal Analysis, vol IT-38 n°2, 713–718, 1992.

8. J-P CONZE, A. RAUGI, "Fonctions harmoniques pour un opérateur de transition et applications," *Bull. Soc. Math. de France*, 118, n° 2, 101-138, 1990.

9. J-P CONZE, "Sur la régularité des solutions d'une équation fonctionnelle", IRMAR res. rep., IRMAR, Université de Rennes I. 1990.

10. J-P CONZE, course in preparation, IRMAR, Université de Rennes I.

11. R.E. CROCHIERE, L.R. RABINER, *Multirate digital signal processing*, Engle-wood Cliffs, NJ: Prentice Hall, 1983.

12. I. DAUBECHIES, "Orthonormal bases of compactly supported wavelets," *Communications on Pure and Applied Math.*, vol. 91, pp. 909–996, 1988.

13. I. DAUBECHIES, *Ten Lectures on Wavelets*, SIAM, Philadelphia, Pennsylvania, 1992.

14. I. DAUBECHIES, "The wavelet transform, time-frequency localization and signal analysis," *IEEE Trans. on Information Theory*, vol 36, 961–1005, Sept. 1990.

15. I. DAUBECHIES, J. LAGARIAS, "Two–scale difference equations, II local regularity, infinite products of matrices, and fractals," preprint AT & T Bell Laboratories, 1989.

16. P. DEWILDE, H. DYM, "Lossless chain scattering matrices and optimum linear prediction: the vector case", *Circuit Theory and Applications*, vol 9, 135-175, 1980.

17. P. DEWILDE, H. DYM, "Lossless inverse scattering, digital filters, and estimation theory", *IEEE Trans. on IT*, IT 30, 644-661, 1984.

18. H. DYM, *J-contractive matrix functions, reproducing kernel Hilbert spaces and interpolation*, Regional conference series in mathematics, vol. 71, Amer. Math. Soc., Providence, R-I, 1989.

19. D. ESTEBAN, C. GALAND, "Application of quadrature mirror filters to split–band voice coding schemes", Proc. of the ICASSP 1977, 191-195.

20. J-C. FAUVEAU, "Analyse multirésolution par ondelettes non orthogonales et bancs de filtres numériques", Thèse de Doctorat de l'Université de Paris-Sud, 1990.

21. A. FETTWEISS, "Wave digital filters: theory and practice", *Proc. of the IEEE*, vol 74 No 2, 270-327, 1986.

22. W. HACKBUSCH AND U. TROTTENBERG, eds., *Multigrid Methods*, Springer-Verlag, New York, 1982.

23. S.G. MALLAT, "A theory for multiresolution signal decomposition: the wavelet representation," *IEEE Trans. on PAMI,*, 11 No 7, July 1989.

24. S.G. MALLAT, "Multiresolution approximation and wavelets orthonormal bases of $L^2(\mathbf{R})$," *Trans. Amer. Math. Soc.*, 315, No1 , 69-88, 1989.

25. S. MCCORMICK, *Multigrid Methods*, Vol. 3 of the SIAM Frontiers Series, SIAM, Philadelphia, 1987.

26. Y. MEYER, "Le calcul scientifique, les ondelettes et les filtres miroir en quadrature," preprint CEREMADE, Université de Paris IX Dauphine, 1990.

27. Y. MEYER, *Wavelets and operators*, Proceedings of the Special year in modern Analysis, Urbana 1986/87, published by Cambridge University Press, 1989. See also Y. MEYER, *Ondelettes et Opérateurs*, Hermann, Paris, 1990.

28. T.Q. NGUYEN, P.P. VAIDYANATHAN, "Structures for M-channel perfect reconstruction FIR QMF filter banks which yield linear phase analysis filters", *IEEE-ASSP 38 No 3*, 433-446, 1990.

29. M.J. SMITH AND T.P. BARNWELL, "Exact reconstruction techniques for tree-structured subband coders," *IEEE Trans. on ASSP*, vol. 34, pp. 434–441, 1986.

30. P.P. VAIDYANATHAN, "Quadrature mirror filter banks, M-band extensions and perfect–reconstruction techniques,"*IEEE ASSP Magazine 4 No 3*, 4-20, 1987.

31. P.P. VAIDYANATHAN, "Theory and design of M-channel maximally decimated quadrature mirror filters with arbitrary M, having perfect reconstruction property", *IEEE ASSP-35*, 476-492, 1987.

32. M. VETTERLI, "Wavelets and Filter Banks : Theory and Design", *IEEE Trans. on Signal Processing*, vol SP-40 n°9, 2207–2232, 1992.

33. M.V. WICKERHAUSER, "Acoustic signal compression with wave-packets", Dept. of math, Yale University, 1990.

The Design of Frequency Sampling Filters

Peter A. Stubberud
University of Nevada, Las Vegas

Cornelius T. Leondes
University of California, San Diego

Abstract

Under certain conditions, a frequency sampling filter can implement a
linear phase filter more efficiently than an equivalent filter implemented by a
direct convolution structure. However, the system function of a frequency
sampling filter requires pole-zero cancellations on the unit circle. For
practical implementations, finite word length effects usually prevent exact
pole-zero cancellation which can result in filter instability. To prevent instability, the poles and zeros on the unit circle can be moved to a circle of radius
r where $0 < r < 1$ by replacing z^{-1} with rz^{-1} in the filter's system function.
The resulting filter is guaranteed to be stable; however, the filter's magnitude
and phase characteristics are affected by the choice of r. This chapter develops a method for determining optimal coefficients which minimizes a linear
combination of the mean square error in the stopband and passband, and the
sum of square error of the impulse response symmetry subject to passband and
stopband constraints and a fixed value of $r < 1$. The optimization problem is
solved using the Lagrange multiplier optimization method which results in a
set of linear equations, the solution of which determines the filter's
coefficients.

I. Introduction

Many digital signal processing systems require linear phase filtering.
Digital linear phase filters designed by either the window design method[1] or
the optimal filter design method[1] are generally implemented by the direct
convolution method which uses the filter's impulse response as filter coefficients. If a linear phase filter has a finite impulse response (FIR) of length N,
then the filter's impulse response has the form

$$h(n) = \begin{cases} h(N-1-n) & 0 \le n \le N-1 \\ 0 & \text{otherwise} \end{cases}$$

CONTROL AND DYNAMIC SYSTEMS, VOL. 68
Copyright © 1995 by Academic Press, Inc.
All rights of reproduction in any form reserved.

and a direct convolution implementation of the filter requires $(N+1)/2$ multiplies and N-1 adds per output sample when N is odd and $N/2$ multiplies and N-1 adds per output sample when N is even [1]. As a filter's passband or transition band narrows or its stopband requirements become more stringent, the length of the filter's impulse response increases. Therefore, if the filter is implemented using direct convolution, the filter's computational requirements also increase.

Unlike direct convolution implementations which use the filter's impulse response as coefficients in the filter's implementation, frequency sampling filters use frequency samples, which are specific values from the filter's frequency response, as coefficients in the filter's implementation. The frequency sampling filter design technique discussed in this chapter interpolates a frequency response from a set of N evenly spaced samples from the filter's frequency response. Although frequency sampling filters interpolate a frequency response through a set of frequency samples, the frequency response may not be well behaved between samples. References [2; 3; 4; 5; 6; 7; 8; 9; [10; 11; 12] describe some of the design methods currently used to control the interpolation errors between frequency samples. When using a frequency sampling filter to implement a frequency selective filter, the frequency samples that lie in the filter's stopband can be set to zero. Thus, all of the non-zero frequency samples lie in the filter's passband and transition band. Therefore, as a frequency sampling filter's passband and transition band narrow and the filter's stopband increases, the computational requirements of a frequency sampling filter can decrease. Thus, a frequency sampling filter has the potential to implement narrowband frequency selective filters more efficiently than a direct convolution filter.

Unlike a direct convolution implementation which is nonrecursive and thus inherently stable, a frequency sampling filter uses a recursive structure which places poles on the unit circle to cancel zeros on the unit circle. In practical implementations, finite word length effects typically prevent exact pole-zero cancellation. An uncancelled pole on the unit circle will cause a frequency sampling filter to be unstable. To prevent instability, the poles and zeros on the unit circle can be moved to a circle of radius r where $0 < r < 1$ by replacing z^{-1} with rz^{-1} in the filter's system function. As a result, the filter is guaranteed to be stable, but the frequency response of the resultant frequency sampling filter is different from that of the original frequency sampling filter. The methods in references [2; 3; 4; 5; 6; 7; 8; 9; 10; 11; 12] design frequency sampling filters assuming $r = 1$, and then a value of r close to 1 is chosen so that the modified filter's frequency response does not differ much from the frequency response designed for $r = 1$. However, references [13; 14; 15; 16] show that the filter's output roundoff noise decreases as r decreases, and reference [13] also show that the filter's signal to noise ratio (SNR) improves as r decreases. As a result, there exists a need for a design technique which determines a value of r which produces an acceptable roundoff noise level while satisfying various frequency response design constraints. In this chapter, such a design method is developed. This design method minimizes a linear combination of the mean square error between the desired and actual frequency responses in the passband and stopband, and the sum of square error of the impulse response symmetry while constraining r to a fixed value and the stopband frequency samples to zero. Additional constraints in the passband may be used to approximate a desired passband response. This results in a constrained optimization problem which can be solved by using the Lagrange multiplier optimization method. The method applies to both Type 1 and Type 2 frequency sampling filters.

II. Frequency Sampling Filters

A. Type 1 Frequency Sampling Filters

Consider a FIR filter which has an impulse response, h(n), of length N and a frequency response, $H(e^{j\omega})$. Suppose we approximate $H(e^{j\omega})$ by a set of N values taken from the frequency response. Let H(k) for $k \in D$ where $D = \{0, 1, 2, \ldots, N-1\}$ represent this set of values. For a Type 1 frequency sampling filter, we select the set H(k) for $k \in D$ so that

$$H(k) = H(e^{j\omega})\Big|_{\omega = \frac{2\pi}{N}k}$$

where H(k) can be written as

$$H(k) = |H(k)|e^{j\theta(k)}.$$

Thus, for a Type 1 frequency sampling filter, the set, $H(k)^1$ for $k \in D$, is chosen so that it represents a set of N evenly spaced samples taken from the filter's frequency response for $0 \le \omega < 2\pi$. The impulse response, h(n), of length N which interpolates a frequency response through the set of frequency samples, H(k) for $k \in D$, can be determined from the inverse discrete Fourier transform (IDFT),

$$h(n) = \frac{1}{N}\sum_{k=0}^{N-1} H(k)e^{j\frac{2\pi}{N}kn}, \tag{1}$$

and the set of frequency samples, H(k) for $k \in D$, can be determined from the filter's impulse response by the discrete Fourier transform (DFT),

$$H(k) = \sum_{n=0}^{N-1} h(n)e^{-j\frac{2\pi}{N}kn}. \tag{2}$$

If we let H(z) represent the z transform of h(n), then the system function, H(z), of the filter which interpolates a frequency response through the set of frequency samples, H(k) for $k \in D$, is

---

1. H(k) represents the system function, H(z), evaluated at $z = e^{j(2\pi/N)k}$. Although this notation is a mathematical faux pas, it is commonly used throughout the literature and this chapter.

$$H(z) = \sum_{n=-\infty}^{\infty} h(n)z^{-n} = \sum_{n=0}^{N-1} h(n)z^{-n} \tag{3}$$

By substituting Equation (1) into Equation (3), interchanging the order of summation and summing over the n index, H(z) becomes

$$H(z) = \frac{1 - z^{-N}}{N} \sum_{k=0}^{N-1} \frac{H(k)}{1 - e^{j\frac{2\pi}{N}k}z^{-1}} \tag{4}$$

Equation (4) has the form of the Lagrange interpolating formula. The complex function, H(z), interpolates a polynomial through the set of points H(k) for $k \in D$ when $z = e^{j(2\pi/N)k}$ for $k \in D$ so that

$$H(z)\big|_{z=e^{j(2\pi/N)k}} = H(k)$$

Thus as desired, the frequency response passes through the set of N evenly spaced samples.

    Equation (4) can be expressed in a computationally more efficient form if we constrain the filter to have a real impulse response in which case the frequency response satisfies

$$H(e^{j\omega}) = \overline{H}(e^{-j\omega})$$

where $\overline{H}(e^{-j\omega})$ is the complex conjugate of $H(e^{-j\omega})$. This implies that the frequency samples, H(k) for $k \in D$, which span the frequency response for $0 \le \omega < 2\pi$ will have the form

$$H(k) = \overline{H}(N - k)$$

where $\overline{H}(N-k)$ is the complex conjugate of H(N-k). This implies that

$$|H(k)| = |H(N - k)|$$

and

$$\theta(k) = -\theta(N - k).$$

Substituting this frequency response constraint into Equation (4), H(z) can be written as

$$H(z) = \frac{1 - r^N z^{-N}}{N} \left[\frac{H(0)}{1 - rz^{-1}} \right.$$

$$\left. + \sum_{k=1}^{\frac{N-1}{2}} \frac{2|H(k)|\left[\cos(\theta(k)) - rz^{-1}\cos\left(\theta(k) - \frac{2\pi}{N}k\right)\right]}{1 - 2\cos\left(\frac{2\pi}{N}k\right)rz^{-1} + r^2 z^{-2}} \right] \tag{5}$$

for N odd and

$$H(z) = \frac{1 - r^N z^{-N}}{N} \left[\frac{H(0)}{1 - rz^{-1}} + \frac{H\left(\frac{N}{2}\right)}{1 + rz^{-1}} \right.$$

$$\left. + \sum_{k=1}^{\frac{N}{2}-1} \frac{2|H(k)|\left[\cos(\theta(k)) - rz^{-1}\cos\left(\theta(k) - \frac{2\pi}{N}k\right)\right]}{1 - 2\cos\left(\frac{2\pi}{N}k\right)rz^{-1} + r^2 z^{-2}} \right] \tag{6}$$

for N even where r = 1 [4]. Equation (5) can then be realized by the structure shown in Figure 1A where Figure 1B describes the structure of the kth resonator[2]. Equation (6) can also be realized by the structure in Figure 1 if a structure realizing

$$\frac{H\left(\frac{N}{2}\right)}{1 + rz^{-1}}$$

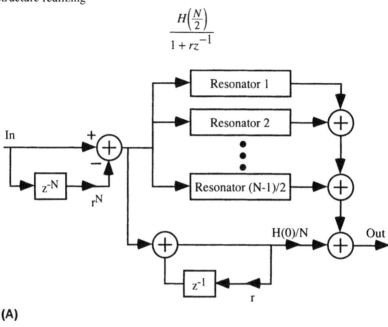

(A)

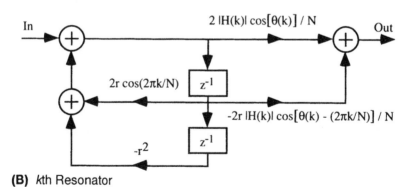

(B) kth Resonator

Figure 1. Structure for a Type 1 Frequency Sampling Filter with a Real Finite Impulse Response of Length N Where N is Odd.

---

2. The term *resonator* is used in this chapter to denote a system which has either a single pole or a complex conjugate pair of poles on or near the unit circle.

is placed in parallel with the resonators. This type of frequency sampling filter is called a Type 1 frequency sampling filter. When implementing this filter, the zeros created by the term $(1-z^{-N})$ exactly cancel the poles in the resonator sections created by the terms $(1-2\cos(2\pi k/N)z^{-1}+z^{-2})$ for $k \in D$. However, when the filter is implemented with finite precision arithmetic, this cancellation usually does not occur. Therefore to ensure stability, z^{-1} is replaced by $rz^{-1}, r < 1$. It should be noted that the frequency sampling filter represented by Equations (5) and (6) does not necessarily have linear phase.

B. Type 2 Frequency Sampling Filters

A second type of frequency sampling filter can be designed by interpolating a frequency response through a set of N evenly spaced frequency samples starting at $\omega = \pi/N$ instead of $\omega = 0$. This type of frequency sampling filter is called a Type 2 frequency sampling filter.

To develop Type 2 frequency sampling filters, we relate the system function and the frequency samples, H(k) for $k \in D$, as follows

$$H(k) = |H(k)| e^{j\theta(k)} = H(z)\Big|_{z=e^{j\frac{2\pi}{N}\left(k+\frac{1}{2}\right)}}$$

A transformation between H(k) and h(n) can be established by defining a modified DFT as

$$H(k) = \sum_{n=0}^{N-1} h(n)e^{-j\frac{2\pi}{N}\left(k+\frac{1}{2}\right)n} \tag{7}$$

and a modified IDFT as

$$h(n) = \frac{1}{N}\sum_{k=0}^{N-1} H(k)e^{j\frac{2\pi}{N}\left(k+\frac{1}{2}\right)n}$$

Using the modified IDFT expression for h(n), the z transform of h(n) can be written as

$$H(z) = \sum_{n=0}^{N-1}\left[\frac{1}{N}\sum_{k=0}^{N-1} H(k)e^{j\frac{2\pi}{N}\left(k+\frac{1}{2}\right)n}\right]z^{-n}$$

Interchanging the order of the summations and performing the summation over the n index, yields

$$H(z) = \frac{1+z^{-N}}{N}\sum_{k=0}^{N-1}\frac{H(k)}{1-e^{j\frac{2\pi}{N}(k+1/2)}z^{-1}} \cdot \tag{8}$$

Equation (8) can be expressed in a computationally more efficient form if we constrain the filter to have a real impulse response in which case the frequency response will have the form

$$H(e^{j\omega}) = \overline{H}(e^{-j\omega})$$

Thus, the frequency samples, H(k) for $k \in D$, which span the frequency response for $0 \le \omega < 2\pi$ will have the form

$$H(k) = \overline{H}(N-1-k).$$

This implies that

$$|H(k)| = |H(N - 1 - k)|$$

and

$$\theta(k) = -\theta(N - 1 - k).$$

Substituting this constraint into Equation (8), the system function, H(z), for Type 2 frequency sampling filters can be written as

$$H(z) = \frac{1 + r^N z^{-N}}{N} \left[\frac{H\left(\frac{N-1}{2}\right)}{1 + rz^{-1}} \right.$$

$$\left. + \sum_{k=0}^{\frac{N-3}{2}} \frac{2|H(k)|\left\{\cos(\theta(k)) - rz^{-1}\cos\left[\theta(k) - \frac{2\pi}{N}\left(k + \frac{1}{2}\right)\right]\right\}}{1 - 2\cos\left[\frac{2\pi}{N}\left(k + \frac{1}{2}\right)\right]rz^{-1} + r^2 z^{-2}} \right] \qquad (9)$$

for N odd and

$$H(z) = \frac{1 + r^N z^{-N}}{N} \left[\sum_{k=0}^{\frac{N}{2}-1} \frac{2|H(k)|\left\{\cos(\theta(k)) - rz^{-1}\cos\left[\theta(k) - \frac{2\pi}{N}\left(k + \frac{1}{2}\right)\right]\right\}}{1 - 2\cos\left[\frac{2\pi}{N}\left(k + \frac{1}{2}\right)\right]r z^{-1} + r^2 z^{-2}} \right]$$

$$(10)$$

for N even where r = 1 [4]. When implementing a Type 2 frequency sampling filter with finite precision arithmetic, the value of r in Equations (9) and (10) is chosen less than 1 to guarantee stability. Equation (10) can be realized by the structure shown in Figure 2A where Figure 2B shows the realization of the *k*th resonator. Equation (9) can also be realized by the structure shown in Figure 2 if a structure realizing

$$\frac{H\left(\frac{N-1}{2}\right)}{N}$$
$$\frac{}{1 + rz^{-1}}$$

is placed in parallel with the resonators. Note that the frequency sampling filter represented by Equations (9) and (10) does not necessarily have linear phase.

C. Computational Advantage of Frequency Sampling Filters

When most of the frequency sampling filter's frequency samples, the H(k)'s, are exactly zero, most of the frequency sampling filter's resonators do not need to be realized. Therefore, in the case of a narrow band filter where only a small number of the filter's frequency samples are non-zero, the resulting structure may require fewer arithmetic operations than the direct convolution structure.

Each of the frequency sampling filters described by Equations (5), (6), (9) and (10) requires 4 multiplies and 3 adds per resonator. If only K of the frequency samples are non-zero, then the frequency sampling structure requires no more than 4K+1 multiplies and 3K+1 adds per output sample. If a filter with a finite impulse response of length N was implemented using the direct convolution method, it would require approximately N/2 multiplies and approximately N adds per output sample. Because multiplications typically require more time to compute and are more complex to implement than adds, a frequency sampling structure becomes computationally more efficient than a direct convolution implementation when it uses fewer multiplies. Therefore, if we wish to implement a linear phase filter which has an impulse response of length of N, a frequency sampling filter described by any of the Equations (5), (6), (9) or (10) can be implemented more efficiently (in the sense of fewer multiplies) than a direct convolution filter when 4K < N/2 or K < N/8.

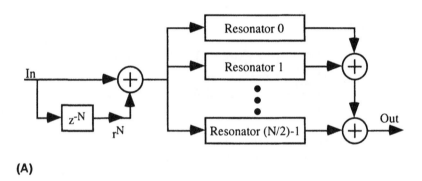

(A)

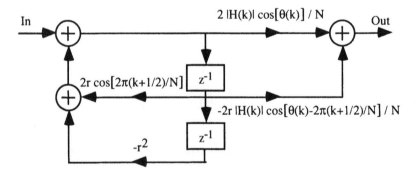

(B) kth Resonator

Figure 2. Structure for a Type 2 Frequency Sampling Filter with a Real Finite Impulse Response of Length N Where N is Even.

III. A Frequency Sampling Filter Design Method Which Accounts for Finite Word Length Effects

In this section, the frequency sampling filter design problem for $r < 1$ is developed as a constrained optimization problem which is solved by the method of Lagrange multipliers.

It was shown in section II of this chapter that a frequency sampling filter approximates a desired frequency response by interpolating a frequency response through a set of frequency samples. Although the frequency response of a frequency sampling filter passes through the frequency samples, the frequency response may not be well behaved between the specific samples. The design method developed in this chapter controls the interpolation errors between frequency samples by minimizing a weighted mean square error between the desired and actual frequency responses in the stopband and passband subject to passband constraints.

If we design a linear phase filter which minimizes the mean square error between a desired frequency response, $H_d(e^{j\omega})$, and the filter's frequency response, $H(e^{j\omega})$, then the design method would determine the function, $H(e^{j\omega})$, so that $H(e^{j\omega})$ minimizes the quantity

$$\int_{-\pi}^{\pi} \left| H(e^{j\omega}) - H_d(e^{j\omega}) \right|^2 d\omega$$

A filter's frequency response is generally specified in terms of passband and stopband requirements, but the transition band usually has no requirements. For these types of filter specifications, the mean square error criterion is overly restrictive because it requires the mean square error to be minimized in the transition band. The transition band error is minimized at the cost of further improvement in the filter's passband and stopband performance. Thus, in this design technique, the mean square error criterion is not used in the transition band.

For a frequency sampling filter to be computationally efficient, the frequency samples in the stopband are constrained to be identically zero. Because of the large number of these stopband constraints, we are usually prevented from constraining $h(n) = h(N-1-n)$ for $n = 0, 1, ..., N-1$, which is the necessary and sufficient condition for a FIR filter to have linear phase. To approximate linear phase, this design technique minimizes the sum squared error of the impulse response symmetry.

The design method developed in this section determines the optimal coefficients which minimizes a linear combination of the mean square error in the stopband and passband, and the sum of square error of the impulse response symmetry subject to passband and stopband constraints and a fixed value of $r < 1$. The optimization problem is solved using the Lagrange multiplier optimization method which results in a set of linear equations, the solution of which determines the filter's coefficients. The design method applies to both Type 1 and Type 2 frequency sampling filters.

The frequency response of a frequency sampling filter which has a finite impulse response, h(n), of length N can be expressed as

$$H(e^{j\omega}) = \sum_{n=0}^{N-1} h(n)e^{-j\omega n} = \mathbf{h}^T \mathbf{s}(e^{j\omega}) = \mathbf{s}^T(e^{j\omega})\mathbf{h} \qquad (11)$$

where

$$\mathbf{h} = \begin{bmatrix} h(0) \\ h(1) \\ h(2) \\ \vdots \\ h(N-1) \end{bmatrix}, \quad \mathbf{s}(e^{j\omega}) = \begin{bmatrix} 1 \\ e^{-j\omega} \\ e^{-j2\omega} \\ \vdots \\ e^{-j(N-1)\omega} \end{bmatrix}$$

and the superscript T denotes transpose.

If we let J_{sb} represent the mean square error over the stopband frequencies, then

$$J_{sb} = \frac{1}{m(\omega_{sb})} \int_{\omega \in \omega_{sb}} \left| H(e^{j\omega}) - H_d(e^{j\omega}) \right|^2 d\omega \qquad (12)$$

where ω_{sb} is the set of stopband frequencies, $m(\omega_{sb})$ is the linear measure of the set ω_{sb}, and $H_d(e^{j\omega})$ is the desired frequency response. Because $H_d(e^{j\omega})$ is equal to zero in the stopband, Equation (12) can be written as

$$J_{sb} = \frac{1}{m(\omega_{sb})} \int_{\omega \in \omega_{sb}} \left| H(e^{j\omega}) \right|^2 d\omega \qquad (13)$$

Substituting Equation (11) into Equation (13) and restricting $\mathbf{h}$ to be real

$$J_{sb} = \frac{1}{m(\omega_{sb})} \int_{\omega \in \omega_{sb}} \mathbf{h}^T \mathbf{s}(e^{j\omega})\bar{\mathbf{s}}^T(e^{j\omega})\mathbf{h} \ d\omega$$

where $\bar{\mathbf{s}}(e^{j\omega})$ is the complex conjugate of $\mathbf{s}(e^{j\omega})$. If we let $\mathbf{W}(e^{j\omega}) = \mathbf{s}(e^{j\omega})\bar{\mathbf{s}}^T(e^{j\omega})$ then

$$\mathbf{h}^T \mathbf{s}(e^{j\omega})\bar{\mathbf{s}}^T(e^{j\omega})\mathbf{h} = \mathbf{h}^T \mathbf{W}(e^{j\omega})\mathbf{h}$$

where

$$\mathbf{W}(e^{j\omega}) = \begin{bmatrix} 1 & e^{j\omega} & e^{j2\omega} & \cdots & e^{j(N-1)\omega} \\ e^{-j\omega} & 1 & e^{j\omega} & \cdots & e^{j(N-2)\omega} \\ e^{-j2\omega} & e^{-j\omega} & 1 & \cdots & e^{j(N-3)\omega} \\ \vdots & \vdots & \vdots & \ddots & \vdots \\ e^{-j(N-1)\omega} & e^{-j(N-2)\omega} & e^{-j(N-3)\omega} & \cdots & 1 \end{bmatrix}$$

It can be shown[17] that quadratic expressions of the form $\mathbf{h}^T \mathbf{W}(e^{j\omega}) \mathbf{h}$ can be written as

$$\mathbf{h}^T \mathbf{W}(e^{j\omega})\mathbf{h} = \mathbf{h}^T \left(\frac{\mathbf{W}(e^{j\omega}) + \mathbf{W}^T(e^{j\omega})}{2} \right)\mathbf{h} + \mathbf{h}^T \left(\frac{\mathbf{W}(e^{j\omega}) - \mathbf{W}^T(e^{j\omega})}{2} \right)\mathbf{h}$$

$$= \mathbf{h}^T \left(\frac{\mathbf{W}(e^{j\omega}) + \mathbf{W}^T(e^{j\omega})}{2} \right)\mathbf{h}$$

If we define

$$q(\omega) = \frac{W(e^{j\omega}) + W^T(e^{j\omega})}{2}$$

then

$$h^T W(e^{j\omega})h = h^T q(\omega)h$$

where $q(\omega)$ is a symmetric matrix. Substituting this expression into Equation (13),

$$J_{sb} = \frac{1}{m(\omega_{sb})} h^T \int_{\omega \in \omega_{sb}} q(\omega)\, d\omega \quad h$$

where

$$q(\omega) = \begin{bmatrix} 1 & \cos(\omega) & \cdots & \cos[(N-1)\omega] \\ \cos(\omega) & 1 & \cdots & \cos[(N-2)\omega] \\ \vdots & \vdots & \ddots & \cdots \\ \cos[(N-1)\omega] & \cos[(N-2)\omega] & \cdots & 1 \end{bmatrix}$$

Now let

$$Q(\omega) = \int q(\omega)\, d\omega$$

then

$$Q(\omega) = \begin{bmatrix} \omega & \sin(\omega) & \cdots & \dfrac{\sin[(N-1)\omega]}{N-1} \\ \sin(\omega) & \omega & \cdots & \dfrac{\sin[(N-2)\omega]}{N-2} \\ \vdots & \vdots & \ddots & \cdots \\ \dfrac{\sin[(N-1)\omega]}{N-1} & \dfrac{\sin[(N-2)\omega]}{N-2} & \cdots & \omega \end{bmatrix}$$

where $Q(\omega)$ is a symmetric matrix. If we let $Q_{ic}(\omega)$ represent the element in the ith row and the cth column of the matrix, $Q(\omega)$, then $Q(\omega)$ can be written as

$$Q_{ic}(\omega) = \begin{cases} \omega & i = c \\ \dfrac{\sin(c-i)\omega}{(c-i)} & i \neq c \end{cases}$$

where $i, c = 0, 1, \ldots , N\text{-}1$. If we define

$$Q_s = \int_{\omega \in \omega_{sb}} q(\omega)\, d\omega$$

where Q_s is generally calculated by evaluating $Q(\omega)$ at the appropriate stop-band limits, then J_{sb} can be written as

$$J_{sb} = \frac{1}{m(\omega_{sb})} h^T Q_s h$$

A desired passband performance can be obtained by minimizing the mean square error between the desired and actual amplitudes of the frequency responses over the set of passband frequencies, ω_{pb}. To simplify the expression for the mean square error over the set, ω_{pb}, we approximate the amplitude of $H(e^{j\omega})$ by a real function. If a FIR filter has linear phase, then its impulse response will have the form $h(n) = h(N\text{-}1\text{-}n)$, and

$H(e^{j\omega}) = H_r(\omega)e^{-j[(N-1)/2]\omega}$ where

$$H_r(\omega) = \begin{cases} \displaystyle\sum_{n=0}^{\frac{N}{2}-1} 2h(n)\cos\left[\omega\left(\frac{N-1}{2}-n\right)\right] & \text{N even} \\[4mm] h\left(\frac{N-1}{2}\right) + \displaystyle\sum_{n=0}^{\frac{N-1}{2}-1} 2h(n)\cos\left[\omega\left(\frac{N-1}{2}-n\right)\right] & \text{N odd} \end{cases}$$

for all $\omega \in \Re$ (real numbers). For our design technique, the passband frequency response should have a phase characteristic which is as close as possible to ideal linear phase characteristics, that is $h(n) \approx h(N-1-n)$. Under these conditions, the passband of the zero phase frequency response, $H_r(\omega)$, of the filter can be approximated by

$$H_r(\omega) = \begin{cases} \displaystyle\sum_{n=0}^{\frac{N}{2}-1} 2\frac{h(n)+h(N-1-n)}{2}\cos\left[\omega\left(\frac{N-1}{2}-n\right)\right] & \text{N even} \\[4mm] h\left(\frac{N-1}{2}\right) + \displaystyle\sum_{n=0}^{\frac{N-1}{2}-1} 2\frac{h(n)+h(N-1-n)}{2}\cos\left[\omega\left(\frac{N-1}{2}-n\right)\right] & \text{N odd} \end{cases}$$

for $\omega \in \omega_{pb}$. If we let

$$\mathbf{x}(\omega) = \begin{bmatrix} \cos\left(\frac{N-1}{2}\omega\right) \\ \cos\left(\frac{N-3}{2}\omega\right) \\ \vdots \\ 1 \\ \vdots \\ \cos\left(\frac{N-3}{2}\omega\right) \\ \cos\left(\frac{N-1}{2}\omega\right) \end{bmatrix}_{N \text{ odd}} \quad , \quad \mathbf{x}(\omega) = \begin{bmatrix} \cos\left(\frac{N-1}{2}\omega\right) \\ \cos\left(\frac{N-3}{2}\omega\right) \\ \vdots \\ \cos\left(\frac{1}{2}\omega\right) \\ \cos\left(\frac{1}{2}\omega\right) \\ \vdots \\ \cos\left(\frac{N-3}{2}\omega\right) \\ \cos\left(\frac{N-1}{2}\omega\right) \end{bmatrix}_{N \text{ even}}$$

then

$$H_r(\omega) = \mathbf{h}^T \mathbf{x}(\omega) = \mathbf{x}^T(\omega)\mathbf{h} \qquad (14)$$

for $\omega \in \omega_{pb}$.

   If we let J_{pb} represent the mean square error over the passband frequencies, then

$$J_{pb} = \frac{1}{m(\omega_{pb})}\int_{\omega \in \omega_{pb}}\left[H_r(\omega) - H_d(\omega)\right]^2 d\omega \qquad (15)$$

where ω_{pb} is the set of passband frequencies, $m(\omega_{pb})$ is the linear measure of the set ω_{pb}, and $H_d(\omega)$ is the amplitude of the desired frequency response. Substituting Equation (14) into Equation (15), J_{pb} can be written as

$$J_{pb} = \frac{1}{m(\omega_{pb})}\left[\mathbf{h}^T \int_{\omega\in\omega_{pb}} \mathbf{x}(\omega)\mathbf{x}^T(\omega)\,d\omega\,\mathbf{h} \right.$$

$$\left. -2\int_{\omega\in\omega_{pb}} H_d(\omega)\mathbf{x}^T(\omega)\,d\omega\,\mathbf{h} + \int_{\omega\in\omega_{pb}} H_d^2(\omega)d\omega \right] \quad (16)$$

If we let $\mathbf{Y}(\omega) = \mathbf{x}(\omega)\,\mathbf{x}^T(\omega)$, then

$$Y_{ic}(\omega) = \cos\left[\left(i - \frac{N-1}{2}\right)\omega\right]\cos\left[\left(c - \frac{N-1}{2}\right)\omega\right]$$

where $Y_{ic}(\omega)$ represents the element in the ith row and the cth column of the matrix, $\mathbf{Y}(\omega)$. If we let

$$\mathbf{Z}(\omega) = \int \mathbf{Y}(\omega)\,d\omega$$

then the element in the ith row and the cth column of the matrix, $\mathbf{Z}(\omega)$ can be written as

$$Z_{ic}(\omega) = \begin{cases} \omega & i = c = \dfrac{N-1}{2} \\[3ex] \dfrac{\omega}{2} + \dfrac{\sin\left[2\left(i - \dfrac{N-1}{2}\right)\omega\right]}{4\left(i - \dfrac{N-1}{2}\right)} & i = c \neq \dfrac{N-1}{2} \text{ and } i = -c + N - 1 \\[4ex] \dfrac{\sin[(i+c-N+1)\omega]}{2(i+c-N+1)} + \dfrac{\sin[(i-c)\omega]}{2(i-c)} & \text{otherwise} \end{cases}$$

where i, c = 0, 1, ... , N-1 for N odd and

$$Z_{ic}(\omega) = \begin{cases} \dfrac{\omega}{2} + \dfrac{\sin\left[2\left(i - \dfrac{N-1}{2}\right)\omega\right]}{4\left(i - \dfrac{N-1}{2}\right)} & i = c \text{ and } i = -c + N - 1 \\[4ex] \dfrac{\sin[(i+c-N+1)\omega]}{2(i+c-N+1)} + \dfrac{\sin[(i-c)\omega]}{2(i-c)} & \text{otherwise} \end{cases}$$

where i, c = 0, 1, ... , N-1 for N even. If we define

$$\mathbf{Z}_p = \int_{\omega\in\omega_{pb}} \mathbf{Y}(\omega)\,d\omega$$

where $\mathbf{Z}_p$ is generally calculated by evaluating $\mathbf{Z}(\omega)$ at the appropriate passband limits, then the term

$$\mathbf{h}^T \int_{\omega_{pb}} \mathbf{x}(\omega)\mathbf{x}^T(\omega)\,d\omega\,\mathbf{h}$$

in Equation (16) can be written as

$$\mathbf{h}^T\mathbf{Z}_p\mathbf{h}.$$

If we also define the following terms

$$\mathbf{R}^T(\omega) = \int H_d(\omega)\mathbf{x}^T(\omega)\,d\omega \qquad \mathbf{R}_p^T = \int_{\omega \in \omega_{pb}} H_d(\omega)\mathbf{x}^T(\omega)\,d\omega$$

$$\gamma(\omega) = \int |H_d(\omega)|^2\,d\omega \qquad \gamma_p = \int_{\omega \in \omega_{pb}} |H_d(\omega)|^2\,d\omega$$

then Equation (16) can be written as

$$J_{pb} = \frac{1}{m(\omega_{pb})}\left[\mathbf{h}^T\mathbf{Z}_p\mathbf{h} - 2\mathbf{R}_p^T\mathbf{h} + \gamma_p\right] \tag{17}$$

If $H_d(\omega)$ is equal to a constant in the passband, then without loss of generality, we can let $H_d(\omega) = 1$, and Equation (17) becomes

$$J_{pb} = \frac{1}{m(\omega_{pb})}\left[\mathbf{h}^T\mathbf{Q}_p\mathbf{h} - 2\mathbf{R}_p^T\mathbf{h}\right] + 1$$

where

$$\mathbf{R}_p^T = \int_{\omega \in \omega_{pb}} \mathbf{x}^T(\omega)\,d\omega$$

If we define

$$\mathbf{R}^T(\omega) = \int \mathbf{x}^T(\omega)\,d\omega$$

then

$$
\mathbf{R}(\omega) \;=\;
\begin{bmatrix}
\frac{2}{N-1}\sin\left(\frac{N-1}{2}\omega\right) \\
\frac{2}{N-3}\sin\left(\frac{N-3}{2}\omega\right) \\
\vdots \\
\omega \\
\vdots \\
\frac{2}{N-3}\sin\left(\frac{N-3}{2}\omega\right) \\
\frac{2}{N-1}\sin\left(\frac{N-1}{2}\omega\right)
\end{bmatrix}_{N\ \text{odd}}
,\qquad
\begin{bmatrix}
\frac{2}{N-1}\sin\left(\frac{N-1}{2}\omega\right) \\
\frac{2}{N-3}\sin\left(\frac{N-3}{2}\omega\right) \\
\vdots \\
2\sin\left(\frac{1}{2}\omega\right) \\
2\sin\left(\frac{1}{2}\omega\right) \\
\vdots \\
\frac{2}{N-3}\sin\left(\frac{N-3}{2}\omega\right) \\
\frac{2}{N-1}\sin\left(\frac{N-1}{2}\omega\right)
\end{bmatrix}_{N\ \text{even}}
$$

Thus, for filters which approximate constant values in their passbands, $\mathbf{R}_p$ can be calculated by evaluating $\mathbf{R}(\omega)$ at the appropriate passband limits eliminating the need for integrating $H_d(\omega)\mathbf{x}^T(\omega)$.

An approximation of a desired passband performance can also be obtained by imposing constraints on the amplitude of the passband frequency response and its derivatives at particular frequencies. Using the approximation of the passband amplitude, $H_r(\omega) = \mathbf{h}^T\mathbf{x}(\omega) = \mathbf{x}^T(\omega)\mathbf{h}$, the passband constraints can be written as

$$H_r(\omega)\big|_{\omega=\omega_0} = K_{00} \qquad \frac{dH_r(\omega)}{d\omega}\bigg|_{\omega=\omega_0} = K_{10} \quad \cdots \quad \frac{d^n H_r(\omega)}{d\omega^n}\bigg|_{\omega=\omega_0} = K_{n0}$$

$$\vdots \qquad\qquad\qquad \vdots \qquad\qquad\qquad\qquad \vdots$$

$$H_r(\omega)\big|_{\omega=\omega_m} = K_{0m} \qquad \frac{dH_r(\omega)}{d\omega}\bigg|_{\omega=\omega_m} = K_{1m} \quad \cdots \quad \frac{d^n H_r(\omega)}{d\omega^n}\bigg|_{\omega=\omega_m} = K_{nm}$$

where

$$\frac{d^n H_r(\omega)}{d\omega^n} = \frac{d^n \mathbf{x}^T(\omega)}{d\omega^n}\mathbf{h}.$$

If we let

$$\mathbf{C}_p^T = \left[\mathbf{x}(\omega_0) \quad \mathbf{x}(\omega_1) \quad \cdots \quad \frac{d\mathbf{x}(\omega)}{d\omega}\bigg|_{\omega=\omega_0} \quad \cdots \quad \frac{d^n\mathbf{x}(\omega)}{d\omega^n}\bigg|_{\omega=\omega_m} \right]$$

and

$$\mathbf{K}_p^T = \begin{bmatrix} K_{00} & K_{01} & \cdots & K_{10} & \cdots & K_{nm} \end{bmatrix},$$

the passband constraints can be written in the matrix form,

$$\mathbf{C}_p \mathbf{h} = \mathbf{K}_p.$$

Recall that one advantage of the frequency sampling structure is the fact that it can be an extremely efficient realization of a FIR filter when most of the filter's frequency samples are exactly zero. To set frequency samples in the stopband equal to zero for an ideal Type 1 frequency sampling filter with $r = 1$, we would constrain the frequency response to be zero at $\omega = 2\pi k/N$ when $k \in D$ and $2\pi k/N \in \omega_{sb}$. To get frequency samples equal to zero for an ideal Type 2 frequency sampling filters with $r = 1$, we would constrain the frequency response to be zero at $\omega = 2\pi(k+1/2)/N$ when $k \in D$ and $2\pi(k+1/2)/N \in \omega_{sb}$. However, when $r < 1$, the frequency samples no longer correspond to the frequency response evaluated at $\omega = 2\pi k/N$ for Type 1 frequency sampling filters and $\omega = 2\pi(k+1/2)/N$ for Type 2 frequency sampling filters. When $r < 1$ there exists a different relationship between the frequency response and the frequency samples.

The frequency response of a frequency sampling filter which has a finite impulse response, $h(n)$, of length N can be expressed as

$$H(e^{j\omega}) = \sum_{n=0}^{N-1} h(n)e^{-j\omega n} \tag{18}$$

When z^{-1} is replaced by rz^{-1}, the frequency sampling filter is modified by moving any unit circle poles and zeros off the unit circle and onto a circle of radius r where $r < 1$. The frequency response, $M(e^{j\omega})$, of this modified frequency sampling filter can be calculated by substituting $re^{-j\omega}$ for $e^{-j\omega}$ in Equation (18).

$$M(e^{j\omega}) = \sum_{n=0}^{N-1} h(n)\left(re^{-j\omega}\right)^n \qquad (19)$$

$$= \sum_{n=0}^{N-1} h(n)r^n e^{-j\omega n}$$

$$= \sum_{n=0}^{N-1} m(n)e^{-j\omega n}$$

The impulse response of $M(e^{j\omega})$ is $m(n)$ where $m(n) = h(n)\, r^n$. To compensate for this effect, $h(n)$ must be premultiplied by the sequence r^{-n}. If we call this set of coefficients $g(n)$, then $g(n) = h(n)\, r^{-n}$. The frequency samples, $G(k)$, which correspond to the impulse response $g(n)$ are determined from $G(e^{j\omega})$ where

$$G(e^{j\omega}) = \sum_{n=0}^{N-1} g(n)e^{-j\omega n} \qquad (20)$$

When these frequency samples are put into the modified frequency sampling filter, the filter's impulse response will be $h(n)$ and the filter will have the frequency response, $H(e^{j\omega})$. This can be shown by calculating the frequency response of the modified filter when the frequency samples, $G(k)\ k \in D$, are used instead of the $H(k)$'s. A frequency sampling filter that uses the frequency samples, $G(k)$, will have a frequency response given by Equation (20). If the frequency sampling filter is modified by replacing z^{-1} with rz^{-1}, the frequency response, $M(e^{j\omega})$, of this modified frequency sampling filter can be calculated by substituting $re^{-j\omega}$ for $e^{-j\omega}$ in Equation (20).

$$M(e^{j\omega}) = \sum_{n=0}^{N-1} g(n)\left(re^{-j\omega}\right)^n$$

$$= \sum_{n=0}^{N-1} h(n)r^{-n}r^n e^{-j\omega n}$$

$$= \sum_{n=0}^{N-1} h(n)e^{-j\omega n}$$

$$= H(e^{j\omega})$$

Thus, when the frequency samples, $G(k)\ k \in D$, are put into the modified frequency sampling filter, the filter's impulse response will be $h(n)$ and the filter will have frequency response $H(e^{j\omega})$. Therefore, to set the frequency samples in the stopband to zero, constrain

$$G(e^{j\omega})\Big|_{\omega=\frac{2\pi}{N}k} = 0 \quad k \in F \quad F = \left\{ k : k \in D \cap \frac{2\pi}{N}k \in \omega_{sb} \right\}$$

for Type 1 frequency sampling filters and

$$G(e^{j\omega})\Big|_{\omega=\frac{2\pi}{N}\left(k+\frac{1}{2}\right)} = 0 \quad k \in F \quad F = \left\{ k : k \in D \cap \frac{2\pi}{N}\left(k+\frac{1}{2}\right) \in \omega_{sb} \right\}$$

for Type 2 frequency sampling filters. Because $G(e^{j\omega})$ is a complex function, each of the stopband constraints will in general require two mathematical constraints to maintain **h** as a real variable. For example, let a Type 1 frequency sampling filter have the constraint,

$$G(e^{j\omega})\Big|_{\omega=\frac{2\pi}{N}k_0} = 0.$$

Therefore, if we specify **h** to be real, the above constraint requires that

$$\text{Re}\left[G(e^{j\omega})\right]\Big|_{\omega=\frac{2\pi}{N}k_0} = 0$$

and

$$\text{Im}\left[G(e^{j\omega})\right]\Big|_{\omega=\frac{2\pi}{N}k_0} = 0 \qquad\qquad (21)$$

If a constrained frequency sample occurs at $\omega_o = 0$ or $\omega_o = \pi$,

$$\text{Im}\left[G(e^{j\omega})\right]\Big|_{\omega=\omega_o} = 0$$

because $G(e^{j\omega})$ evaluated at ω_0 is a real function. Thus, the constraint in Equation (21) is redundant when $\omega_0 = 0$ or $\omega_0 = \pi$ and should be omitted.

To describe $G(e^{j\omega})$ in matrix form, we let

$$\mathbf{P} = \begin{bmatrix} 1 & 0 & 0 & \cdots & 0 \\ 0 & r^{-1} & 0 & \cdots & 0 \\ 0 & 0 & r^{-2} & \cdots & 0 \\ \vdots & \vdots & \vdots & \ddots & 0 \\ 0 & 0 & 0 & 0 & r^{-(N-1)} \end{bmatrix}$$

then

$$G(e^{j\omega}) = \mathbf{h}^T \mathbf{P} \mathbf{s}(e^{j\omega}) = \mathbf{s}^T(e^{j\omega})\mathbf{P}\mathbf{h}$$

and the stopband constraints can be written in the matrix form

$$\mathbf{C}_s \mathbf{h} = \mathbf{0}$$

where

$$
\mathbf{C}_s = \begin{bmatrix}
\mathrm{Re}\!\left[\mathbf{s}^T(e^{j\omega})\right]\Big|_{\omega=\omega_0} \\[4pt]
\mathrm{Im}\!\left[\mathbf{s}^T(e^{j\omega})\right]\Big|_{\omega=\omega_0} \\[4pt]
\mathrm{Re}\!\left[\mathbf{s}^T(e^{j\omega})\right]\Big|_{\omega=\omega_1} \\[4pt]
\mathrm{Im}\!\left[\mathbf{s}^T(e^{j\omega})\right]\Big|_{\omega=\omega_1} \\[4pt]
\vdots \\[4pt]
\mathrm{Re}\!\left[\mathbf{s}^T(e^{j\omega})\right]\Big|_{\omega=\omega_M} \\[4pt]
\mathrm{Im}\!\left[\mathbf{s}^T(e^{j\omega})\right]\Big|_{\omega=\omega_M}
\end{bmatrix} \mathbf{P}
$$

The passband and stopband constraints can be combined into a single equation and written as

$$
\mathbf{Ch} = \mathbf{K} \tag{22}
$$

where

$$
\mathbf{C} = \begin{bmatrix} \mathbf{C}_p \\ \hline \mathbf{C}_s \end{bmatrix} \quad \text{and} \quad \mathbf{K} = \begin{bmatrix} \mathbf{K}_p \\ \hline \mathbf{0} \end{bmatrix}.
$$

If most of the filter's frequency samples are exactly zero, the stopband constraints will consume most of the design's N degrees of freedom provided by the variable $\mathbf{h}$ because in general each of the frequency samples constrained to zero requires two constraints. As a result, there are not enough degrees of freedom left to constrain $h(n) = h(N\text{-}1\text{-}n)$ for $n = 0, 1, ..., N\text{-}1$, which is the necessary and sufficient condition for an FIR filter to have linear phase. Instead, we will approximate linear phase by minimizing the sum square error of the impulse response symmetry. If we denote the sum square error of the impulse response symmetry as J_{ph}, then

$$
J_{ph} = \begin{cases}
\displaystyle\sum_{n=0}^{\frac{N-1}{2}} [h(n) - h(N-1-n)]^2 & \text{N odd} \\[20pt]
\displaystyle\sum_{n=0}^{\frac{N}{2}-1} [h(n) - h(N-1-n)]^2 & \text{N even}
\end{cases}
$$

To express these error terms in matrix form, we let

$$
\mathbf{V} = \begin{bmatrix}
1 & 0 & 0 & 0 & \cdots & 0 & 0 & 0 & -1 \\
0 & 1 & 0 & 0 & \cdots & 0 & 0 & -1 & 0 \\
\vdots & \vdots & \ddots & \ddots & \vdots & \cdot\cdot & \cdot\cdot & \vdots & \vdots \\
0 & 0 & \cdots & 1 & 0 & -1 & \cdots & 0 & 0
\end{bmatrix}
$$

for N odd, and

$$V = \begin{bmatrix} 1 & 0 & 0 & \cdots & \cdots & 0 & 0 & -1 \\ 0 & 1 & 0 & \cdots & \cdots & 0 & -1 & 0 \\ \vdots & \vdots & \ddots & & & \ddots & \ddots & \vdots \\ 0 & 0 & \cdots & 1 & -1 & \cdots & 0 & 0 \end{bmatrix}$$

for N even. Then

$$Vh = \begin{bmatrix} h(0) - h(N-1) \\ h(1) - h(N-2) \\ h(2) - h(N-3) \\ \vdots \end{bmatrix}$$

and

$$J_{ph} = h^T V^T Vh$$

where the appropriate matrix, V, is used depending upon whether N is odd or even. If we define

$$A = V^T V$$

then J_{ph} can be written as

$$J_{ph} = h^T Ah$$

where

$$A = \begin{bmatrix} 1 & 0 & 0 & \cdots & 0 & \cdots & 0 & 0 & -1 \\ 0 & 1 & 0 & \cdots & 0 & \cdots & 0 & -1 & 0 \\ \vdots & \vdots & \ddots & \ddots & \vdots & \cdot^{\cdot^{\cdot}} & \cdot^{\cdot^{\cdot}} & \vdots & \vdots \\ 0 & 0 & \cdots & 1 & 0 & -1 & \cdots & 0 & 0 \\ 0 & 0 & \cdots & 0 & 0 & 0 & \cdots & 0 & 0 \\ 0 & 0 & \cdots & -1 & 0 & 1 & \cdots & 0 & 0 \\ \vdots & \vdots & \cdot^{\cdot^{\cdot}} & \cdot^{\cdot^{\cdot}} & \vdots & \ddots & \ddots & \vdots & \vdots \\ 0 & -1 & 0 & \cdots & 0 & \cdots & 0 & 1 & 0 \\ -1 & 0 & 0 & \cdots & 0 & \cdots & 0 & 0 & 1 \end{bmatrix}$$

for N odd, and

$$A = \begin{bmatrix} 1 & 0 & 0 & \cdots & \cdots & 0 & 0 & -1 \\ 0 & 1 & 0 & \cdots & \cdots & 0 & -1 & 0 \\ \vdots & \ddots & \ddots & 0 & 0 & \cdot^{\cdot^{\cdot}} & \cdot^{\cdot^{\cdot}} & \vdots \\ 0 & \cdots & 0 & 1 & -1 & 0 & \cdots & 0 \\ 0 & \cdots & 0 & -1 & 1 & 0 & \cdots & 0 \\ \vdots & \cdot^{\cdot^{\cdot}} & \cdot^{\cdot^{\cdot}} & 0 & 0 & \ddots & \ddots & \vdots \\ 0 & -1 & 0 & \cdots & \cdots & 0 & 1 & 0 \\ -1 & 0 & 0 & \cdots & \cdots & 0 & 0 & 1 \end{bmatrix}$$

for N even.

Assuming that the number of non-redundant constraints in Equation (22) is less than the number of elements in the vector h, the design problem can now be stated as follows. Minimize the error function

$$J(\mathbf{h}) = \alpha J_{pb} + \beta J_{sb} + (1 - \alpha - \beta)J_{ph} \qquad (23)$$

where $\alpha \geq 0$, $\beta \geq 0$ and $\alpha + \beta \leq 1$ subject to the constraints

$$\mathbf{Ch = K}.$$

The scalar terms α and β in Equation (23) allow the designer to weight the relative importance between the mean square error of the passband amplitude, the mean square error in the stopband and the sum square error of the impulse response symmetry.

This problem can be solved using the method of Lagrange multipliers by defining a Lagrange multiplier vector as

$$\boldsymbol{\lambda} = \begin{bmatrix} \lambda_{00} \\ \lambda_{01} \\ \vdots \\ \lambda_{nm} \\ \lambda_{SB1} \\ \lambda_{SB2} \\ \vdots \end{bmatrix}$$

and minimizing the augmented cost function,

$$J_a(\mathbf{h}, \boldsymbol{\lambda}) = \alpha J_{pb} + \beta J_{sb} + (1 - \alpha - \beta)J_{ph} + \boldsymbol{\lambda}^T(\mathbf{Ch - K}). \qquad (24)$$

Substituting the appropriate expressions for J_{pb}, J_{sb} and J_{ph} into Equation (24) yields,

$$J_a(\mathbf{h}, \boldsymbol{\lambda}) = \frac{\alpha}{m(\omega_{pb})}\left[\mathbf{h}^T\mathbf{Z}_p\mathbf{h} - 2\mathbf{R}_p^T\mathbf{h} + \gamma_p\right]$$

$$+ \frac{\beta}{m(\omega_{sb})}\mathbf{h}^T\mathbf{Q}_s\mathbf{h} + (1 - \alpha - \beta)\mathbf{h}^T\mathbf{Ah} + \boldsymbol{\lambda}^T(\mathbf{Ch - K})$$

The necessary conditions for an optimal solution are

$$\frac{\partial J_a(\mathbf{h}, \boldsymbol{\lambda})}{\partial \mathbf{h}} = 0 \qquad (25)$$

and

$$\frac{\partial J_a(\mathbf{h}, \boldsymbol{\lambda})}{\partial \boldsymbol{\lambda}} = 0 \qquad (26)$$

Because $\mathbf{Z}_p$, $\mathbf{Q}_s$ and $\mathbf{A}$ are symmetric matrices, Equation (25) becomes

$$\frac{2\alpha}{m(\omega_{pb})}\left[\mathbf{Z}_p\mathbf{h} - \mathbf{R}_p\right] + \frac{2\beta}{m(\omega_{sb})}\mathbf{Q}_s\mathbf{h} + 2(1 - \alpha - \beta)\mathbf{Ah} + \mathbf{C}^T\boldsymbol{\lambda} = 0 \qquad (27)$$

Equation (26) implies

$$\frac{\partial J_a(\mathbf{h}, \boldsymbol{\lambda})}{\partial \boldsymbol{\lambda}} = \mathbf{Ch - K} = 0 \qquad (28)$$

Now Equations (27) and (28) can be written in the following matrix form

$$\begin{bmatrix} \dfrac{2\alpha}{m(\omega_{pb})}\mathbf{Z}_p + \dfrac{2\beta}{m(\omega_{sb})}\mathbf{Q}_s + 2(1 - \alpha - \beta)\mathbf{A} & \vdots & \mathbf{C}^T \\ \cdots\cdots\cdots\cdots\cdots\cdots\cdots\cdots\cdots\cdots\cdots\cdots & \vdots & \cdots \\ \mathbf{C} & \vdots & \mathbf{0} \end{bmatrix}\begin{bmatrix} \mathbf{h} \\ \cdots \\ \boldsymbol{\lambda} \end{bmatrix} = \begin{bmatrix} \dfrac{2\alpha}{m(\omega_{pb})}\mathbf{R}_p \\ \cdots\cdots \\ \mathbf{K} \end{bmatrix} \qquad (29)$$

A solution to Equation (29) will exist when **C** has full rank which occurs when **C** contains no redundant or trivial constraints.

When Equation (29) is solved, the impulse response is available in the vector **h**. Once the impulse response values have been calculated, the frequency samples for a Type 1 frequency sampling filter can be calculated from the DFT of g(n),

$$H(k) = \sum_{n=0}^{N-1} g(n)e^{-j\frac{2\pi}{N}kn} = \mathbf{h}^T \mathbf{Ps}(e^{j\omega})\Big|_{\omega=\frac{2\pi}{N}k} \qquad (30)$$

where $g(n) = r^{-n}h(n)$, and the frequency samples for a Type 2 frequency sampling filter can be calculated from the modified DFT of g(n),

$$H(k) = \sum_{n=0}^{N-1} g(n)e^{-j\frac{2\pi}{N}\left(k+\frac{1}{2}\right)n} = \mathbf{h}^T \mathbf{Ps}(e^{j\omega})\Big|_{\omega=\frac{2\pi}{N}\left(k+\frac{1}{2}\right)} \qquad (31)$$

where $g(n) = r^{-n}h(n)$.

IV. Examples

A. Example 1. A Type 1 Frequency Sampling Filter for N odd

In this example, we will design a filter which approximates the frequency response,

$$\left|H_d(e^{j\omega})\right| = \begin{cases} 1 & 0 \le \omega \le \omega_s \\ 0 & \omega_s \le \omega \le \pi \end{cases}$$

$$\arg\left[H_d(e^{j\omega})\right] = -\omega\left(\frac{N-1}{2}\right)$$

using a Type 1 frequency sampling filter which has an impulse response of length N, where N = 101. For this example, we will let $\omega_s = 12\pi/N$ and r = 0.953125 = 0.111101_2, and will approximate the passband of $|H_d(e^{j\omega})|$ with a maximally flat frequency response at $\omega = 0$. To attain this type of passband approximation, we will not minimize the mean square error in the passband which implies that $\alpha = 0$ and that Equation (29) can be written as

$$\begin{bmatrix} \dfrac{2\beta}{m(\omega_{sb})}\mathbf{Q}_s + 2(1-\beta)\mathbf{A} & \mathbf{C}^T \\ \mathbf{C} & \mathbf{0} \end{bmatrix}\begin{bmatrix} \mathbf{h} \\ \lambda \end{bmatrix} = \begin{bmatrix} \mathbf{0} \\ \mathbf{K} \end{bmatrix}. \qquad (32)$$

To determine the filter's coefficients, we calculate **h** from Equation (32) where $\omega_{sb} = \pi - \omega_s$, $\mathbf{Q}_s = \mathbf{Q}(\pi) - \mathbf{Q}(\omega_s)$, and the matrix **C** and the vector **K** are determined from the type of passband approximation desired and stopband constraints. The stopband constraints are determined by setting $G(e^{j\omega}) = 0$ for $\omega = 2\pi k/N$, k = k_0, k_0+1, ..., (N-1)/2 where k_0 is chosen so that it is the first integer for which $2\pi k_0/N \ge \omega_s$. The passband constraints are selected by the type of passband approximation that is desired. In this example, we have decided to approximate the passband with a maximally flat frequency response at $\omega = 0$. This can be accomplished by constraining

$H_r(\omega)|_{\omega=0} = 1$ and the first L derivatives of $H_r(\omega)|_{\omega=0}$ to zero. The odd order derivatives of a linear phase filter evaluated at $\omega = 0$ are zero, thus constraining them is redundant, and they are omitted The even order derivatives evaluated at $\omega = 0$ can be expressed as

$$\frac{d^n H_r(\omega)}{d\omega^n}\bigg|_{\omega=0}$$

$$= \left[(-1)^{\frac{n}{2}}\left(\frac{N-1}{2}\right)^n \quad \cdots \quad (-1)^{\frac{n}{2}}1^n \quad 0 \quad (-1)^{\frac{n}{2}}1^n \quad \cdots \quad (-1)^{\frac{n}{2}}\left(\frac{N-1}{2}\right)^n \right]\mathbf{h}$$

when n is even. For $L = 6$, the matrix $\mathbf{C}$ can be expressed as

$$C_{ic} = [(N-1)/2 - m]^{2i} \qquad\qquad i = 0, 1, 2, 3 \qquad\qquad c = 0, 1,..., N-1$$

$$C_{ic} = r^{-c}\cos\left[c\frac{2\pi}{N}\left(\frac{i}{2}+4\right)\right] \qquad i = 4, 6, ..., N-9 \qquad c = 0, 1,..., N-1$$

$$C_{ic} = r^{-c}\sin\left[c\frac{2\pi}{N}\left(\frac{i-1}{2}+4\right)\right] \qquad i = 5, 7,..., N-8 \qquad c = 0, 1,..., N-1$$

where C_{ic} is the element in the ith row and the cth column of the matrix $\mathbf{C}$. The column vector $\mathbf{K}$ is $\mathbf{K} = [1\ 0\ 0\ \cdots\ 0]$. When solved, the solution vector, $\mathbf{h}$, of Equation (32) is substituted into Equation (30) to determine the filter's frequency samples.

Figure 3A shows the magnitude responses of filters designed for $L = 6$, $\alpha = 0$ and $\beta = 0.25, 0.9, 0.999$. Figure 3B shows the passbands in detail, and Figure 4 shows the phase deviations from the ideal phase of -50ω. Table 1 gives the non-zero frequency samples for this example when $L = 6$, $\alpha = 0$ and $\beta = 0.9$. Figure 5A shows the magnitude responses of filters designed for $\alpha = 0$, $\beta = 0.9$ and $L = 2, 6, 10$. Figure 5B shows the passbands in detail, and Figure 6 shows the phase deviations from the ideal phase of -50ω.

B. Example 2. A Type 1 Frequency Sampling Filter for N odd

In this example, we will design a Type 1 frequency sampling filter which minimizes the mean square error of the desired frequency response,

$$\left|H_d(e^{j\omega})\right| = \begin{cases} \dfrac{4\omega}{\sin(4\omega)} & 0 \le \omega \le \omega_p \\ 0 & \omega_s \le \omega \le \pi \end{cases}$$

where $\omega_p < \omega_s$, and approximates a linear phase filter. We will use a Type 1 frequency sampling filter which has an impulse response of length N, where N is odd.

To determine the impulse response of the frequency sampling filter, we calculate the vector $\mathbf{h}$ by solving Equation (29),

$$\left[\begin{array}{c|c} \dfrac{2\alpha}{m(\omega_{pb})}\mathbf{Z}_p + \dfrac{2\beta}{m(\omega_{sb})}\mathbf{Q}_s + 2(1-\alpha-\beta)\mathbf{A} & \mathbf{C}^T \\ \hline \mathbf{C} & \mathbf{0} \end{array} \right] \left[\begin{array}{c} \mathbf{h} \\ \hline \lambda \end{array} \right] = \left[\begin{array}{c} \dfrac{2\alpha}{m(\omega_{pb})}\mathbf{R}_p \\ \hline \mathbf{K} \end{array} \right]$$

where $m(\omega_{pb}) = \omega_p - 0$, $\mathbf{Z}_p = \mathbf{Z}(\omega_p) - \mathbf{Z}(0)$, $m(\omega_{sb}) = \pi - \omega_s$, $\mathbf{Q}_s = \mathbf{Q}(\pi) - \mathbf{Q}(\omega_s)$, $\mathbf{K} = 0$, $\mathbf{R}_p^T = \displaystyle\int_{\omega=0}^{\omega_p} \frac{4\omega}{\sin(4\omega)} \mathbf{x}^T(\omega)\, d\omega$ and

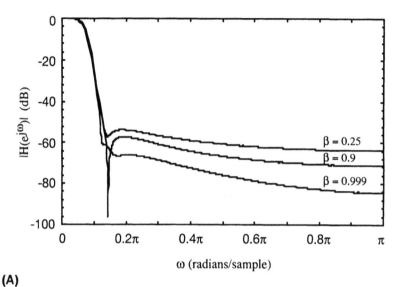

(A)

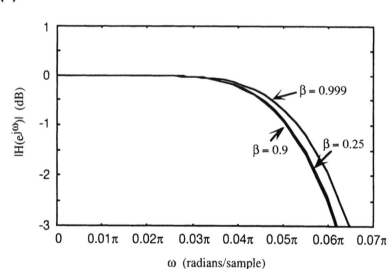

(B)

Figure 3. Magnitude of Frequency Response for Example 1 where
$N = 101$, $\omega_s = 12\pi/N$, $r = 0.953125$, $\alpha = 0$ and $L = 6$.

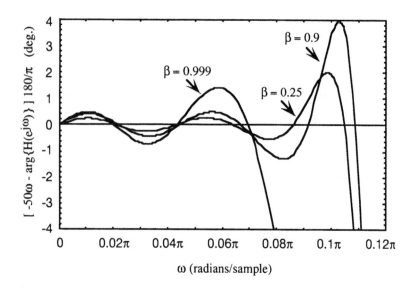

Figure 4. Difference Between Ideal Linear Phase of -50ω and the
Phase of the Frequency Response for Example 1 where N = 101,
ω_s = 12π/N, r = 0.953125, α = 0 and L = 6.

Table 1

Non-zero Frequency Samples for Example 1 when L = 6 and β = 0.9.

k	|H(k)|	θ(k)
0	11.38995857	0
1	10.68186574	-3.11928414
2	11.68441972	0.04974796
3	10.07855019	2.84593402
4	4.40865063	-0.86062435
5	0.77142629	1.50833959

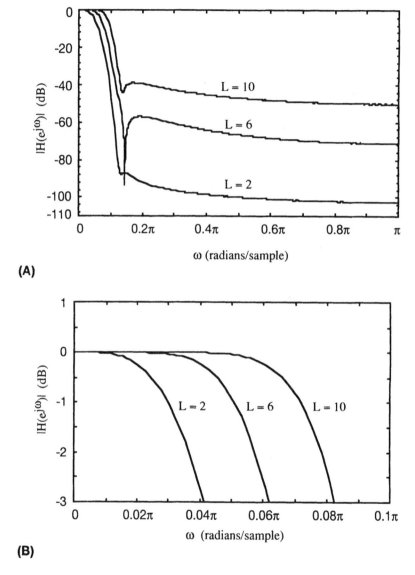

(A)

(B)

Figure 5. Magnitude of the Frequency Response for Example 1 where
N = 101, ω_s = 12π/N, r = 0.953125, α = 0 and β = 0.9.

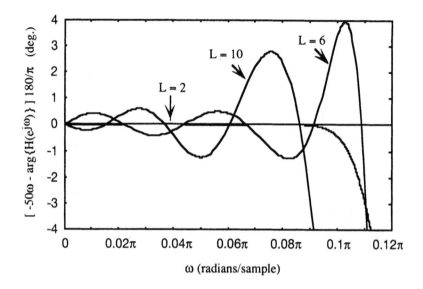

Figure 6. Difference Between Ideal Linear Phase of -50ω and the Phase of the Frequency Response for Example 1 where N = 101, ωₛ = 12π/N, r = 0.953125, α = 0 and β = 0.9.

$$
\mathbf{C} = \begin{bmatrix}
\operatorname{Re}\left[\mathbf{s}^{T}(e^{j\omega})\right]\Big|_{\omega=\frac{2\pi}{N}k_0} \\[2mm]
\operatorname{Im}\left[\mathbf{s}^{T}(e^{j\omega})\right]\Big|_{\omega=\frac{2\pi}{N}k_0} \\[2mm]
\operatorname{Re}\left[\mathbf{s}^{T}(e^{j\omega})\right]\Big|_{\omega=\frac{2\pi}{N}(k_0+1)} \\[2mm]
\operatorname{Im}\left[\mathbf{s}^{T}(e^{j\omega})\right]\Big|_{\omega=\frac{2\pi}{N}(k_0+1)} \\[2mm]
\vdots \\[2mm]
\operatorname{Re}\left[\mathbf{s}^{T}(e^{j\omega})\right]\Big|_{\omega=\frac{2\pi}{N}(k_0+M)} \\[2mm]
\operatorname{Im}\left[\mathbf{s}^{T}(e^{j\omega})\right]\Big|_{\omega=\frac{2\pi}{N}(k_0+M)}
\end{bmatrix} \mathbf{P}
$$

The value of k_0 in the matrix $\mathbf{C}$ is an integer which is chosen so that the first stopband frequency sample, $H(k_0)$, is the first frequency sample greater than ω_s. M is chosen so that the last frequency sample, $H(k_0+M)$, is the last stopband frequency sample less than π. Substituting these values into Equation (29), we can calculate the vector, $\mathbf{h}$, which contains the impulse response, h(n). Letting $g(n) = r^{-n}h(n)$, we can determine the frequency samples from Equation (30) which is

$$H(k) = \sum_{n=0}^{N-1} g(n)e^{-j\frac{2\pi}{N}nk} \text{ for } k = 0, 1, ..., (N-1)/2.$$

or in matrix form,

$$H(k) = \mathbf{h}^T \mathbf{Ps}(e^{j\omega})\Big|_{\omega=\frac{2\pi}{N}k} \text{ for } k = 0, 1, ..., (N-1)/2.$$

Figure 7A shows the magnitude of the frequency response for this example when $N = 75$, $\omega_p = 0.08\pi$, $\omega_s = 0.16\pi$, $r = 0.953125 = 0.111101_2$, $k_0 = 6$, $k_0+M = (N-1)/2 = 37$, $\alpha = 0.02$ and $\beta = 0.96$. Figure 7B shows in detail the passband of $|H(e^{j\omega})|$ shown in Figure 7A. Figure 8 shows the filter's phase deviation from the ideal phase of -37ω, and Table 2 gives the values of the non-zero frequency samples, $H(0)$, $H(1)$, ..., $H(5)$. This filter can be realized by the Type 1 frequency sampling filter structure illustrated in Figure 1 where only the resonators for $k = 0, 1,..., 5$ are realized.

C. Example 3. A Type 2 Frequency Sampling Filter for N even

For this example, we will design a Type 2 lowpass frequency sampling filter which approximates a linear phase filter with the following magnitude specification

$$\left|H_d(e^{j\omega})\right| = \begin{cases} 1 & 0 \le \omega \le \omega_p \\ 0 & \omega_s \le \omega \le \pi \end{cases}$$

where $\omega_p < \omega_s$. We will use a Type 2 frequency sampling filter which has an impulse response of length N, where N is even. To attain the desired passband approximation in this example, we will minimize the mean square error in the passband while constraining $H_r(\omega)|_{\omega=0} = 1$ and the first L derivatives of $H_r(\omega)|_{\omega=0}$ to zero

To determine the impulse response of the frequency sampling filter, we calculate the vector $\mathbf{h}$ by solving Equation (29),

$$\left[\begin{array}{c|c} \dfrac{2\alpha}{m(\omega_{pb})}\mathbf{Z}_p + \dfrac{2\beta}{m(\omega_{sb})}\mathbf{Q}_s + 2(1-\alpha-\beta)\mathbf{A} & \mathbf{C}^T \\ \hline \mathbf{C} & \mathbf{0} \end{array}\right] \left[\begin{array}{c} \mathbf{h} \\ \hline \lambda \end{array}\right] = \left[\begin{array}{c} \dfrac{2\alpha}{m(\omega_{pb})}\mathbf{R}_p \\ \hline \mathbf{0} \end{array}\right]$$

where $m(\omega_{pb}) = \omega_p - 0$, $m(\omega_{sb}) = \pi - \omega_s$, $\mathbf{Z}_p = \mathbf{Z}(\omega_p) - \mathbf{Z}(0)$, $\mathbf{Q}_s = \mathbf{Q}(\pi) - \mathbf{Z}(\omega_s)$, and $\mathbf{R}_p = \mathbf{R}(\omega_p) - \mathbf{R}(0)$. If $L = 4$, the passband constraint matrix, $\mathbf{C}_p$, can be expressed as

$$C_{pic} = [(N-1)/2 - m]^{2i} \qquad i = 0, 1, 2 \qquad c = 0, 1,..., N-1$$

and the stopband constraint matrix, $\mathbf{C}_s$, can be expressed as

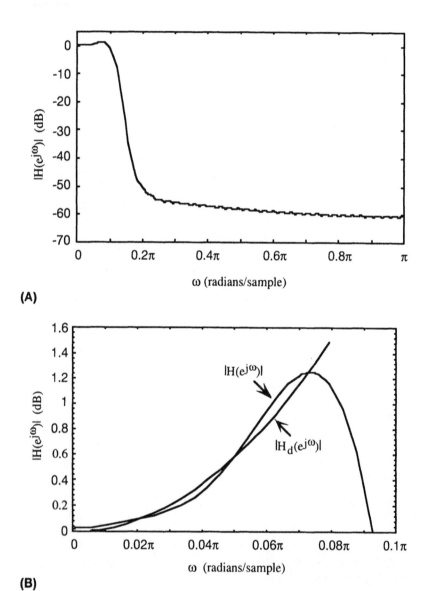

(A)

(B)

Figure 7. Magnitude of the Frequency Response for Example 2 where $N = 75$, $\omega_p = 0.08\pi$, $\omega_s = 0.16\pi$, $r = 0.953125$, $\alpha = 0.02$ and $\beta = 0.96$.

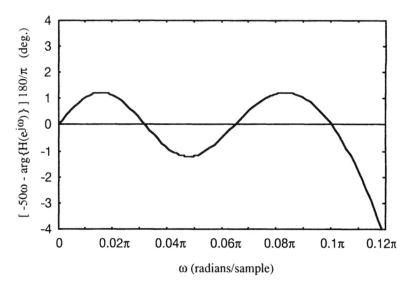

Figure 8. Difference Between Ideal Linear Phase of -37ω and the
Phase of the Frequency Response for Example 2 where N = 75.

Table 2

Non-zero Frequency Samples for Example 2 where N = 75

k	|H(k)|	θ(k)
0	6.16144177	0
1	5.74058525	-3.11655094
2	6.46345261	0.20005815
3	7.43710583	-3.10480369
4	4.66398723	-0.33715578

$$
C_s = \begin{bmatrix}
\mathrm{Re}\left[s^T(e^{j\omega})\right]\Big|_{\omega=\frac{2\pi}{N}(k_0+1/2)} \\[2ex]
\mathrm{Im}\left[s^T(e^{j\omega})\right]\Big|_{\omega=\frac{2\pi}{N}(k_0+1/2)} \\[2ex]
\vdots \\[2ex]
\mathrm{Re}\left[s^T(e^{j\omega})\right]\Big|_{\omega=\frac{2\pi}{N}(k_0+1/2+M)} \\[2ex]
\mathrm{Im}\left[s^T(e^{j\omega})\right]\Big|_{\omega=\frac{2\pi}{N}(k_0+1/2+M)}
\end{bmatrix} P
$$

The value of k_0 in the matrix C is an integer which is chosen so that the first stopband frequency sample, $H(k_0)$, is the first frequency sample greater than ω_s. M is chosen so that the last stopband frequency sample, $H(k_0+M)$, is the last stopband frequency sample less than π. Substituting these values into Equation (29), we can calculate the vector, **h**, which contains the impulse response, h(n). Letting $g(n) = r^{-n}h(n)$, we can determine the frequency samples from Equation (31) which is

$$
H(k) = \sum_{n=0}^{N-1} g(n)e^{-j\frac{2\pi}{N}(k+1/2)n} \quad \text{for } k = 0, 1, ..., N/2 - 1
$$

or in matrix form,

$$
H(k) = h^T P s(e^{j\omega})\Big|_{\omega=\frac{2\pi}{N}(k+1/2)} \quad \text{for } k = 0, 1, ..., N/2 - 1.
$$

Figure 9A shows the magnitude of the frequency response for this example when N = 100, ω_p = 0.07π, ω_s = 0.13π, r = 0.953125 = 0.111101$_2$, k_0 = 6, k_0+M = N/2 - 1 = 49, α = 0.005 and β = 0.99. Figure 9B shows in detail the passband of $|H(e^{j\omega})|$ shown in Figure 9A. Figure 10 shows the filter's phase deviation from the ideal phase of -99ω/2, and Table 3 gives the values of the non-zero frequency samples, H(0), H(1), ..., H(5). This filter can be realized by the Type 2 frequency sampling filter structure illustrated in Figure 2 where only the resonators for k = 0, 1,..., 5 are realized.

IV. Summary and Conclusions

In this chapter, we presented a frequency sampling filter design technique which accounts for finite word length effects. The design method uses the Lagrange multiplier optimization method to design a frequency sampling filter which minimizes the mean square error in the stopband, the mean square error of the passband amplitude and the sum square error of the impulse response symmetry subject to constraints on the passband and stopband and a fixed value of r < 1. As the examples demonstrated, this design method is easily adaptable for both Type 1 and Type 2 frequency sampling filters, and is well suited for designing narrow band near linear phase frequency sampling filters.

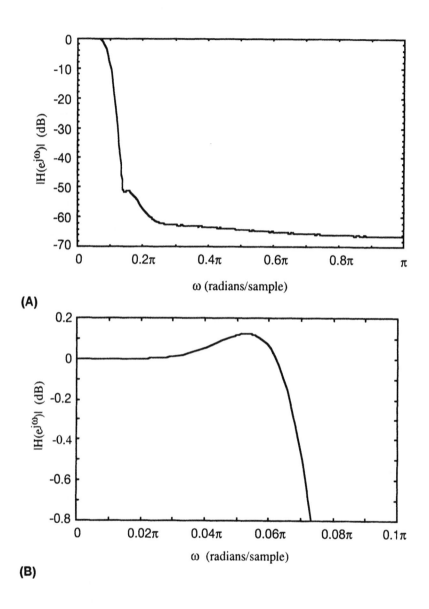

Figure 9. Magnitude of the Frequency Response for Example 3 where
$N = 100$, $\omega_p = 0.07\pi$, $\omega_s = 0.13\pi$, $r = 0.953125$, $k_0 = 6$, $k_0+M = 49$,
$\alpha = 0.005$ and $\beta = 0.99$.

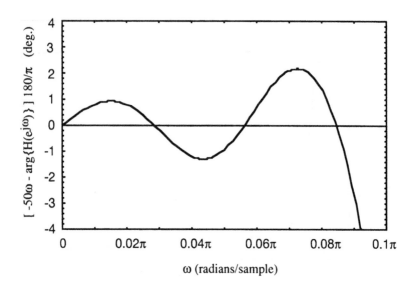

Figure 10. Difference Between Ideal Linear Phase of -99ω/2 and the
Phase of the Frequency Response for Example 3 where N = 100.

Table 3

Non-zero Frequency Samples for Example 3.

k	|H(k)|	θ(k)
0	11.00184860	-1.58770651
1	10.18982374	1.61629299
2	11.47032920	-1.41258967
3	12.00331799	1.47094957
4	6.82793503	-2.12235917
5	1.40670737	0.47008988

V. References

[1]   A. V. Oppenheim and R. W. Schafer, *Discrete-Time Signal Processing*, Prentice Hall, New Jersey (1989).

[2]   L. R. Rabiner and B. Gold, *Theory and Application of Digital Signal Processing*, Prentice Hall, New Jersey (1975).

[3]   P. A. Stubberud and C. T. Leondes, "The Design of Frequency Sampling Filters by the Method of Lagrange Multipliers," *IEEE Transactions on Circuits and Systems*, vol. 40, no. 1, pp. 51-54 (1993).

[4]   L. R. Rabiner and R. W. Schafer, "Recursive and Nonrecursive Realizations of Digital Filters Designed by Frequency Sampling Techniques," *IEEE Transactions on Audio and Electroacoustics*, Vol. AU-19, No. 3, pp. 200-207 (1971).

[5]   L. R. Rabiner and R W. Schafer, "Correction to "Recursive and Nonrecursive Realizations of Digital Filters Designed by Frequency Sampling Techniques"," *IEEE Transactions on Audio and Electroacoustics*, vol. 20, No 1, pp. 104-105 (1972).

[6]   B. Gold and K. L. Jordan Jr., "A Direct Search Procedure for Designing Finite Impulse Response Filters," *IEEE Transactions on Audio and Electroacoustics*, vol. AU-17, No 1, pp. 33-36 (1969).

[7]   R. E. Bogner, "Frequency Sampling Filters - Hilbert Transformers and Resonators," *The Bell System Technical Journal*, vol. 48, No. 3, pp. 501-510 (1969).

[8]   L. R. Rabiner, "Linear Program Design of Finite Impulse Response (FIR) Digital Filters," *IEEE Transactions on Audio and Electroacoustics*, vol. 20, No 4, pp. 280-288. (1972).

[9]   L. R. Rabiner, "The Design of Finite Impulse Response Digital Filters Using Linear Programming Techniques," *The Bell System Technical Journal*, vol. 51, No. 6, pp.1177-1198 (1972).

[10]   T. J. Mc Creary, "On Frequency Sampling Digital Filters," *IEEE Transactions on Audio and Electroacoustics*, vol. 20, No 3, pp. 222-223 (1972).

[11]   B. N. S. Babu and R. Yarlagadda, "A Direct Approach to the

Frequency Sampling Design of Digital Filters," *Int. J. Electronics*, vol. 47, No. 2, pp. 123-138 (1979).

[12]   D. W. Burlage, R. C. Houts and G. L. Vaughn, "Time-Domain Design of Frequency-Sampling Digital Filters for Pulse Shaping Using Linear Programming Techniques," *IEEE Transactions on Acoustics, Speech and Signal Processing*, Vol. ASSP-22, No. 3, pp. 180-185 (1974).

[13]   J. W. Adams, R. Cotner and H. Samueli, "On the Roundoff Noise Problem in Frequency Sampling Filters," *Proceedings of the IEEE Midwest Symposium on Circuits and Systems*, Lincoln, NE (1986).

[14]   C. Weinstein, "Quantization Effects in Frequency Sampling Filters," *Nerem Record*, pp. 222-223 (1968).

[15]   A. V. Oppenheim and C. J. Weinstein, "Effects of Finite Register Length in Digital Filtering and the Fast Fourier Transform," *Proceedings of the IEEE*, Vol. 60, pp. 957-976 (1972).

[16]   P. A. Stubberud, *Frequency Sampling Digital Filters*, Ph.D. dissertation, University of California, Los Angeles (1990).

[17]   W. L. Brogan, *Modern Control Theory*, Third Edition, Prentice Hall, New Jersey, pp 264 - 266 (1991).

Low-Complexity Filter-Banks for Adaptive and Other Applications

Mukund Padmanabhan
IBM T. J. Watson Research Center
P. O. Box 704
Yorktown Heights, NY 10598

Ken Martin
Electrical Engineering Department
University of Toronto
Toronto, Ontario, Canada M5S 1A4

I Introduction

Typically, the term filter-bank is used to refer to a single-input, multiple-output structure, and such structures find use in a wide variety of applications such as subband coding [1], frequency domain adaptive filtering [2, 3], communication systems, frequency estimation [4], transform computations [5], etc. Consequently, a great deal of attention has been focussed on their design and properties. Most of these developments, however, do not place a great deal of importance on the issues involved in implementing the filter-bank, and it is only recently that this aspect has started getting some notice [6, 7, 8, 9, 10, 11, 12]. This issue is however no less important, because in the complex systems of today, the filtering operation constitutes only a small part of the functions of the overall system, and the addition of a dedicated DSP or a computer to the system, just for the task of filtering, would make its cost, and size prohibitive. One way to solve this problem is to take advantage of the advances in VLSI to make application specific integrated circuits to implement the filter-bank; hence the entire filter-bank could be implemented on a single chip, which occupies very little space, and is also relatively cheap. However, this solution makes it necessary to take several additional issues into account when designing the filter-bank : if the filter-bank is to be amenable to VLSI implementation, it should have a low hardware complexity (multipliers and adders), its data flow path should be regular, it should have good properties under finite precision arithmetic, it should be pipelineable, etc. In this chapter, we will describe a filter-bank that has all these properties, and examine some of its applications.

The applications, and the theoretical aspects presented here are developed in the context of the special filter-bank structure described in this chapter, rather than as a globally applicable theory that can be applied to any filter-bank. Also, applications such as frequency estimation, transform computation, etc., have been emphasized, rather than applications such as

Copyright © 1995 by Academic Press, Inc.
All rights of reproduction in any form reserved.

subband-coding in multirate situations. (In fact, the problem of applying the filter-bank structure in multirate situations is still in the research stage.) We refer readers more interested in subband-coding applications to [9], which contains an excellent review of developments in filter-banks, in the context of multirate systems and subband coding.

A   Organization

This chapter will be organized as follows: In Section. II, the basic filter structure is developed, some of its properties are examined, and some useful transfer functions associated with it, are derived. The filter structure has a set of internal nodes, such that the transfer function from the input to these nodes, is a set of bandpass transfer functions with non-overlapping center frequencies. Hence, by taking the outputs from these internal nodes, the structure may be used as a filter-bank. In Section. III, the first application of the filter-bank is examined. In this application the filter is used as an adaptive line enhancer, for the application of sinusoidal frequency estimation and enhancement. The developed line enhancer has the good properties of linear hardware complexity, and also provides improved convergence properties, as compared to most other IIR adaptive filters. In Section. IV, the filter-bank structure is used to compute trigonometric transforms, such as the DFT, DCT, DST, etc., in a time-recursive manner. This structure has the advantages of linear complexity and amenability to VLSI implementation, and provides a possible solution to the problem of implementing a complete transformer on a single chip. Finally, in Section. V, the filter structure is used to synthesize low complexity filter-banks, related to "Lerner" filters, that exhibit very good magnitude and phase response characteristics. Some other applications of these filter-banks are also examined.

II   The basic filter-bank structure

The filter-bank structure described here takes on the form of digital resonators in a feedback loop, and can be arrived at using four separate approaches. These approaches actually lead to a single input single output digital filter structure - the resonator-based digital filter structure [13, 14, 15]; however, the structure has a set of internal nodes such that the transfer functions from the input to these nodes are a set of bandpass transfer functions with different center frequencies. Hence, by taking the outputs from these internal nodes, the filter structure can be used as a single input multiple output filter-bank. The development and some properties of the filter-bank are briefly described next. As this chapter will concern itself chiefly with applications of the filter-bank, we will only describe those properties that are relevant and necessary to these applications. For a

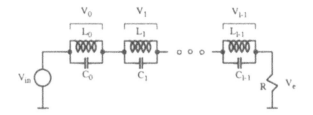

Figure 1: (a) Singly terminated L-C ladder.

more detailed description of the filter properties, we will refer the interested reader to [13, 14, 15].

A   Background

The first approach leading to the resonator-based filter structure is based on [16], where the DFT of an input signal is obtained by matching the input signal to the weighted sum of reference oscillator outputs, the weights being updated by the LMS algorithm. Drawing on the analysis of [17], [18, 19] obtained the transfer function from the input to the ith weighted oscillator output, and the form of this transfer function was used to suggest the resonator-based filter structure.

The independent approach of [13, 20], was based on [21], and started off with structures to obtain certain transforms of the given signal (such as the DFT). A state space model for the signal was assumed, and an identity observer [21] was used to reconstruct the states of the model, from which the desired transforms of the signal were obtained. As the signal was modelled as the sum of the outputs of a bank of resonators, the identity observer turned out to have a 'resonator-in-feedback-loop' structure, and the potential of this structure for general filtering applications was later realized and reported in [13].

The third approach is based on a tone-decoder originally used in the telecommunications industry [22]. This tone-decoder was an active RC realization of the singly-terminated ladder filter shown in Fig. 1a. A signal-flow-graph simulation of this filter is shown in Fig. 1b, and is seen to take the form of resonators in a feedback loop. The resonator-based filter structure may be arrived at by transforming the analog signal-flow-graph to the digital domain, using the bilinear transform, and modifying it to eliminate delay-free loops.

The fourth approach is based on using a recursive structure to compute transforms such as the DFT. The time-recursive computation of transforms may be done by using an FIR filter-bank [23], and to save hardware, these FIR filters can be implemented using frequency-sampling structures [24, 25]. Unfortunately, this does not represent a very practical solution because frequency-sampling structures are known to have poor finite precision

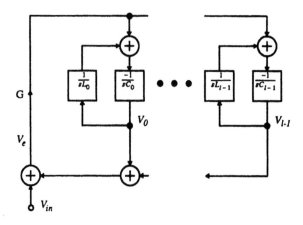

Figure 1: (b) Signal-flow graph of singly terminated L-C ladder.

properties. Alternatively, a feedback structure may be used to implement the transformer (Section IV). This structure may be used to implement any transform; however, for trigonometric transforms such as the DFT, DCT, DST, etc., it reduces to the resonator-based filter-bank structure.

B   Development

Our development of the filter-bank structure will follow the third approach outlined above. Consider the transfer function from V_{in} to V_e of Fig. 1a. This transfer function is given by

$$H_e(s) = \frac{V_e}{V_{in}} = \frac{R}{R + \sum_{i=0}^{l-1} \frac{s/L_i}{s^2 + 1/L_i C_i}} . \tag{1}$$

Transforming this analog transfer function into the digital domain, using the bilinear transform, $s \to \frac{1-z^{-1}}{1+z^{-1}}$, we get

$$
\begin{aligned}
H_e(z) &= \frac{R}{R + \sum_{i=0}^{l-1} \frac{L_i}{1+L_i C_i} \frac{1-z^{-2}}{1 - 2(\frac{L_i C_i - 1}{L_i C_i + 1})z^{-1} + z^{-2}}} , \\
&= \frac{1}{1 + \frac{1}{R} \sum_{i=0}^{l-1} K_i \frac{1-z^{-2}}{1 - 2a_i z^{-1} + z^{-2}}} , \tag{2}
\end{aligned}
$$

where $K_i = \frac{L_i}{1+L_i C_i}$, and $a_i = \frac{1 - 1/L_i C_i}{1 + 1/L_i C_i}$, with $|a_i| < 1$. The form of (2) suggests that it might be possible to obtain this transfer function by embedding

l biquads, with the transfer functions

$$\frac{K_i}{R} \frac{1 - z^{-2}}{1 - 2a_i z^{-1} + z^{-2}} , \tag{3}$$

in a feedback loop. However, these second transfer functions have a delay-free forward path; hence, the filter structure will have a delay-free loop, and thus cannot be implemented.

It is well known from classical circuit and filter theory [26], that this delay-free loop problem arises because the signal-flow-graph of Fig. 1b models the voltage and current variables of Fig. 1a, and one way to solve the problem is to do the modelling in terms of wave variables. Here however, we will get around the delay-free loop problem not by using this technique, but by other means. First express the transfer function (3) as

$$\frac{K_i}{R} + \frac{K_i}{R} \frac{2a_i z^{-1} - 2z^{-2}}{1 - 2a_i z^{-1} + z^{-2}} . \tag{4}$$

Here $\frac{K_i}{R}$ represents the delay-free part of the biquad transfer function, and the second part of the transfer function has no delay-free forward path. The transfer function $H_e(z)$ may now be written as

$$
\begin{aligned}
H_e(z) &= \frac{1}{1 + \sum_{i=0}^{l-1} \frac{K_i}{R} + \sum_{i=0}^{l-1} \frac{K_i}{R} \frac{2a_i z^{-1} - 2z^{-2}}{1 - 2a_i z^{-1} + z^{-2}}} , \\
&= \frac{C}{1 + \sum_{i=0}^{l-1} G_i \frac{2a_i z^{-1} - 2z^{-2}}{1 - 2a_i z^{-1} + z^{-2}}} , \tag{5}
\end{aligned}
$$

where $C = \frac{1}{1 + \sum_{i=0}^{l-1} \frac{K_i}{R}}$ and $G_i = \frac{K_i C}{R} = \frac{K_i}{R + \sum_{i=0}^{l-1} K_i}$ are constants. Further, $\sum_{i=0}^{l-1} K_i < 1$.

Ignoring the scale factor C, the filter structure now takes on the form of biquads in a feedback loop as shown in Fig. 2a. The transfer function of the ith biquad is given by

$$H_{fb,i}(z) = G_i \frac{2a_i z^{-1} - 2z^{-2}}{1 - 2a_i z^{-1} + z^{-2}} . \tag{6}$$

It may be seen that the poles of the $H_{fb,i}(z)$ lie on the unit circle, hence, it represents a digital resonator, with the resonator frequency, ω_i, being related to a_i as $a_i = \cos(\omega_i)$. Consequently, the structure takes on the form of l digital resonators in a feedback loop. As these resonators arise from tank circuits of 1a, they are of second order; however, it is also possible to have inductors or capacitors in place of the tank circuits in Fig. 1a. The corresponding resonators now become first order, with their transfer functions taking on the forms $K_i \frac{z^{-1}}{1 - z^{-1}}$ or $-K_i \frac{z^{-1}}{1 + z^{-1}}$. (These transfer functions may also be obtained by substituting $a_i = \pm 1$, and adding an additional factor of 0.5 to (6)).

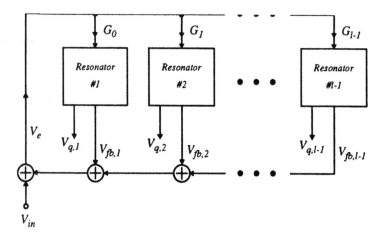

Figure 2: (a) General form of filter-bank.

The structure now takes on the form of l digital resonators in a feedback loop, with the transfer function of the resonators being given by

$$H_{fb,i}(z) = G_i p(a_i) \frac{2a_i z^{-1} - 2z^{-2}}{1 - 2a_i z^{-1} + z^{-2}} , \qquad (7)$$

where $p(a_i) = 0.5$ if $a_i = \pm 1$, and $p(a_i) = 1$ otherwise.

These second order resonators may be implemented by using any biquad structure such as the LDI biquad [27], the direct-form biquad, the coupled-form biquad, etc [1]. The LDI-based and coupled-form based biquads are shown in Figs. 2b and 2c respectively, and the overall structure is shown in Fig. 3, with LDI-based biquads being used to implement the second order resonators. Alternatively, these biquads could be replaced by the structure of Fig. 2c to give a coupled-form-based filter-bank structure. Also, in order to scale the internal nodes of the filter, the multiplier G_i of (7) is split up into two multipliers, $\sqrt{G_i}$, as shown in Fig. 3. Further, assuming that of the l resonators, l_2 are biquads, and the remainder $l - l_2$ are first order resonators, and using N to denote the order of the filter, we have

$$N = 2l_2 + l - l_2 . \qquad (8)$$

In order to motivate the applications of the filter structure, we will next look at the transfer functions from the input, V_{in}, to some of its internal nodes. For instance, the transfer function from V_{in} to V_e is given by (5), without the scale factor C, and may be seen to be a transfer function with multiple notches, the notch frequencies being the same as the

---

[1]For the best performance [14], the biquad whose resonant frequency is least sensitive to coefficient truncation should be used - hence, for low resonant frequencies, LDI biquads should be used, for frequencies near $f_s/4$, direct-form biquads should be used.

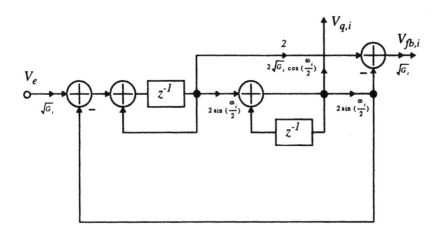

Figure 2: (b) LDI based biquad.

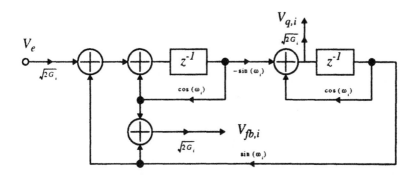

Figure 2: (c) Coupled-form biquad.

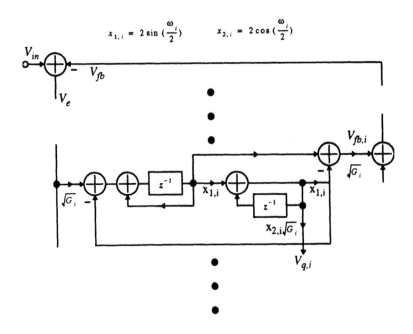

Figure 3: Filter-bank Structure.

resonator frequencies. This transfer function will turn out to motivate the development of the adaptive application of the filter-bank. Further, the structure also yields multiple bandpass transfer functions from V_{in} to $V_{fb,i}$ of Fig. 3. The ith such transfer function is given by

$$H_{fb,i}^{f}(z) = \frac{V_{fb,i}}{V_{in}} = \frac{G_i p(a_i) \frac{2a_i z^{-1} - 2z^{-2}}{1 - 2a_i z^{-1} + z^{-2}}}{1 + \sum_{j=0}^{l-1} G_j p(a_j) \frac{2a_j z^{-1} - 2z^{-2}}{1 - 2a_j z^{-1} + z^{-2}}}, \tag{9}$$

and may be seen to have a value of unity at the frequency of the ith resonator, and zeros at the frequencies of all other resonators; hence, the transfer functions $H_{fb,i}^{f}(z)$, $i = 0, \cdots, l-1$, represent bandpass transfer functions with non-overlapping center frequencies. Unfortunately, however, the characteristics of these bandpass transfer functions are not very good, because they feature high sidelobes (the main sidelobe is down by only 13 dB, and the sidelobes fall off to only around 21 dB), and the filter structure cannot be used in this form as a filter-bank. However, by grouping the $V_{fb,i}$ appropriately to form composite channels, it turns out to be possible to vastly improve the transfer characteristics of the resulting filter-bank. This is discussed in Section V.

Another transfer function of interest is the transfer function from V_{in}

to an internal node, $V_{q,i}$, of the ith resonator, given by

$$H^f_{f_{q,i}}(z) = \frac{V_{q,i}}{V_{in}} = \frac{G_i \frac{2\sqrt{1-a_i^2}z^{-1}}{1-2a_iz^{-1}+z^{-2}}}{1 + \sum_{j=0}^{l-1} G_j p(a_j) \frac{2a_jz^{-1}-2z^{-2}}{1-2a_jz^{-1}+z^{-2}}} . \tag{10}$$

This output is also shown in Fig. 3, and represents the quadrature outputs of the filter-bank; i.e., at the frequency ω_i, the transfer function $H^f_{q,i}(z)$ has the same magnitude as $H^f_{f_{b,i}}(z)$, and exactly $90°$ phase shift. Note, however, that this output is not available for the first order resonators.

1 Choice of resonator frequencies

The question that arises next is: how do we choose the frequencies of the resonators? The answer to this question is dependent on the application being addressed. For the spectral estimation application, the resonator frequencies are actually adapted till they are equal to the frequencies of a multi-sinusoidal input. For the application of transform computation or subband coding, the resonator frequencies are chosen to lie at the N roots of 1 or -1. If the filter structure is used to implement an arbitrary transfer function, then the resonator frequencies are determined by the poles of the unknown transfer function [13, 14].

Further, for all applications other than the last one mentioned above, it turns out to be unnecessary to use different G_i for different resonators, i.e., $G_i = G \quad \forall \quad i$. For this reason, the symbols G_i and G are used interchangeably in the following text.

C Implementation issues

So far we have developed the structure of the resonator-based filter-bank, however, it remains to justify the initial claim that the structure is amenable to VLSI implementation. Two of the requirements for VLSI amenability are low complexity, and a regular data flow path. From Fig. 3, the filter-bank structure can be seen to have a linear hardware complexity, and a regular data flow path; hence, both these requirements are met. Further, as a VLSI implementation would mean that only a finite number of bits are available for representing the coefficients and the internal word storage of the filter structure, the behaviour of the filter-bank structure under finite precision conditions is also important.

1 Finite wordlength effects

When considering a finite-precision implementation of the filter-structure, there are three main issues that need to be considered. The first is that of linear stability, i.e., will the quantization of the filter coefficients result in the poles of the structure moving outside the unit circle? Secondly, assuming that the filter remains stable after coefficient quantization, will the actually implemented transfer function differ significantly from the desired one ? Thirdly, when both the coefficients and internal words are quantized, the wordlength quantization introduces nonlinearities into the filter structure, that could lead to the existence of limit cycle oscillations even when the filter has zero input; hence, we need to verify that the filter structure is incapable of sustaining such oscillations.

Let us first consider the issue of linear stability. When considering this issue, we will assume infinite internal wordlength, and look at only the effects of finite coefficient precision. For the sake of simplicity, we will also assume that there are no first order resonators in the loop i.e., $l = l_2$.

We saw earlier that the transfer function of a second order resonator is given by (6). Assuming that after coefficient truncation, the poles of the resonators are still retained on the unit circle (this is true if LDI or direct-form biquads, rather than the coupled-form biquads are used to implement the resonators), the transfer function of the resonator after coefficient quantization may be written as

$$H_{fb,i}(z) = G \frac{2\hat{a}_i z^{-1} - 2z^{-2}}{1 - 2\hat{a}_i z^{-1} + z^{-2}} \; . \tag{11}$$

The transfer function from V_{in} to V_e is now given by

$$H_e(z) = \frac{V_e}{V_{in}} = \frac{1}{1 + \sum_{j=0}^{l-1} G_j \frac{2\hat{a}_j z^{-1} - 2z^{-2}}{1 - 2\hat{a}_j z^{-1} + z^{-2}}} \; . \tag{12}$$

Further, from (5), we know that this transfer function relates, through the bilinear transform, to an analog LCR prototype as shown in Fig. 1a, with $\hat{a}_j = \frac{L_j C_j - 1}{L_j C_j + 1}$, and as long as

$$\sum_{j=0}^{l-1} G_j < 1 \, , \quad \text{and} \quad |\hat{a}_j| < 1 \, , \tag{13}$$

the analog LCR prototype is stable, and as the bilinear transform maps a stable analog transfer function to a stable digital transfer function, we can conclude that the digital filter structure is stable under finite coefficient precision.

If coupled-form biquads are used to implement the second order resonators in Fig. 3, then assuming magnitude truncation for the coefficients, the poles of the resonators are no longer retained on the unit circle after coefficient truncation. For this case also, it is possible to show that the overall filter structure remains stable (for a proof of this see [14, 15]).

We will next consider the issue of sensitivity of the transfer function to changes in the filter coefficients. The resonator-based filter-structure falls under the category of "orthogonal" filter-structures [28], i.e., the transfer functions from the input of the filter to its states are orthonormal. Further, if the output is taken at V_{fb} of Fig. 3, the transfer functions from the states to the output are also orthonormal. If the state-space representation of the filter structure is

$$X(n+1) = A\,X(n) + B\,V_{in}(n) \quad V_{fb}(n) = C\,X(n) + D\,V_{in}(n)\,, \quad (14)$$

where A, B, C, and D are respectively $N\mathrm{x}N$, $N\mathrm{x}1$, $1\mathrm{x}N$, and $1\mathrm{x}1$ matrices (further, $D = 0$). Now, the above constraints of orthonormality are equivalent [29] to stating that the matrix R defined as

$$R = \begin{bmatrix} A & B \\ C & D \end{bmatrix}, \quad (15)$$

is orthonormal, i.e., $R^*R = I_{N+1}$. From [30], such a network can always be related to a doubly terminated analog lossless prototype network. The desired transfer function now corresponds to the energy transferred from the source to the sink of the analog prototype. From the classical arguments of [31], the first order sensitivity of the transfer function, to changes in the element values of the prototype, is zero, at the frequencies of maximum power transfer in the passband. This low sensitivity property of the analog prototype translates into low passband sensitivity of the digtal filter [2]. Several examples are given to demonstrate this low passband sensitivity in [15].

Finally, we will briefly mention that as far as zero input stability under finite internal wordlength and finite coefficient precision is concerned, it can be shown that if coupled-form biquads are used, and if magnitude truncation is used at the input to the delay elements, then the structure will not sustain zero input limit cycles (for a proof of this statement, see [15]). Unfortunately, it has not yet been possible to show that limit cycles will be suppressed if other biquad forms such as the LDI and direct-form are used (to implement the resonators), though this is conjectured to be the case.

2 Pipelineability

Another issue of some importance is the pipelineability of the structure. From Fig. 3, it is clear that the critical delay of the filter structure arises from the additions that have to be performed in the feedback path, because, before a new data sample can be processed, the signal V_e should be

---

[2]The reader may be a little confused by this statement, because in Section. II B, the filter structure was related to a singly terminated LCR ladder, which is known to have *poor* sensitivity properties. This seeming contradiction may be resolved by noting that the manipulation of the signal-flow-graph of the bilinear-transformed filter, (to get rid of the delay-free-loop), destroys the 1-1 correspondence between the analog singly-terminated ladder and the final digital filter structure. Hence, the poor sensitivity properties of the analog filter are not carried over to the digital domain.

computed. And in order to compute V_e, all the fedback outputs $V_{fb,i}$ have to be added, and then subtracted from V_{in}. Hence, the rate at which input data can be accepted is determined by the time taken to compute $l+1$ additions, and turns out to be the critical delay of the filter. This critical delay is quite large and constrains the maximum clock rate of the system, thus making it unsuitable for very high speed applications. For the case where the resonator frequencies lie at the roots of 1, or -1, however, this problem may be overcome by using transformation techniques, similar to the one outlined in [32, 33]. This issue will be discussed further in Section IV.

III   Adaptive Line Enhancement $(ALE)^{\dagger}$

We will next look at some of the applications of the filter-bank. A common problem that arises in several situations is that of estimating and isolating the frequency components of a multi-sinusoidal-input signal, over time-varying conditions. One common solution to this problem is the use of an Adaptive Line Enhancer (ALE), an adaptive filter which tracks and isolates the sinusoidal components of the input over time-varying conditions. Typically, these ALE's turn out to be notch filters (IIR or FIR), and the adaptation process changes the notch frequencies till they are equal to the input frequencies, at which point, the error is minimized. Such formulations have been done using FIR filters [17, 34], and IIR filters [35, 36, 37, 38], with the IIR filters providing the advantage of low bias, but having the disadvantage of poor convergence properties (when a gradient-based strategy is used for the adaptation). Further, most IIR ALE's have the disadvantages of large complexity ($O(l^2)$), require complicated stability monitoring during the adaptation process, and also do not provide enhanced components that are in phase with their corresponding components in the input (which could be important for applications such as carrier recovery).

In this chapter, we will first address the problem of developing an IIR ALE that does not have the above-mentioned disadvantages, and still provides the advantages associated with IIR ALE's. The ALE is based on the filter-bank structure of Fig. 3, and a filter-bank with l second order resonators in a feedback loop can be used to track l or fewer sinusoids in the input. In the steady-state, the enhanced and isolated sinusoids are available at the $V_{fb,i}$ outputs. It is also possible to replace the LDI biquads of Fig. 3 with other biquad forms such as the direct-form.

Consider the transfer function H_e given by (5). This transfer function has zeros at the frequencies of all the resonators, and the adaptive algorithm is based on varying the resonator frequencies till they are equal to the frequencies of the sinusoidal components of the input. This is equivalent to minimizing the energy of the signal V_e. Now, the simplest approach to

---

† Figs 4a, 4b, 5, 6, 7a, 7b and 7c of this Section reproduced with permission from [4].

doing this is to use a gradient-based adaptive algorithm, i.e., the gradient of the error surface $\langle V_e^2 \rangle$ is estimated with respect to the ith resonator frequency, and the ith resonator frequency is adapted in a direction opposite this gradient. However, this simple solution has problems arising from the fact that the adapted filter is an IIR filter. As is well known [39], the error surface for IIR filters is invariably multi-modal, and the use of a gradient-based strategy could cause the filter to settle in some undesirable local minimum. Another disadvantage of using the gradient-based strategy is that the generation of the gradient signals requires a complexity of $O(l^2)$.

In order to overcome these problems, a "pseudo-gradient" based adaptive algorithm is introduced next for this specific filter structure. The pseudo-gradient based ALE avoids the problems of multi-modality of the error surface, and also has the advantage of linear complexity. It also provides enhanced components that are in phase with the corresponding components in the input.

A Development of the adaptive algorithm

In the adaptive application, the resonator frequencies are changed by the addition of a correcting quantity at every iteration. Now, intuitively, this correcting quantity can be expected to have certain properties to ensure convergence. Let us assume that we have a single input sinusoid of frequency ω_{in}, and that the current resonator frequencies are given by ω_i. A correcting quantity has to be generated for each resonator, so there are a total of l correcting quantities. To ensure convergence, we would like the ith correcting quantity to have the following properties :

i) If $\omega_{in} < \omega_i$, then the frequency of the ith resonator must be lowered to bring it closer to ω_{in}; hence, the correcting quantity should be negative.

ii) If $\omega_{in} > \omega_i$, then the frequency of the ith resonator must be increased to bring it closer to ω_{in}; hence, the correcting quantity should be positive.

iii) If ω_{in} is equal to any one of the resonator frequencies, ω_j, then the correcting quantity should be zero, as the adaptive system is already in the desired state.

$$(16)$$

Further, the biquad implementations make it possible to change the resonator frequency only by changing some coefficient of the biquad, and in order to keep the adaptation process as simple as possible, we would like to use biquads that have their resonant frequency controlled by a *single* coefficient. This coefficient will henceforth be referred to as k_i; for instance, for LDI-biquads, $k_i = 2\sin(\omega_i/2)$, and for direct-form biquads, $k_i = \cos(\omega_i)$. Coupled-form biquads, however, cannot be used in this application because they have two coefficients, $\cos(\omega_i)$, and $\sin(\omega_i)$, that are dependent on the resonant frequency.

1   The Gradient

As mentioned earlier, one possible choice for the ith correcting quantity is the negative of the gradient of the error power, $\langle V_e^2 \rangle$, with respect to the ith resonator frequency. We will first investigate this choice, and see if it satisfies the conditions (16). Now, the power of the error signal is given in the frequency domain by

$$E[V_e^2] = \frac{1}{2\pi j} \oint H_e(z) H_e(z^{-1}) V_{in}(z) V_{in}(z^{-1}) z^{-1} dz , \qquad (17)$$

where $H_e(z)$ is the transfer function from the input to the error, and is given by (5). Instead of differentiating this with respect to the ith resonator frequency, ω_i, we will differentiate this quantity with respect to $a_i = \cos(\omega_i)$. As a_i is related to ω_i in a monotonic fashion, this preserves the validity of the correcting quantity. Hence, we have the ith component of the gradient of the error surface as

$$\nabla_i = \frac{d}{da_i} E[V_e^2] , \qquad (18)$$

$$= \frac{1}{2\pi j} \oint V_{in}(z^{-1}) V_{in}(z)$$
$$\left[H_e(z^{-1}) \frac{d}{da_i} H_e(z) + H_e(z) \frac{d}{da_i} H_e(z^{-1}) \right] z^{-1} dz . \qquad (19)$$

Defining a sensitivity transfer function as

$$H_{s,i}(z) = \frac{d}{da_i} H_e(z) , \qquad (20)$$

the expression for the gradient may now be written as

$$\nabla_i = \frac{1}{2\pi j} \oint V_{in}(z^{-1}) V_{in}(z)$$
$$[H_e(z) H_{s,i}(z^{-1}) + H_e(z^{-1}) H_{s,i}(z)] dz , \qquad (21)$$

$$= \frac{2}{2\pi j} \oint V_e(z) V_{s,i}(z^{-1}) z^{-1} dz , \qquad (22)$$

i.e., the gradient is obtained by correlating the error output and the output of the sensitivity filter $H_{s,i}(z)$. Hence the adaptive algorithm changes the resonator frequencies in a direction opposite the gradient, till the gradient eventually goes to zero, i.e., when the correlation between the signal V_e and $V_{s,i}$, the outputs of the 'sensitivity' filters, are all equal to zero.

Let us now see if the conditions (16) are satisfied by the gradient ∇_i. Assuming that the amplitude and frequency of the input sinusoid is A, and

ω_{in}, respectively, we can write

$$\begin{aligned}
\nabla_i(\omega_{in}) &= \frac{A^2}{2} \operatorname{Re}\left\{H_e^*(e^{j\omega_{in}})H_{s,i}(e^{j\omega_{in}})\right\} \\
&= \frac{A^2}{2} \operatorname{Re}\left\{H_e(z^{-1})H_{s,i}(z)|_{z=e^{j\omega_{in}}}\right\} .
\end{aligned} \qquad (23)$$

From (5) and (20), we can write

$$H_{s,i}(z) = -2G_i A_i(z)A_i(z)z^{-1}(1 - z^{-2}) , \qquad (24)$$

where

$$A_i(z) = \frac{\frac{1}{1-2a_i z^{-1}+z^{-2}}}{1 + \sum_{j=0}^{l-1} G_j \frac{2a_j z^{-1}-z^{-2}}{1-2a_j z^{-1}+z^{-2}}} . \qquad (25)$$

Also, writing $H_e(z^{-1})$ as $A_i(z^{-1})(1 - 2a_i z + z^2)$, (23) becomes

$$\begin{aligned}
\nabla_i(\omega_{in}) &= -2G_i \\
&\quad \operatorname{Re}[A_i(z)A_i(z^{-1})\, z^{-1}(1 - 2a_i z + z^2) \\
&\quad (1 - z^{-2})\, A_i(z)\,|_{z=e^{j\omega_{in}}}] , \qquad (26) \\
&= -2G_i\ \operatorname{Re}[|A_i(e^{j\omega_{in}})|^2\, [2\cos(\omega_{in}) - 2a_i]\, 2\sin(\omega_{in}) \\
&\quad e^{j(\pi/2-\omega_{in})}\, |A_i(e^{j\omega_{in}})|e^{j\,\phi(\omega_{in})}] , \qquad (27) \\
&= 2G_i\, |A_i(e^{j\omega_{in}})|^2\, [2a_i - 2\cos(\omega_{in})]\, 2\sin(\omega_{in}) \\
&\quad \cos(\pi/2 + \phi(\omega_{in}) - \omega_{in})\, |A_i(e^{j\omega_{in}})| , \qquad (28)
\end{aligned}$$

where $A_i(z)|_{z=e^{j\omega_{in}}} = |A_i(e^{j\omega_{in}})|\, e^{j\,\phi(\omega_{in})}$. Now, $|A_i(e^{j\omega_{in}})|$ is a positive quantity that has zeros at all resonator frequencies except the ith resonator frequency. Further, as $\cos(\omega_i) = a_i$, and as $\cos(\omega_{in})$ is a monotonically decreasing function in $[0, \pi]$, we have

$$2a_i - 2\cos(\omega_{in}) \left\{ \begin{array}{lll} < & 0 \ , & \omega_{in} < \omega_i \\ > & 0 \ , & \omega_{in} > \omega_i \\ = & 0 \ , & \omega_{in} = \omega_i \end{array} \right\} . \qquad (29)$$

From (28), the first two terms in the expression for $\nabla_i(\omega)$ grouped together provide the properties that we would like the correcting quantity to have. Further the third term $\sin(\omega_{in})$, is positive for $\omega_{in} \in [0, \pi]$; however, the $\cos()$ term in (28) causes the gradient to lose the desired properties. $\phi(\omega_{in})$ in this last term is given by

$$\begin{aligned}
\phi(\omega_{in}) &= \omega_{in} - \arctan\left[\frac{2G_i \sin(\omega_{in})}{1 - \sum_{j=0}^{l-1} G_j}K\right] \qquad \omega_{in} < \omega_i , \quad (30) \\
&= \omega_{in} - \arctan\left[\frac{2G_i \sin(\omega_{in})}{1 - \sum_{j=0}^{l-1} G_j}K\right] + \pi \qquad \omega_{in} > \omega_i , \quad (31)
\end{aligned}$$

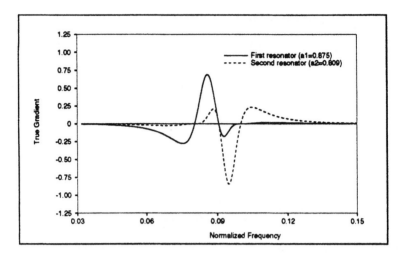

Figure 4: (a) True gradient for a two resonator system with a1 and a2 fixed. (©1991 IEEE)

where

$$K = \sum_{i=0}^{l-1} \frac{1}{2\cos(\omega_{in}) - 2a_i}.$$ (32)

When $\cos(\pi/2 - \omega_{in} + \phi(\omega_{in}))$ is evaluated, it is seen that the quantity becomes negative in regions between the resonant frequencies, implying that the gradient cannot be used as the correcting quantity.

A plot of the quantity (28) (as a function of ω_{in}) is shown in Fig. 4a, for a two resonator system with the two resonator frequencies fixed at $\omega_1 = 0.804$ Hz, and $\omega_2 = 0.1$ Hz. It may be seen that the gradient changes sign between the two resonator frequencies.

A further disadvantage of using the true gradient is that a separate filter-bank is needed to generate each sensitivity output, so the complexity of generating the gradients is $O(l^2)$.

2   The Pseudo-gradient

We saw in the last section that the true gradient of the error surface could not be used as the correcting quantity because it does not meet the requirements (16). We also saw that it is the cos() term in (28), that causes the true gradient to lose the desired properties. Consequently, if we could generate a pseudo-gradient, which retains the first few terms of the true gradient, but not the cos() term, then this pseudo-gradient could be used as a valid correcting quantity. Now, the true gradient was generated by correlating the output of a 'sensitivity' filter, $H_{s,i}(z)$, and the error output, V_e. It would seem a logical inference that the desired pseudo-gradient could also be generated in a similar way, i.e., by correlating the error output with the output of a pseudo-sensitivity filter (to be defined), and it remains now to find the transfer function of the pseudo-sensitivity filter.

By inspection, from (24) and (28), we can deduce that the transfer function of the ith pseudo-sensitivity filter should be

$$H_{ps1,i}(z) = -2GA_i(z)z^{-1}(1 - z^{-2}) . \tag{33}$$

The correlation between the output of the pseudo-sensitivity filter and the error output for the case of a single input sinusoid can now be obtained, as in (28), to be

$$\nabla_i^{ps}(\omega_{in}) = 2G |A_i(e^{j\omega_{in}})|^2 [2a_i - 2\cos(\omega_{in})] 2\sin(\omega_{in}) , \tag{34}$$

which can be seen to satisfy the conditions (16). A further major advantage of using this pseudo-gradient is that all the l pseudo-sensitivity outputs are available at the internal nodes of the filter structure, consequently, no extra hardware is needed to generate the pseudo-gradient. Hence, the complexity of the adaptive system is $O(l)$, as compared to $O(l^2)$ for most other ALE's.

It actually turns out that there are several possibilities for the ith pseudo-sensitivity transfer function, that generate a pseudo-gradient with the desirable conditions (16). All these choices yield almost identical performance in the absence of noise; however, in the presence of noise, their performance can be quite different. In order to narrow down our choice of the pseudo-gradient, we can now take this factor also into account.

One of the primary considerations in the noise performance of the ALE is the existence of a bias in the final solution. Consider the case of a single resonator tracking a single input sinusoid, in white noise. If (33) is used as the pseudo-sensitivity transfer function, then the final converged value of the resonator frequency will be different from the frequency of the input sinusoid, leading to a bias in the final solution. As this is an undesirable characteristic, we would like to find a pseudo-sensitivity transfer function that does not have this characteristic. It turns out that such a choice does exist, and the transfer function of the pseudo-sensitivity filter is given by

$$H_{ps,i}(z) = -2G_iA_i(z)z^{-1} . \tag{35}$$

The form of the pseudo-gradient can now be derived, similar to (28), as

$$\nabla_i^{ps}(\omega_{in}) = 2G_i |A_i(e^{j\omega_{in}})|^2 [2a_i - 2\cos(\omega_{in})] . \tag{36}$$

This quantity satisfies the conditions (16), and it is shown in Section III B 1 that this choice of $H_{ps,i}(z)$ produces an unbiased estimate in the presence of noise. Further, the pseudo-sensitivity output is obtained simply by scaling the lowpass output, $V_{q,i}$, of Fig. 3, and does not add to the complexity of the system.

Thus, in adapting the filter, one simply changes the ith resonator frequency to drive the correlation between the error signal and the ith lowpass

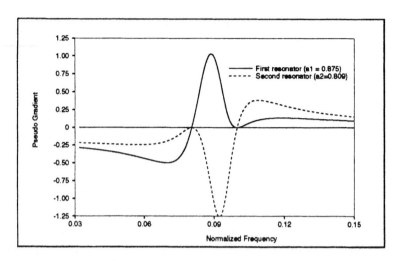

Figure 4: (b) Pseudo-gradient for a two resonator system with a1 and a2 fixed. (©1991 IEEE)

output to zero. This particular pseudo-gradient appears to be very similar to the frequency update used in an earlier proposed method for frequency estimation [40]. However, the update used here is felt to be more robust and less complicated than the method of [40] [3].

A plot of the pseudo-gradient for a two resonator system is shown in Fig. 4b. After the resonators have converged to their steady states, the frequency of the input sinusoid may be obtained using the functional relationship between the coefficient k_i and the resonant frequency of the biquad ω_i. To avoid the computational overhead of evaluating this function, lookup tables may be used. A point worth noting is that the relation between k_i and ω_i is independent of the other k_j and also of G_i, the damping of the loop. Hence, only one table of values need be maintained which is valid for all values of the damping.

---

[3] The work of [40] was based on matching the weighted sum of reference oscillator outputs to the input signal, and it was shown in [41] that this approach was equivalent to a static filter-bank for fixed resonator frequencies. If the resonator frequencies are themselves being adapted, then the weights of the filter have to converge much faster than the resonator frequencies themselves, for the equivalence to hold. Assuming that this is the case, the signal Im (Y_m) of [40] corresponds to the pseudo-sensitivity output used here. Hence, it would appear that the error signal used to modify the resonator frequencies of this filter-bank structure are very similar to the error signal used to adapt the resonator frequencies of [40]. However, it should be kept in mind that in [40], both the reference frequencies and the weights have to be adapted, and the convergence of the system depends on the weights converging to their steady state much faster than the reference frequencies. These problems are not present in the adaptive system presented here, in fact, there is no need to adapt any weights at all, only the resonator frequencies need be adapted.

B Properties of the ALE

1 Bias in the resonator frequency

In this section, it is shown that for the case of a single resonator tracking a single input sinusoid in additive white Gaussian noise, the resonator frequency converges to exactly the frequency of the input sinusoid without any bias. Assuming the input to be a sinusoid of frequency ω_{in} and amplitude A, and the noise to be additive white Gaussian with variance σ^2, the correlation between the pseudo-sensitivity output and the error output has an additional term due to the noise (cross correlation terms between the signal and noise are assumed to be equal to zero because of the assumption of whiteness) and now becomes

$$\nabla(\omega_{in}) = \nabla_1^{ps}(\omega_{res}, \omega_{in}) + \frac{2\sigma^2}{2\pi} \, \text{Re}\left\{ \int_0^\pi H_e^*(\omega) H_{ps,1}(\omega) \, d\omega \right\} ,$$

where $H_{ps,1}$ denotes the pseudo-sensitivity filter, $\nabla_1^{ps}()$ indicates the component of the correlation due to the signal alone, and ω_{res} indicates the frequency of the resonator. (As the resonator is adapting, ∇_1^{ps} becomes a function of the resonator frequency also.) The resonator now adapts till $\nabla(\omega_{in}) = 0$ and finally settles at a frequency ω_1 such that

$$\nabla_1^{ps}(\omega_1, \omega_{in}) = -\frac{2\sigma^2}{2\pi} \, \text{Re}\left\{ \int_0^\pi H_e^*(\omega) H_{ps,i}(\omega) \, d\omega \right\} . \qquad (37)$$

Hence, the bias in the steady state solution is given by $\omega_{in} - \omega_1$. This bias may be obtained by evaluating the integral in (37) and dividing it by a linear approximation to $\nabla_1^{ps}(\omega)$ near the resonant frequency [38].

By invoking Parsevals theorem in the z domain and the Residue theorem, we can show that

$$\frac{\sigma^2}{2\pi j} \oint H_e(z^{-1}) H_{ps,1}(z) z^{-1} d z \; = \; 0 .$$

Hence, from (37), we have $\nabla_1^{ps}(\omega_1, \omega_{in}) = 0$. This implies that $\omega_1 = \omega_{in}$, i.e., the resonator frequency in the steady state is exactly equal to the input sinusoidal frequency for the case of a single sinusoid. The generalization to the multi-sinusoidal case is an area of future research.

2 SNR Enhancement

The SNR enhancement is defined as the ratio of the SNR of the enhanced output to the SNR of the input. As the signal part of the enhanced output is exactly identical to the signal part of the input, the SNR enhancement simply is the ratio of the noise power at the input, to the noise power at the enhanced output. For the single sinusoidal case, the transfer function from the input of the ALE to the enhanced output (the fed back output) is given

---

Subsections III.B, III.B.2, III.C.1–III.C.3 reprinted with permission from *IEEE Transactions on Circuits and Systems-II*, vol. CAS-38, No. 10, pp. 1145–1159, Oct. 1991. © 1991 IEEE.

by $H_{fb,1}$. Therefore, the noise power at the enhanced output, assuming that the input is additive white gaussian noise, is

$$\overline{n_{enh}^2} = \frac{\sigma^2}{2\pi j} \oint H_{fb,1}(z) H_{fb,1}(z^{-1}) z^{-1}\, dz \,, \qquad (38)$$

where σ^2 is the power of the input additive white Gaussian noise. Evaluating the integral for the one-resonator case using the Residue theorem, we obtain

$$\overline{n_{enh}^2} = \frac{\sigma^2 G}{(1-G)} \,, \qquad (39)$$

and the SNR enhancement ratio as

$$\text{SNR enh.} = \frac{\sigma^2}{\sigma^2 \frac{G}{1-G}} = \frac{1-G}{G} \,. \qquad (40)$$

C   Simulations of the Adaptive Line Enhancer

The performance of the ALE is next examined through simulations. The filter-bank structure of Fig. 3 was used for these simulations, i.e., LDI biquads were used to implement the resonators. The value of G_i was also set to the same value for all i.

Stability Monitoring :   As the adapted filter is an IIR filter, it is necessary to monitor its stability during the adaptation process. From Section II C 1, the filter-bank is stable if $\sum_{j=0}^{l-1} G_j < 1$, and $|a_i| < 1$. For the case of LDI biquads, the condition $|a_i| < 1$ translates as $0 < k_i < 2$, where k_i is the actual coefficient used in the biquad. Also, as the G_i are all equal to some constant, G, they need to be chosen such that $G < 1/l$. The stabilti check for the filter simply becomes

$$0 < k_i < 2 \quad i = 0, \cdots, l-1 \,. \qquad (41)$$

Also, the actual pseudo-sensitivity used in the algorithm is a scaled version of the pseudo-sensitivity of (35), and is simply the signal $V_{ps,i}$ of Fig. 3.

1   Comparison of performance of true and pseudo gradients

Simulations of the performance of a two resonator system, when the true-sensitivity and pseudo-sensitivity filters are used, are shown in Fig. 5. The simulated system consists of two resonators, one with a resonant frequency of 0.1 Hz which is fixed, and with the other resonator adapting to notch out an input sinusoid at a frequency of 0.09545. A μ value of 0.0001 is used to adapt the resonators for both cases, and the damping factor G is set equal to 1/12. It can be seen that when the true sensitivities are used, convergence is not obtained.

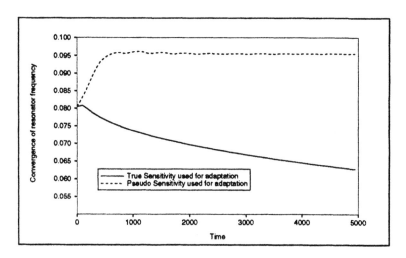

Figure 5: Convergence of resonator frequency of a two resonator system for true and pseud-gradient updates, with one resonator fixed. (©1991 IEEE)

2   Bias in resonator frequency

To confirm the derivations made in Sect. 1, a plot of the percentage bias in the resonator frequency as obtained from simulations is compared to the theoretical value of zero in Fig. 6, for an input SNR of 10 dB and 0 dB for $G = 1/6$. The simulated system consists of a single resonator in a feedback loop, and the coefficient k_1 of the resonator is adapted using an instantaneous approximation to the gradient as

$$k_1(n+1) = k_1(n) - \mu V_e(n) V_{ps,1}(n) ,$$

where $e(n)$ is the error and $V_{ps,1}(n)$ is the pseudo-sensitivity output. The bias shown in the simulations is seen to be negligibly small, which is in accordance with the theory. This condition of zero bias is valid only for the case of one resonator tracking a single input frequency; for multiple resonators, a bias is expected to exist but as of date, it has not been possible to derive theoretical expressions for this general case.

3   Simulations of the tracking

Simulations of the performance of the adaptive line enhancer are shown next. The resonator coefficients are adapted using the power normalized update [38]

$$k_i(n+1) = k_i(n) - \mu \frac{V_{ps,i}(n) V_e(n)}{||V_{ps,i}||^2 + P_{min}} , \tag{42}$$

where $V_{ps,i}(n)$ is the ith pseudo-sensitivity output and $e(n)$ is the error output. Here, an instantaneous approximation has been used for the pseudo-

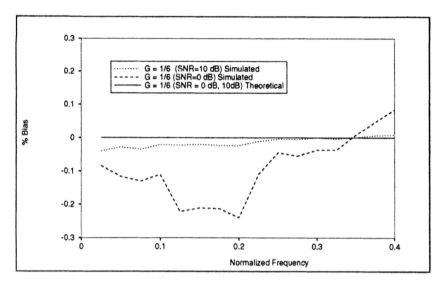

Figure 6: Bias plot for one resonator system. (©1991 IEEE)

gradient, and the power normalization is done in order to speed up convergence when the resonator coefficients are far away from their steady state values.

For the simulations presented, G is set equal to $1/12l$, where l is the number of resonators, P_{min} is set equal to 0.002, and to prevent the tracking of resonators from being identical, the μ value of the different resonators are separated from each other by 10% of the largest μ value. Also, the input signal is normalized so as to have unit power.

The first example (see Fig. 7a) shows the convergence of the coefficients of a system of three resonators, with the resonator frequencies initialized at 0.08, 0.10, and 0.12, and the input being the sum of three sinusoids in white noise, with frequencies 0.1, 0.12, and 0.14 respectively, all with an SNR of 10 dB. A μ value of 0.001 [4] is used for the first resonator.

The second example (see Fig. 7b) shows convergence when the resonator frequencies are initialized far away from the steady state values. The input consists of two sinusoids with frequencies 0.125 and 0.13 with an SNR of 3 dB each. The resonator frequencies were initialized at 0.001 and 0.002 respectively, and a μ value of 0.0001 is used for the first resonator.

The third example (see Fig. 7c) shows convergence when a two-resonator system tracks two input sinusoids with different powers. The resonator

---

[4] In the examples presented, the μ value was chosen arbitrarily. However, for the one-resonator case, it is possible to interpret the tracking loop as a phase-locked-loop, with a second order loop filter. This should enable the μ to be chosen so as to obtain any desired transient response. This idea has been used in [42, 14]. However, as the generalization for the multiple resonator case is not straightforward, it is not clear at this time, how the μ values can be chosen in a systematic fashion to get a desired transient response.

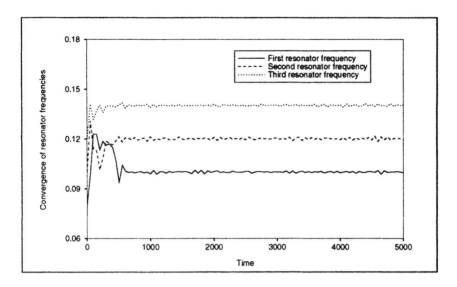

Figure 7: (a) Convergence of a three resonator system. (©1991 IEEE)

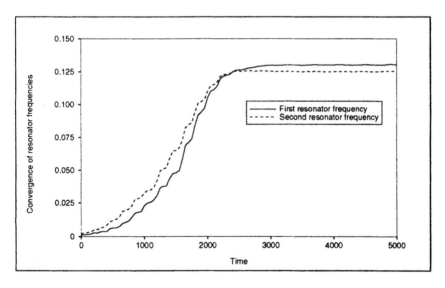

Figure 7: (b) Convergence of a two resonator system. (©1991 IEEE)

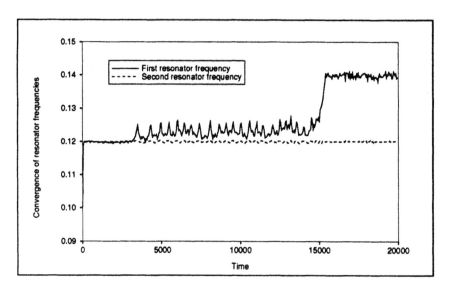

Figure 7: (c) Convergence of a two resonator system for input frequency components with different powers. (©1991 IEEE)

frequencies are initialized at 0.1 and 0.11, and the input consists of two sinusoids with frequencies 0.12 and 0.14 at an SNR of 20 dB and 0 dB respectively. This simulation used a μ value of 0.001 for the first resonator.

IV   Computation of Transforms[‡]

The next application that we will study is that of computing transforms. Transforms are used frequently in coding systems to achieve data compression [43, 1], and also in other applications such as transform domain adaptive filtering [2], in order to improve the convergence speed. Now, the computation of the transform adds quite a great deal to the complexity of the system, and conventionally, this complexity has been minimized by using fast algorithms to compute the transform [44, 45]. These fast algorithms are typically order recursive; i.e., an N point transform is broken down into smaller size transforms in a recursive manner to obtain computational savings. Though these fast algorithms do bring down the computational cost of the transform, they are unfortunately not very suitable for VLSI implementation because of the extensive data reordering that is needed to implement them. In contrast, time-recursive solutions [25, 46, 24, 20],[47, Ch 6] offer better possibilities for VLSI implementation. Here, the input data is assumed to arrive sequentially in time, and at any instant, the

---

[‡] Figs. 9, 10, 11, 12 and Tables. 1, 2, 3, 4 of this Section reproduced with permission from [5].

transform of the last N data samples is available, thus providing a "sliding" transform of the data. This is equivalent to passing the data through an FIR filter-bank, and sampling the outputs of the filter-bank to get the transform values. For the case of trigonometric transforms, this FIR filter-bank can be implemented with a linear hardware complexity using either a frequency-sampling structure, or the resonator-based filter-bank structure. Both these solutions are described next; however, due to its poor finite precision properties, the frequency-sampling structure is not very suitable for VLSI implementation, and the resonator-based filter-bank turns out to be the best solution.

A   Filter-bank interpretation of "sliding" transforms

Consider a transform with the basis matrix

$$\underline{\underline{H}} = \begin{bmatrix} h_{0,0} & \cdots & h_{0,N-1} \\ \vdots & \vdots & \vdots \\ h_{N-1,0} & \cdots & h_{N-1,N-1} \end{bmatrix}, \tag{43}$$

with the input data available sequentially in time. A "sliding" transformer provides the transform of the last N data samples at any point of time, hence it can be thought of as a system that outputs a set of N linear combinations of the last N data samples at any point of time. This is equivalent to saying that the transform values may be obtained by passing the data through a bank of N FIR filters, $H_{bp,i}(z)$, $i = 0, \cdots, N - 1$, and sampling the outputs, $V_{f,i}$, the filter-bank. Also, the transfer function of the ith filter is related to the elements of the ith row of the basis matrix [48, 23] in the following manner :

$$H_{bp,i}(z) = h_{i,N-1}z^{-1} + \ldots + h_{i,0}z^{-N} . \tag{44}$$

This set of transfer functions may be implemented with a linear hardware complexity, either using a frequency-sampling structure, or using a feedback structure, as described in Sections IV B, C.

B   Frequency-sampling structure

For illustration purposes, we will first start with the DFT. The $(l, m)^{th}$ element of the DFT basis matrix is given by $e^{j\frac{2\pi lm}{N}}$, $l, m = 0, \cdots, N - 1$. Hence, the transfer function, $H_{bp,i}(z)$, of (44), becomes

$$H_{bp,i}(z) = e^{j\frac{2\pi(N-1)}{N}}z^{-1} + e^{j\frac{2\pi(N-2)}{N}}z^{-2} + \cdots + z^{-N} ,$$

$$= [1 - z^{-N}]\frac{z_i z^{-1}}{1 - z_i z^{-1}} , \tag{45}$$

where $z_i = e^{j\frac{2\pi i}{N}}$, and the properties of geometric series have been used to simplify the sum in (45). We now need only one complex multiplier, z_i, to implement each channel of the filter-bank; further, as the term $1 - z^{-N}$ is common to all the channels, a single block with this transfer function can be shared between all the channels. This can be simplified even further if the filter-bank outputs were designed to be $V_{f,i} + V_{f,N-i}$ and $V_{f,i} - V_{f,N-i}$, instead of $V_{f,i}$ and $V_{f,N-i}$. The transfer function from the input to these two outputs now becomes

$$H_{bp,i\oplus N-i}(z) = [1 - z^{-N}] \left[\frac{z_i z^{-1}}{1 - z_i z^{-1}} + \frac{z_{N-i} z^{-1}}{1 - z_{N-i} z^{-1}} \right] ,$$

$$= [1 - z^{-N}] \frac{2a_i z^{-1} - 2z^{-2}}{1 - 2a_i z^{-1} + z^{-2}} \quad i = 1, \cdots, \lfloor \frac{N+1}{2} \rfloor , \quad (46)$$

$$H_{bp,i\ominus N-i}(z) = [1 - z^{-N}] \left[\frac{z_i z^{-1}}{1 - z_i z^{-1}} - \frac{z_{N-i} z^{-1}}{1 - z_{N-i} z^{-1}} \right] ,$$

$$= j[1 - z^{-N}] \frac{2b_i z^{-1}}{1 - 2a_i z^{-1} + z^{-2}} \quad i = 1, \cdots, \lfloor \frac{N+1}{2} \rfloor , \quad (47)$$

where $a_i = \frac{z_i + z_{N-i}}{2} = \cos(\frac{2\pi i}{N})$, and $b_i = \frac{z_i - z_{N-i}}{2j} = \sin(\frac{2\pi i}{N})$. The main advantage of this implementation is that all the internal arithmetic can be carried out in the real domain (assuming real input data), and it is only necessary to use complex numbers at the very last stage. The complete filter-bank is shown in Fig. 8, with coupled form biquads being used to implement the transfer function (46, 47). The desired outputs $V_{f,i}$ and $V_{f,N-i}$ can be obtained simply by adding and subtracting $V_{f,i\oplus N-i}$ and $V_{f,i\ominus N-i}$ respectively.

Hence, the number of multipliers needed to implement the entire filter-bank is only $2N$, and its data flow path is also very regular. Unfortunately, however, these nice features are offset by the poor finite precision behaviour of the structure, which makes it unsuitable for VLSI implementation. The reasons for this are examined next.

1 Finite precision effects

Stability: The transfer function of the biquads in (46, 47) is seen to have its poles on the unit circle; hence it represents a digital resonator, and the filter-bank is stable and FIR because this pole is exactly cancelled out by a pair of the zeros of the common block $1 - z^{-N}$. Under finite precision conditions, the cancellation will not occur, and the filter-bank will have a pole on or very close to the unit circle, thus representing a marginally stable structure.

Minimality: The transfer functions of the different channels of the filter-bank are of Nth order (44); however, there are $2N$ delays being used to implement the filter-bank of Fig. 8. Now under conditions of finite internal wordlength, assuming that the quantization is done just before the delays, the effect of finite internal wordlength can be modelled by adding a white noise source just before the delay elements. The contribution of these

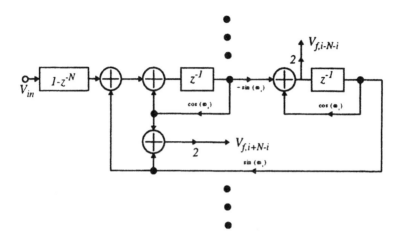

Figure 8: Frequency Sampling Structure.

noise sources at the output is determined by the transfer function from the
input of the delay elements, to the output. This transfer function can be
seen to have its poles on the unit circle; consequently, the additive noise
due to quantization could get blown up very badly at the output.

For these reasons, the frequency-sampling structure cannot be used in
practical situations. However, it does suggest a possible way to obtain
structures that do not have the same problems, and have the same hardware
complexity. These structures make use of a feedback loop to overcome the
problems mentioned above, and essentially reduce to the resonator-based
filter-bank structure.

C   Feedback Structure

We will next describe a feedback transformer to implement the filter-bank
of (44). The structure contains certain "basis generating" IIR filters and
a "compensating" filter in a feedback loop, and its capability to perform
transformations is dependent on choosing the "compensating" filter coefficients in such a way that the overall structure resembles an FIR filter with
all its poles at 0. The term "basis-generating" is used because the filters
can be considered as a cascade of an IIR filter, $\frac{1}{1\pm z^{-K}}$ (K is an integer
that is dependent on the transform to be implemented), and an FIR filter
$H_{bp,i}(z)$; the IIR filter can be thought of as producing an impulse every K

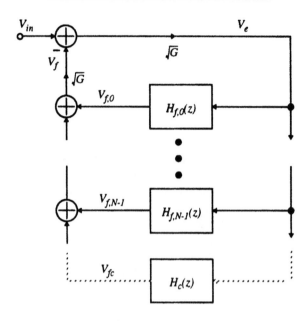

Figure 9: Generalized Feedback Transformer Structure. (©1993 IEEE)

sample periods, that excites the FIR filter, and the impulse response of the filter is hence a periodic sequence related to the ith row of the basis matrix.

The overall filter structure is shown in Fig. 9, with the filter $H_{f,i}(z)$ representing the ith "basis-generating" filter, and $V_{f,i}$ being related to the ith output of the filter-bank through a scale factor. The transfer functions of $H_{f,i}(z)$ and $H_c(z)$ are given by

$$H_{f,i}(z) = \frac{V_{f,i}}{V_e} = \frac{H_{bp,i}(z)}{1 + \eta z^{-K}} \quad i = 0, \cdots, N-1 . \tag{48}$$

$$H_c(z) = \frac{V_{fc}}{V_e} = \frac{\left[h_{c,1}z^{-1} + \cdots + h_{c,N}z^{-N} \right]}{1 + \eta z^{-K}} \quad \text{for} \quad K \leq N , \quad \text{or} \tag{49}$$

$$= \frac{\left[h_{c,1}z^{-1} + \cdots + h_{c,K-1}z^{-(K-1)} \right]}{1 + \eta z^{-K}} , \quad \text{for} \quad K > N \tag{50}$$

where $\eta = \pm 1$. The following derivations assume that $K \leq N$; they may be easily extended to the case where $K > N$. The loop gain of the structure shown in Fig. 9 is given by

$$\delta(z) = G \left[H_c(z) + \sum_{i=0}^{N-1} H_{f,i}(z) \right] ,$$

$$= \frac{G}{1 + \eta z^{-K}} \sum_{j=1}^{N} \left[h_{c,j} + \sum_{i=0}^{N-1} h_{i,N-j} \right] z^{-j} . \tag{51}$$

If we choose the coefficients $h_{c,j}$ and $\sqrt{G}$ such that the following conditions are satisfied,

$$
\left\{
\begin{array}{l}
h_{c,j} = -\sum_{i=0}^{N-1} h_{i,N-j} \quad j = 1,..,N \ , \ j \neq K \\
h_{c,K} = 0 \\
G = \frac{-\eta}{\sum_{i=0}^{N-1} h_{i,N-K}}
\end{array}
\right\}
, \qquad (52)
$$

then the loop gain becomes

$$
\delta(z) = \frac{-\eta z^{-K}}{1 + \eta z^{-K}} , \qquad (53)
$$

and we obtain the transfer function from the input to the error V_e, and to the fedback signal V_f as

$$
H_e(z) = \frac{V_e}{V_{in}} = \frac{\sqrt{G}}{1 + \delta(z)} = \sqrt{G}(1 + \eta z^{-K}) , \qquad (54)
$$

$$
\text{and} \qquad H_{fb}(z) = \frac{V_f}{V_{in}} = \frac{\delta(z)}{1 + \delta(z)} = -\eta z^{-K} . \qquad (55)
$$

Hence, the overall filter behaves like an FIR filter [5] . If the basis matrix $\underline{H}$ is such that all column sums except the $(N - K)^{th}$ are equal to zero, then from (52), $h_{c,j} = 0$, and no compensating filter is required. This condition holds true for the basis matrices of transforms such as the DFT and the WHT for $K = N$; however for some special cases such as the DCT and DST, even though these conditions do not hold true, we do not need to include a separate compensating filter in the feedback loop, as is shown later.

If the input data is available sequentially in time, and if $h_{c,j}$ and $\sqrt{G}$ are chosen as in (52), from (48) and (54), $V_{f,i} = H_{f,i} H_e V_{in} = \sqrt{G} H_{bp,i} V_{in}$, i.e., the transfer function from the input to the ith filter-bank output, scaled by $1/\sqrt{G}$ is equal to $H_{bp,i}$, where $H_{bp,i}$ is as given in (44). The desired transformation can now be obtained by scaling the $V_{f,i}$ outputs by $1/\sqrt{G}$, which due to the finite memory of the overall filter, will yield the transform of the last N data samples.

---

[5] If the $H_{f,i}(z)$ are implemented using N delays, then it is always possible to choose the $h_{c,j}$ such that the first two conditions of (52) are satisfied. Further, if finite precision constrains the actually implemented value of $\sqrt{G}$ to be equal to $\gamma\sqrt{G}$, where $\gamma \leq 1$, then the transfer function $H_e(z)$ may be obtained as $H_e(z) = \frac{\gamma\sqrt{G}(1+\eta z^{-K})}{1+\eta(1-\gamma^2)z^{-K}}$. The poles of the filter structure may now be seen to be evenly distributed around the unit circle at a radius of $(1 - \gamma^2)^{1/K}$ (for infinite precision, $\gamma = 1$, and the poles are all at 0). Hence, the filter remains stable even if finite precision is used to represent $\sqrt{G}$.

1   DFT

From (44), and (48),

$$H_{f,i}(z) = \frac{1-z^{-N}}{1+\eta z^{-K}}\frac{z_i^* z^{-1}}{1-z_i^* z^{-1}} \qquad z_i = w_N^i \quad i = 0, \cdots, N-1 . \qquad (56)$$

Now, if we choose $\eta = -1$, and $K = N$, the expression for the $H_{f,i}(z)$ becomes $\frac{z_i^* z^{-1}}{1-z_i^* z^{-1}}$, and this may be implemented with a single delay, and the overall structure with a canonical number (N) of delays. Further, all column sums of the DFT matrix except the first are seen to sum to zero, hence no compensating filter is needed, and we only need to choose $\sqrt{G} = \frac{1}{\sqrt{N}}$, to obtain FIR behavior. These conditions are summarized in Table 1, and it is seen that the filter structure reduces to the structure of [20]. Further, if the ith and $(N-i)^{th}$ complex first order resonators are grouped together, the structure takes on the form of l_2 real second order resonators, and one or two real first order resonators in a feedback loop. The transfer function of these resonators is the same as (6), with $G_i = 1/N$, and $a_i = \cos(\frac{2\pi i}{N})$. This structure is now seen to be the same as the resonator-based structure of Fig. 3, with the resonator frequencies being the N roots of unity. Further, for this special choice of resonator frequencies, and G, we have

$$V_e = H_e(z)V_{in} = \sqrt{G}[1 - z^{-N}]V_{in} , \qquad (57)$$

$$V_{fb,i} = V_e \left[\frac{z_i z^{-1}}{1 - z_i z^{-1}} + \frac{z_{N-i} z^{-1}}{1 - z_{N-i} z^{-1}} \right]$$

$$= \sqrt{G}[1 - z^{-N}]\frac{2\cos(\frac{2\pi i}{N})z^{-1} - 2z^{-2}}{1 - 2\cos(\frac{2\pi i}{N})z^{-1} + z^{-2}}V_{in} \quad i = 1, \cdots, \left\lfloor \frac{N-1}{2} \right\rfloor (58)$$

$$V_{q,i} = V_e \left[\frac{z_i z^{-1}}{1 - z_i z^{-1}} - \frac{z_{N-i} z^{-1}}{1 - z_{N-i} z^{-1}} \right]$$

$$= \sqrt{G}[1 - z^{-N}]\frac{2\sin(\frac{2\pi i}{N})z^{-1}}{1 - 2\cos(\frac{2\pi i}{N})z^{-1} + z^{-2}}V_{in} \quad i = 1, \cdots, \left\lfloor \frac{N-1}{2} \right\rfloor (59)$$

Hence, $V_{fb,i}$ gives the sum of the desired outputs $V_{bp,i}$, and $V_{bp,N-i}$, and $V_{q,i}$ equals their difference. The desired quantities can now be obtained by taking the sum and difference of $V_{fb,i}/\sqrt{G}$ and $V_{q,i}/\sqrt{G}$ respectively.

We will next consider a related structure that will be useful in the following sections. A related basis matrix may be defined whose elements are given by $\left\{ w_{2N}^{(2l+1)m} \right\}$. The filter-bank associated with this basis matrix takes exactly the same form as the DFT based structure, with the difference that $\eta = 1$, and the z_i are now given by w_{2N}^{2i+1}. The corresponding transfer functions for this structure become

$$V_e = H_e(z)V_{in} = \sqrt{G}[1 + z^{-N}]V_{in} , \qquad (60)$$

Table 1: DFT based Filter-Bank (©1993 IEEE)

		$H_{f,i}(z) = \frac{z_i^* z^{-1}}{1 - z_i^* z^{-1}}$	$z_i = w_N^i$, $w_N = e^{j\frac{2\pi}{N}}$
			$i = 0, \cdots, N - 1$
DFT	$\eta = -1$, $K = N$	$h_{c,j} = 0$	$j = 1, \cdots, N - 1$
		$\sqrt{G} = \frac{1}{\sqrt{N}}$	

$$V_{fb,i} = \sqrt{G}[1 + z^{-N}] \frac{2\cos(\frac{\pi(2i+1)}{N})z^{-1} - 2z^{-2}}{1 - 2\cos(\frac{\pi(2i+1)}{N})z^{-1} + z^{-2}} V_{in} , \qquad (61)$$

$$i = 1, \cdots, \left\lfloor \frac{N}{2} \right\rfloor + 1 ,$$

$$V_{q,i} = \sqrt{G}[1 + z^{-N}] \frac{2\sin(\frac{\pi(2i+1)}{N})z^{-1}}{1 - 2\cos(\frac{\pi(2i+1)}{N})z^{-1} + z^{-2}} V_{in} , \qquad (62)$$

$$i = 1, \cdots, \left\lfloor \frac{N}{2} \right\rfloor + 1 .$$

D   Feedback Transformer for general Trigonometric Transforms - Non-minimal Implementation

Under the category of general trigonometric transforms, we group the various forms of the DCT and the DST [46].

Consider the DCT basis matrix proposed in [43] (DCT-II of [46]) whose elements are given by

$$\left\{ \sqrt{\frac{2}{N}} \, k_l \cos\left[l(m + \frac{1}{2})\frac{\pi}{N}\right] \right\} ,$$

where $\quad k_l = 1 \; l \neq 0, N \quad k_0 = k_N = \frac{1}{\sqrt{2}} \quad l, m = 0, \cdots, N - 1 .$

The $H_{f,i}(z)$ for this case can be derived to be

$$H_{f,i}(z) = -\sqrt{\frac{2}{N}} \, k_i \left[\frac{z^{-1}(1 - z^{-1})\cos(\omega_i/2)}{1 - 2\cos(\omega_i)z^{-1} + z^{-2}}\right] \left[\frac{z^{-N} - e^{j(\pi i)}}{1 + \eta z^{-K}}\right] , \qquad (63)$$

where $\omega_i = \frac{\pi i}{N} \quad i = 0, \cdots, N - 1$. This basis matrix does need a compensating filter, $H_c(z)$, to force the feedback transformer to be FIR. If the $H_{f,i}(z)$ and $H_c(z)$ are implemented as a cascade of a Nth order transfer function $\frac{z^{-N} + (-1)^i}{1 + \eta z^{-K}}$, and a second order section $\frac{z^{-1}(1 - z^{-1})}{1 - 2\cos(\omega_i)z^{-1} + z^{-2}}$, then the transformer could be implemented as shown in Fig. 9, with a non-minimal

Subsections IV.D, IV.E.1, and IV.F reprinted with permission from *IEEE Transactions on Circuits and Systems-II*, vol. 40, No. 1, pp. 41–50, Jan. 1993. © 1993 IEEE.

number of delays in the feedback loop. The loop gain is now given by

$$
\delta(z) = G \sum_{i=0}^{N-1} \left(-\sqrt{\frac{2}{N}} k_i \frac{z^{-1}(1-z^{-1})\cos(\omega_i/2)}{1-2\cos(\omega_i)z^{-1}+z^{-2}} \frac{z^{-N}+(-1)^i}{1+\eta z^{-K}} \right)
$$
$$
+ \frac{h_{c,1}z^{-1} + \cdots + h_{c,N-1}z^{-(N-1)}}{1+\eta z^{-K}} \tag{64}
$$
$$
= \frac{a_1 z^{-1} + \cdots + a_{3N}z^{-3N}}{(1+\eta z^{-K})\prod_{i=0}^{N-1}(1-2\cos(\omega_i)z^{-1}+z^{-2})}, \tag{65}
$$

where $a_1, \cdots, a_{3N}$ are some real numbers. The transfer function $H_e(z)$ now becomes

$$
H_e(z) = \frac{\sqrt{G}}{1+\delta(z)}, \tag{66}
$$

which reduces to

$$
\frac{\sqrt{G}(1+\eta z^{-K})\prod_{i=0}^{N-1}(1-2\cos(\omega_i)z^{-1}+z^{-2})}{(1+\eta z^{-K})\prod_{i=0}^{N-1}(1-2\cos(\omega_i)z^{-1}+z^{-2}) + a_1 z^{-1} + \cdots + a_{3N}z^{-3N}}. \tag{67}
$$

Ideally, of course $\delta(z) = \frac{-\eta z^{-K}}{1+\eta z^{-K}}$, and $\frac{1}{1+\delta(z)} = 1+\eta z^{-K}$. This implies that all the terms $(1-2\cos(\omega_i)z^{-1}+z^{-2})$ in the numerator of (67), which represent zeros lying on the unit circle, cancel corresponding terms in the denominator. Hence, (67) has poles as well as zeros on the unit circle, which cancel exactly under infinite precision arithmetic, to produce a stable FIR transfer function. However, for finite precision, this cancellation will not occur, and the structure turns out to have poles very close to, or on, the unit circle. Hence, if the feedback loop is implemented with a non-minimal number of delays, the structure is as sensitive to coefficient truncation effects as the frequency sampling structure.

E   Feedback Transformer for general Trigonometric Transforms - Minimal Implementation

In order to avoid the problems of Section IV B, D, we need to implement the feedback transformer using a minimal number of delays. In order to do this, we need to separate the $H_{f,i}(z)$ into two groups, depending on whether i is even or odd, and implement each group using a separate filter bank. We will refer to these henceforth as the "even" and "odd" filter-banks and differentiate between them when necessary by using the superscript "even" or "odd" respectively, in all relevant quantities. The even and odd filter-banks have $l = \lfloor \frac{N+1}{2} \rfloor$, and $l = \lfloor \frac{N}{2} \rfloor$ basis generating filters, respectively.

Consider first the even filter-bank. Choosing $K = N$ and $\eta = -1$, we obtain

$$
H_{bp,i}^{even}(z) = \sqrt{\frac{2}{N}} k_i [1 - z^{-N}] \frac{\cos(\omega_i/2)z^{-1}(1-z^{-1})}{1-2\cos(\omega_i)z^{-1}+z^{-2}} \qquad \omega_i = \frac{2\pi i}{N} \quad , \tag{68}
$$

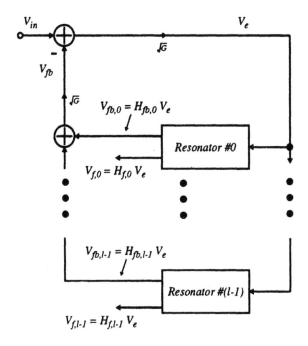

Figure 10: General form of "even" and "odd" filter-banks of Feedback Transformer. (©1993 IEEE)

$$i = 0, ..., \left\lfloor \frac{N-1}{2} \right\rfloor .$$

Comparing this with the transfer function from V_{in} to $V_{fb,i}$ of the DFT-based structure, we see that the two transfer functions differ only in the numerator of the biquadratic term, and in the scale factor. The numerator in (68) can be obtained from the DFT-based structure simply by taking a linear combination of the internal nodes of the ith biquad. Hence, the desired outputs of the "even" filter-bank can be obtained simply from the DFT-based structure. Similarly, the modified-DFT based structure can be used to produce the outputs of the "odd" filter-bank.

The even and odd filter-banks now take the form shown in Fig. 10; *i.e.*, *first or second order resonators in a feedback loop*, which represent minimal realizations, with the transfer function from V_e to the ith feedback output $V_{fb,i}$ being given by $H_{fb,i}(z)$ (see (6)) and the transfer function from V_e to the output $V_{f,i}$ being given by $H_{f,i}(z)$. Also, the final desired outputs can be obtained by interlacing, or taking linear combinations of the outputs of the two filter-banks as indicated in Tables 2, 3.

The same procedure as above can be followed for the other forms of the DCT and DST, and the relevant quantities and transfer functions, together with the definition of the $(i,j)^{th}$ element of the basis matrix, are

summarized in Tables 2, 3. For the case of the DCT-III, DST-III, DCT-IV
and DST-IV it becomes necessary to use complex filters, and consequently,
the implementation is not as computationally efficient as for the other two
versions of the DCT and the DST; hence, only the structures corresponding
to the DCT I-III and DST I-III are described here. In the following,

$$k_j = 1 \ j \neq 0, N \ , \ \ k_0 = k_N = \frac{1}{\sqrt{2}} \ , \tag{69}$$

$$H_{fb,i}(z) = \left\{ \begin{array}{ll} \frac{2\cos(\omega_i)z^{-1}-2z^{-2}}{1-2\cos(\omega_i)z^{-1}+z^{-2}} & \omega_i \neq 0, \pi \\ \frac{\cos(\omega_i)z^{-1}-z^{-2}}{1-2\cos(\omega_i)z^{-1}+z^{-2}} & \omega_i = 0 \text{ or } \pi \end{array} \right\} \ , \ \ i = 0, \cdots, l-1 \ , \tag{70}$$

$$\text{with} \ \left\{ \begin{array}{ll} l-1 = \lfloor \frac{N-1}{2} \rfloor & , \ \omega_i = \frac{2\pi i}{N} \text{ if } \eta = -1 \\ l-1 = \lfloor \frac{N-2}{2} \rfloor & , \ \omega_i = \frac{\pi(2i+1)}{N} \text{ if } \eta = 1 \end{array} \right\} \cdot \tag{71}$$

1 Comparison with other methods

In general, the computation of block transforms arises in applications such
as data compression etc. and several efficient algorithms have been developed to efficiently compute the DCT or DST of a vector [44, 45, 49].
In terms of number of multiplications and additions, these methods are
more computationally efficient than the transformer presented here ([49]
needs only 11 multiplications to compute an 8-pt DCT-II); however, the
transformer does have the merit of being simple, and requires no data
reordering etc. It also provides a more efficient alternative to compute sliding transforms for applications such as transform domain adaptive filtering
[25, 50, 2], as compared to recursive formulae that were developed in [46]
(see Table 4). Further it also provides a much more robust alternative to
the frequency sampling like structure developed in [25] because it shows
very good behavior under finite precision (no stability problems etc), unlike the frequency sampling structure, as shown in the next section. Table 4
indicates the number of multiplications and additions needed to implement
the filter-bank, and this is also equal to the computations necessary to update a "sliding" transform (coupled form biquads (see Fig. 2c) are used to
implement the $H_{fb,i}(z)$, and the signal flow graph of the filter-bank for the
DCT-II, is shown in Fig. 11.) In Table 4,

$$k_1 \overset{\triangle}{=} 4 \left[\lfloor \frac{N-1}{2} \rfloor + \lfloor \frac{N}{2} \rfloor \right] \approx 4N \ ,$$

$$k_2 \overset{\triangle}{=} \log_2 \left[\lfloor \frac{N+3}{2} \rfloor \right] + \log_2 \left[\lfloor \frac{N+4}{2} \rfloor \right] \approx 2\log_2 \left[\frac{N}{2} \right] \ ,$$

$$k_3 \overset{\triangle}{=} \lfloor \frac{N+1}{2} \rfloor + \lfloor \frac{N}{2} \rfloor \approx N \ .$$

Table 2: Feedback Transformer (DCT) (©1993 IEEE)

DCT − I		
	$\left\{ \sqrt{\frac{2}{N}}\, k_i k_j \, \cos\left[\frac{\pi i j}{N}\right] \right\}$	$i,j = 0, \cdots, N-1$
Even	$H_{f,i}^{even}(z) = -\sqrt{\frac{2}{N}}\, k_i\, \frac{z^{-1}(z^{-1}-\cos(\omega_i))}{1-2\cos(\omega_i)z^{-1}+z^{-2}}$ $H_{fb,i}^{even}(z)$- see (70)	$\eta = -1 \; K = N$ $\omega_i = \frac{2\pi i}{N}\;\; i = 0,\cdots,\lfloor \frac{N-1}{2} \rfloor$ $\sqrt{G} = \frac{1}{\sqrt{N}}, \; C = \sqrt{\frac{2}{N}} \left(\frac{1}{\sqrt{2}} - 1 \right)$
Odd	$H_{f,i}^{odd}(z) = \sqrt{\frac{2}{N}}\, \frac{z^{-1}(z^{-1}-\cos(\omega_i))}{1-2\cos(\omega_i)z^{-1}+z^{-2}}$ $H_{fb,i}^{odd}(z)$- see (70)	$\eta = 1 \; K = N$ $\omega_i = \frac{\pi(2i+1)}{N}\;\; i = 0,\cdots,\lfloor \frac{N-2}{2} \rfloor$ $\sqrt{G} = \frac{1}{\sqrt{N}}, \; C = \sqrt{\frac{2}{N}} \left(\frac{1}{\sqrt{2}} - 1 \right)$
	$V_{fb,2i} = V_{f,i}^{even} + C\, k_i\, V_{fbc}^{even} \quad i = 0, \cdots, \lfloor \frac{N-1}{2} \rfloor$ $V_{fb,2i+1} = V_{f,i}^{odd} - C V_{fbc}^{odd} \quad i = 0, \cdots, \lfloor \frac{N-2}{2} \rfloor$	

DCT − II		
	$\left\{ \sqrt{\frac{2}{N}}\, k_i \, \cos\left[i(j+0.5)\frac{\pi}{N}\right] \right\}$	$i,j = 0, \cdots, N-1$
Even	$H_{f,i}^{even}(z) = \sqrt{\frac{2}{N}}\, k_i\, \frac{\cos(\omega_i/2)z^{-1}(1-z^{-1})}{1-2\cos(\omega_i)z^{-1}+z^{-2}}$ $H_{fb,i}^{even}(z)$- see (70)	$\eta = -1 \; K = N$ $\omega_i = \frac{2\pi i}{N}\;\; i = 0,\cdots,\lfloor \frac{N-1}{2} \rfloor$ $\sqrt{G} = \frac{1}{\sqrt{N}}$
Odd	$H_{f,i}^{odd}(z) = -\sqrt{\frac{2}{N}}\, \frac{\cos(\omega_i/2)z^{-1}(1-z^{-1})}{1-2\cos(\omega_i)z^{-1}+z^{-2}}$ $H_{fb,i}^{odd}(z)$- see (70)	$\eta = 1 \; K = N$ $\omega_i = \frac{\pi(2i+1)}{N}\;\; i = 0,\cdots,\lfloor \frac{N-2}{2} \rfloor$ $\sqrt{G} = \frac{1}{\sqrt{N}}$
	$V_{fb,2i} = V_{f,i}^{even} \quad i = 0,\cdots,\lfloor \frac{N-1}{2} \rfloor \qquad V_{fb,2i+1} = V_{f,i}^{odd} \quad i = 0,\cdots,\lfloor \frac{N-2}{2} \rfloor$	

DCT − III		
	$\left\{ \sqrt{\frac{2}{N}}\, k_j \, \cos\left[(i+0.5)j\frac{\pi}{N}\right] \right\}$	$i,j = 0, \cdots, N-1$
Even	$H_{f,i}^{even}(z) = \sqrt{\frac{2}{N}}\, \frac{j z_i z^{-1}}{1 - z_i z^{-1}}$ $H_{fb,i}^{even}(z) = \frac{z_i z^{-1}}{1 - z_i z^{-1}}$	$\eta = j \; K = N$ $z_i = e^{j\frac{(2i-0.5)\pi}{N}}\;\; i = 0,\cdots,N-1$ $\sqrt{G} = \frac{1}{\sqrt{N}}$
Odd	$H_{f,i}^{odd}(z) = \sqrt{\frac{2}{N}}\, \frac{j z_i z^{-1}}{1 - z_i z^{-1}}$ $H_{fb,i}^{odd}(z) = \frac{z_i z^{-1}}{1 - z_i z^{-1}}$	$\eta = -j \; K = N$ $z_i = e^{j\frac{(2i+0.5)\pi}{N}}\;\; j = 0,\cdots,N-1$ $\sqrt{G} = \frac{1}{\sqrt{N}}, \; C = \frac{1}{\sqrt{2}} - 1$
	$V_{fb,2i} = \frac{1}{2}\left[-V_{fb,N-i}^{even} + V_{f,i}^{odd} \right] - C V_{fbc}^{odd} \quad i = 0,\cdots,\lfloor \frac{N-1}{2} \rfloor$ $V_{fb,2i+1} = \frac{1}{2}\left[-V_{f,i+1}^{even} + V_{fb,N-i-1}^{odd} \right] - C V_{fbc}^{odd} \quad i = 0,\cdots,\lfloor \frac{N-2}{2} \rfloor$	

Table 3: Feedback Transformer (DST) (©1993 IEEE)

DST − I

	$\left\{\sqrt{\frac{2}{N}} \sin\left\lceil\frac{\pi(i+1)(j+1)}{N}\right\rceil\right\}$	$i,j = 0, \cdots, N-2$
Even	$H_{f,i}^{even}(z) = \sqrt{\frac{2}{N}} \frac{\sin(\omega_i)z^{-1}}{1-2\cos(\omega_i)z^{-1}+z^{-2}}$ $H_{fb,i}^{even}(z)$- see (70)	$\eta = 1\ \ K = N$ $\omega_i = \frac{\pi(2i+1)}{N}\ \ i = 0,\cdots,\lfloor\frac{N-2}{2}\rfloor$ $\sqrt{G} = \frac{1}{\sqrt{N}}\ \ i = 0,\cdots,\lfloor\frac{N-2}{2}\rfloor$
Odd	$H_{f,i}^{odd}(z) = -\sqrt{\frac{2}{N}} \frac{\sin(\omega_i)z^{-1}}{1-2\cos(\omega_i)z^{-1}+z^{-2}}$ $H_{fb,i}^{odd}(z)$- see (70)	$\eta = -1\ \ K = N$ $\omega_i = \frac{2\pi(i+1)}{N}\ \ i = 0,\cdots,\lfloor\frac{N-3}{2}\rfloor$ $\sqrt{G} = \frac{1}{\sqrt{N}}\ \ i = 0,\cdots,\lfloor\frac{N-1}{2}\rfloor$
	$V_{fb,2i} = V_{f,i}^{even}\quad i = 0,\cdots,\lfloor\frac{N-2}{2}\rfloor \qquad V_{fb,2i+1} = V_{f,i}^{odd}\quad i = 0,\cdots,\lfloor\frac{N-3}{2}\rfloor$	

DST − II

	$\left\{\sqrt{\frac{2}{N}} k_{i+1} \sin\left\lceil\frac{\pi(i+1)(j+0.5)}{N}\right\rceil\right\}$	$i,j = 0, \cdots, N-1$
Even	$H_{f,i}^{even}(z) = \sqrt{\frac{2}{N}} k_{i+1} \frac{\sin(\frac{\omega_i}{2})z^{-1}(1+z^{-1})}{1-2\cos(\omega_i)z^{-1}+z^{-2}}$ $H_{fb,i}^{even}(z)$- see (70)	$\eta = 1\ \ K = N$ $\omega_i = \frac{\pi(2i+1)}{N}\ \ i = 0,\cdots,\lfloor\frac{N-1}{2}\rfloor$ $\sqrt{G} = \frac{1}{\sqrt{N}}\ \ i = 0,\cdots,\lfloor\frac{N-2}{2}\rfloor$
Odd	$H_{f,i}^{odd}(z) = -\sqrt{\frac{2}{N}} k_{i+1} \frac{\sin(\frac{\omega_i}{2})z^{-1}(1+z^{-1})}{1-2\cos(\omega_i)z^{-1}+z^{-2}}$ $H_{fb,i}^{odd}(z)$- see (70)	$\eta = -1\ \ K = N$ $\omega_i = \frac{2\pi(i+1)}{N}\ \ i = 0,\cdots,\lfloor\frac{N-2}{2}\rfloor$ $\sqrt{G} = \frac{1}{\sqrt{N}}\ \ i = 0,\cdots,\lfloor\frac{N-1}{2}\rfloor$
	$V_{fb,2i} = V_{f,i}^{even}\quad i = 0,\cdots,\lfloor\frac{N-1}{2}\rfloor \qquad V_{fb,2i+1} = V_{f,i}^{odd}\quad i = 0,\cdots,\lfloor\frac{N-2}{2}\rfloor$	

DST − III

	$\left\{\sqrt{\frac{2}{N}} k_{j+1} \sin\left\lceil\frac{\pi(i+0.5)(j+1)}{N}\right\rceil\right\}$	$i,j = 0, \cdots, N-1$
Even	$H_{f,i}^{even}(z) = \sqrt{\frac{1}{2N}} \frac{z^{-1}}{1-z_i z^{-1}}$ $H_{fb,i}^{even}(z) = \frac{z_i z^{-1}}{1-z_i z^{-1}}$	$\eta = j\ \ K = N$ $z_i = e^{j\frac{(2i-0.5)\pi}{N}}\ \ i = 0,\cdots,N-1$ $\sqrt{G} = \frac{1}{\sqrt{N}}$
Odd	$H_{f,i}^{odd}(z) = \sqrt{\frac{1}{2N}} \frac{z^{-1}}{1-z_i z^{-1}}$ $H_{fb,i}^{odd}(z) = \frac{z_i z^{-1}}{1-z_i z^{-1}}$	$\eta = -j\ \ K = N$ $z_i = e^{j\frac{(2i+0.5)\pi}{N}}\ \ i = 0,\cdots,N-1$ $\sqrt{G} = \frac{1}{\sqrt{N}}$, $C = \sqrt{\frac{2}{N}}\left(\frac{1}{\sqrt{2}}-1\right)$
	$V_{fb,2i} = V_{fb,N-i}^{even} + V_{f,i}^{odd} + Cz^{-1}V_{in}\quad i = 0,\cdots,\lfloor\frac{N-1}{2}\rfloor$ $V_{fb,2i+1} = -V_{f,i+1}^{even} - V_{fb,N-i-1}^{odd} + Cz^{-1}V_{in}\quad i = 0,\cdots,\lfloor\frac{N-2}{2}\rfloor$	

Table 4: Computational Complexity [6] (©1993 IEEE)

	Filter Bank		From [46]	
	Mult	Add	Mult	Add
DFT	$(N+2)^*$	$(N+log_2 N+1)^*$		
DCT − I	k_1+8	$k_1+k_2+k_3+5$	12N	8N
DCT − II	k_1+k_3+3	$k_1+k_2+k_3+5$	6N	6N
DCT − III	$(2N+2)^*+2$	$(4N+2log_2 N+2)^*$	12N	8N
DST − I	k_1+3	$k_1+k_2+k_3+5$		
DST − II	k_1+k_3+3	$k_1+k_2+k_3+3$		
DST − III	$2(N+1)^*+2$	$(4N+2log_2 N+2)^*$		

$$x_{1,i} = \cos(\theta_i) \qquad c_i = \frac{\sqrt{2}}{2\sin(\frac{\theta_i}{2})\sqrt{N}}$$

$$x_{2,i} = \sin(\theta_i)$$

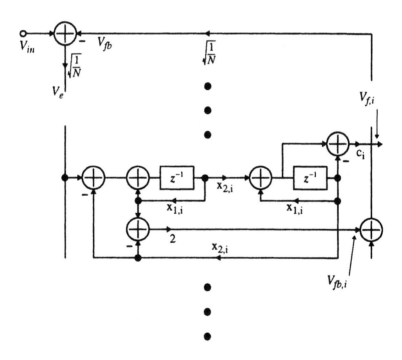

Figure 11: Implementation of either filter-bank of the S-V Structure for DCT-II. (©1993 IEEE)

F   Finite precision effects

The modularity and linear complexity of the transformer structure make it very suitable for implementation using VLSI circuits or DSP's. As the internal wordlength and coefficient wordlength used in these cases is finite, the study of finite precision effects on this structure is of particular interest. Recall from Sect. IV. C. 1, that the even and odd filter-bank's take the form of the resonator-based filter structure. From the results of Section II, we know that this structure has good finite precision properties i.e., if coupled-form biquads (Fig. 2c), or LDI biquads (Fig. 2b) or direct-form biquads are used to implement the second order resonators, then the filter structure remains stable even under coefficient truncation (assuming magnitude truncation); and for the case of coupled-form biquads, it does not even sustain any zero input limit cycles.

Under similar conditions, the poles of the frequency sampling structure also do remain inside the unit circle, however they lie very close to the unit circle, as shown below. Also, the frequency sampling structure may sustain zero input limit cycles.

If infinite coefficient wordlength is used for either the generalized transformer, or the frequency-sampling structure, the impulse response of the transfer function from the input to $V_{f,i}$ is FIR, with length equal to N. Under finite precision however, the poles of either structure move away from zero, and the impulse response becomes IIR. The poles of the generalized transformer however remain close to zero, while the poles of the frequency sampling structure remain close to the unit circle. This is illustrated in Fig. 12, where the error in the impulse response of the third channel of the "even" filter-bank is shown for the transformer structure and the frequency sampling structure corresponding to the DCT-II. This simulation uses a value of N equal to 32, and 9 bits (including one sign bit) are used to represent the internal coefficients. The error in the impulse response dies away to zero much more slowly for the case of the frequency-sampling structure, showing that its poles are closer to the unit circle.

G   Pipelineability of Transformer Structure

The final issue to be considered in this application is the pipelineability of the transformer structure. We have already seen that the linear complexity and good finite precision properties of the structure make it a good candidate for VLSI implementation, however, as explained in Section. II C 2, the critical-path delay of the filter-structure constrains its maximum speed

---

[6]The * denotes complex operations. The coupled form is used to implement the second order $H_{fb,i}(z)$ and $H_{f,i}(z)$, and a tree structure is used to add up all the $V_{fb,i}$ in the feedback loop. Also, as the recursive formulae of [46] require both the DST and DCT to be maintained, and updated in order to compute a sliding transform, the numbers shown for [46], corresponding to the DCT-i in the table, refer to the computation required to obtain both the updated DCT-i and the DST-i values.

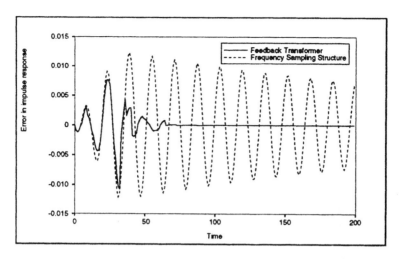

Figure 12: Error in impulse response of generalized Transformer and Frequency Sampling Structure. (©1993 IEEE)

of operation, thus making it unsuitable for high-speed image processing applications [7]. To solve this problem, we will next describe techniques to pipeline the filter-structure, and enable it to be used for high-speed applications. It is important to note however, that this pipelining strategy is only applicable for the case where the resonator frequencies lie at the N roots of 1 or -1.

As explained in Section. II C 2, the critical delay of the filter structure. arises from the additions that have to be performed in the feedback path. Before a new data sample can be processed, the signal V_e (see Fig. 10) should be computed. To compute V_e, all the fedback outputs $V_{fb,i}$ have to be added, and then subtracted from V_{in}. Hence, the rate at which input data can be accepted is determined by the time taken to compute $l + 1$ additions, and turns out to be the critical delay of the filter To reduce this delay, we make use of transformation techniques similar to the ones outlined in [32, 33].

Consider making a modification to the filter structure of Fig. 10. Assume that there is a block with the transfer function κz^{-N} in cascade with all the $H_{fb,i}(z)$ in the feedback loop, where κ is some scalar to be defined.

---

[7] Strictly speaking, if A is the delay associated with a single adder, and b represents the internal wordlength, the delay associated with adding up these l terms is $O(A*(b+log_2l))$ [51], rather than $O(A*b*l)$. For example, consider the case where $l = 4$, and we need to compute $s_i = V_{fb,0} + V_{fb,1}$, $s_2 = V_{fb,2} + V_{fb,3}$, and finally $s_1 + s_2$. It is not necessary to wait for the complete sums s_1 and s_2 to be available before starting the computation $s_1 + s_2$; rather this computation can start as soon as the least significant bits of s_1 and s_2 are available. Hence, the time taken would be only $O(A*(b+log_2l))$. However, this time can be brought down even further to just $O(Ab)$ by following the pipelining procedures outlined here.

The loop gain of the structure now becomes

$$H_{lg}(z) = \frac{-\kappa\eta z^{-2N}}{1 + \eta z^{-N}} \qquad (72)$$

and the transfer function from the input to the ith output, $V_{f,i}$ becomes

$$\frac{1 + \eta z^{-N}}{1 + \eta z^{-N} - \eta\kappa z^{-2N}} H_{f,i}(z) , = , \frac{H'_{bp,i}(z)}{1 + \eta z^{-N} - \kappa\eta z^{-2N}} , \qquad (73)$$

where $H'_{bp,i}(z)$ represents the transfer function from the input, V_{in}, to $V_{f,i}$ of the original structure of Fig. 10 ($H'_{bp,i}(z)$ is an FIR transfer function). If we now cascade a pre-filter before the modified transformer structure with the transfer function

$$H_{prefilt}(z) = 1 + \eta z^{-N} - \kappa\eta z^{-2N} , \qquad (74)$$

then the transfer function from the input of the pre-filter to the output, $V_{f,i}$, is given by $H'_{bp,i}(z)$, which is the desired transfer function.

We now have N additional delays in the feedback loop, and if the signals $V_{fb,i}$ are added using a tree structure, these delays could be placed after each adder in the tree, resulting in the structure of Fig. 13. It may now be seen that as far as the computation in the feedback loop is concerned, it is only necessary to compute a single addition in a clock period, and not $l+1$ additions as before. Hence, the critical delay of the filter structure has now been reduced to the critical delay of the resonators in the loop, which is determined by the biquad structure used to implement the resonators.

The only remaining point to be considered is the performance of the structure under finite precision. It is obvious that the FIR pre-filter is being used to cancel the poles of the modified structure. As we cannot rely on pole-zero cancellation under finite precision conditions, the structure should be such that these poles lie within the unit circle, so that even if they are not cancelled out by the pre-filter, it will not result in an unstable structure.

Consider the pre-filter transfer function (74). Assuming $\eta = -1$, we have

$$H_{prefilt}(z) = (z^{2n} - z^N + \kappa)z^{-2N} . \qquad (75)$$

Substituting $x = z^N$, this becomes a quadratic equation, and solving for the roots of this quadratic, the zeros of the pre-filter may be found to lie at

$$r_1 e^{j\frac{2\pi k}{N}} \quad k = 0, \cdots, N-1 \quad r_2 e^{j\frac{2\pi k}{N}} \quad k = 0, \cdots, N-1 , \qquad (76)$$

where

$$r_1 = \left[\frac{1 + \sqrt{1 - 4\kappa}}{2}\right]^{1/N} \quad \text{and} \quad r_2 = \left[\frac{1 - \sqrt{1 - 4\kappa}}{2}\right]^{1/N} . \qquad (77)$$

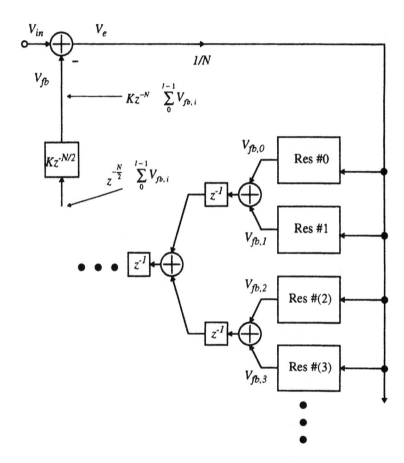

Figure 13: Modified (Pipelineable) Feedback Transformer Structure.

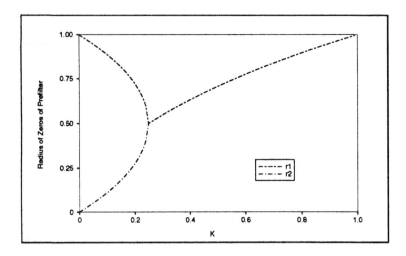

Figure 14: Radius of poles of pipelined transformer as a function of κ.

The quantities $|r_1|^N$ and $|r_2|^N$ are plotted in Fig. 14 for κ values ranging from 0 to 1, and it may be seen that the zeros of the pre-filter, and consequently, the poles of the pipelined transformer structure lie inside the unit circle for this range of values of κ.

V   A "Lerner"-based modification of the filter-bank[†‡]

The final application of the filter structure presented here, is as a multi-output filter-bank. Earlier, we saw that the resonator-based filter-structure could be used for the application of adaptive filtering, and as a filter-bank for the computation of trigonometric transforms. The transfer functions from V_{in} to $V_{fb,i}$ of this filter-bank take on the shape of bandpass filters; unfortunately, the characteristics of these bandpass transfer functions are not very good. They have quite high sidelobes, and their passbands are not very flat, and applications such as frequency-domain adaptive filtering, subband decomposition systems, etc., usually require a much higher suppression of out-of-band energy. To try and meet this requirement, in this Section, we will develop a class of filter-banks based on some classical 'Lerner' filters. The basic idea is to start with a prototype filter-bank with poor characteristics, and then group the channels of this prototype in such a manner that the combination of channels has much better characteristics. Initially a filter-bank with uniformly spaced channels is developed,

---

[7†‡] Figs. 16, 17a, and 20 of this Section reproduced with permission from [52]. Figs. 18a, 18b, and 18c reproduced with permission from [53].

and subsequently, a non-uniform filter-bank is developed by applying an all-pass transformation on the uniform filter-bank. Finally, some applications of the filter-bank are examined.

A Preliminaries

The resonator-based filter structure, with the resonator frequencies set equal to the roots of 1, will be used as the "poor" prototype that starts the design process. This also corresponds to the DFT-based transformer structure of Section IV, i.e., there are $\lfloor \frac{N+2}{2} \rfloor$ resonators in a feedback loop, with the transfer function of thr resonators being given by (70). The band-pass transfer functions of interest are from the input V_{in} to $V_{fb,i}$, denoted $H^f_{fb,i}(z)$, and from the input V_{in} to $V_{q,i}$, denoted $H^f_{fq,i}(z)$ (see Section II). Further, given that the resonator frequencies are the roots of unity, it turns out to be also possible to develop explicit expressions for these transfer functions, which will be used to prove the properties of the filter-bank.

From (57) and (58), we can write

$$H^f_{fb,i}(z) = \frac{V_{fb,i}}{V_{in}} = [1 - z^{-N}] \left[\frac{z_i z^{-1}}{1 - z_i z^{-1}} + \frac{z_i^* z^{-1}}{1 - z_i^* z^{-1}} \right] . \tag{78}$$

Now, recognizing the fact that $z_i^N = z_i^{*N} = 1$, this might be rewritten as

$$
\begin{aligned}
H^f_{fb,i}(z) &= z_i z^{-1} \frac{1 - z_i^N z^{-N}}{1 - z_i z^{-1}} + z_i^* z^{-1} \frac{1 - z_i^{*N} z^{-N}}{1 - z_i^* z^{-1}} , \tag{79} \\
&= (z_i + z_i^*) z^{-1} + \cdots + (z_i^N + z_i^{*N}) z^{-N} , \tag{80} \\
&= 2\cos(\omega_i) z^{-1} + 2\cos(2\omega_i) z^{-2} + \cdots \\
&\quad + 2\cos((N-1)\omega_i) z^{-(N-1)} + z^{-N} . \tag{81}
\end{aligned}
$$

For the first order resonators at dc and $f_s/2$, the outputs are

$$H^f_{fb,0}(z) = z^{-1} + \cdots + z^{-N} , \tag{82}$$

and

$$H^f_{fb,N/2}(z) = -z^{-1} + z^{-2} - \cdots + z^{-N} , \tag{83}$$

respectively. Following the same procedure as for $H^f_{fb,i}(z)$, we may write

$$H^f_{fq,i}(z) = \frac{V_{q,i}}{V_{in}} = \tag{84}$$

$$2 \left[\sin(\omega_i) z^{-1} + \sin(2\omega_i) z^{-2} + \cdots + \sin((N-1)\omega_i) z^{-(N-1)} \right] . \tag{85}$$

for the second order resonators. Note that the $V_{q,i}$ output is not available for the real first order resonators at dc and $f_s/2$.

B   Filter-banks with Lerner Weighted Outputs

The proposed filter-bank takes a linear combination of the resonator outputs $V_{fb,i}$ or $V_{q,i}$ to yield bandpass channels with much better characteristics [52]. The combination of outputs is similar to what has been used for Lerner filters [54]. These filter design methods were originally proposed for realizing continuous-time filter-banks having almost linear-phase bandpass outputs, with good stopband attenuation. In [54], each bandpass filter was realized by a weighted sum of adjacent parallel second order biquadratic filters. The weighting coefficients were ± 1 for adjacent resonators, except the bandpass edge biquads, for which $\pm 1/2$ was used [55]. The + signs were used for all the odd biquads, whereas the − signs were used for all the even biquads, or vice versa. These filter realization techniques were extended to the digital domain using the matched-z transform [56], and the impulse invariant transform [57].

1   Uniform Filter-Bank

In the proposed filter-bank, the outputs of adjacent resonators of the prototype filter-bank are added up with alternating signs to realize each Lerner bandpass output. The more resonator outputs are included in each filter-bank output, the wider the passband will be. Normally, the resonators at the passband edges are weighted with a $\pm 1/2$ rather than a ± 1. This gives a better transition band performance. In addition, the transition band outputs are shared between adjacent channels [55]. For good stopband performance (or equivalently small sidelobes), the total weighting in each bandpass output, of all the resonator outputs having even indices must equal the total weighting of all the resonator outputs having odd indices.

Similar ideas of grouping the bandpass channels of a poor prototype filter-bank, to get a filter-bank with better characteristics have also been explored in [58]. The techniques in [58] formulated the problem as a linear program, and solved it to obtain the weighting coefficient for each resonator output. In contrast, the Lerner groupings introduced here use a predetermined weighting strategy, which appears to give quite an acceptable channel characteristic, and the additional bonus of linear phase. The magnitude response could perhaps be improved on by using the optimization technique of [58], but it is felt that the resulting improvement would be minimal.

A possible weighting scheme for the Lerner based filter-bank is shown in Fig. 15. Note that for this choice of weighting, the first and last Lerner outputs of the filter-bank only use two resonator outputs, whereas all other Lerner outputs have a weighted sum of three resonator outputs. This is because the resonators at dc and $f_s/2$ are only first order resonators, and thus an adjacent biquad with a weighting of 1/2 will cancel their sidelobes. The characteristics of this Lerner-based filter-bank may be summarized as follows:

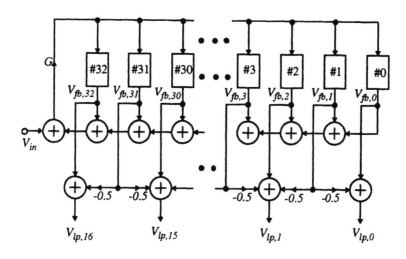

Figure 15: Three resonator Lerner groupings.

(a) The passband is about twice as wide as that of the prototype filter-bank, whereas the stopband attenuation is greatly improved, especially as one gets further away from the passband. In many practical applications, this is preferable to a constant stopband attenuation, This is shown in Fig. 16, where $H_{lp,1}(z) = \frac{V_{lp,1}}{V_{in}}$ of Fig. 15, is graphed along with the transfer function of $H_{fb,2}^{f}(z)$. A flatter passband can be obtained by grouping more than three resonator outputs in the manner described above; the larger the number of resonators used for the Lerner grouping, the flatter the passband, and the larger the passband width. To reduce the passband width, the order of the prototype filter-bank could be increased. A doubling of the order would result in an equal width passband and a further improved stopband at the expense of greater complexity.

(b) The Lerner grouped bandpass channels also have an additional important property. They are all exactly linear phase, and all bandpass channels have the same group delay. This can be shown easily as follows. Consider the simplest kind of Lerner grouping, as in Fig. 15. We now have

$$H_{lp,i}(z) = -0.5H_{fb,2i-1}^{f}(z) + H_{fb,2i}^{f}(z) - 0.5H_{fb,2i+1}^{f}(z) . \qquad (86)$$

Substituting from (81) into (86), we have

$$H_{lp,i}(z) = a_1 z^{-1} + \cdots + a_{N-1} z^{-(N-1)} , \qquad (87)$$
$$\text{where } a_i = -0.5\cos(\omega_{2i-1}) + \cos(\omega_{2i}) - 0.5\cos(\omega_{2i+1}) . \qquad (88)$$

Now noting that

$$\cos((N-1)\omega_i) = \cos\left(\frac{2\pi i}{N}(N-1)\right) = \cos\left(\frac{2\pi i}{N}\right) = \cos(\omega_i) , \qquad (89)$$

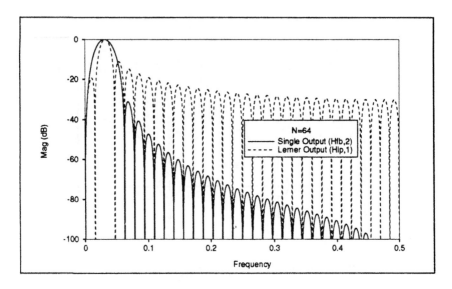

Figure 16: A Lerner grouped output compared to a single output. (©1992 IEEE)

we have

$$a_i = a_{N-i} \, , \tag{90}$$

in (87). Hence, the impulse response corresponding to the transfer function $H_{lp,i}(z)$ has even symmetry. This implies that $H_{lp,i}(z)$ has linear phase, and a group delay given by $z^{-N/2}$. This group delay depends only on the length of the impulse response, and is the same for all channels.

(c) The filter-bank also provides a very useful quadrature Lerner transfer function $H_{lq,i}(z)$, that has almost the same magnitude characteristic as the Lerner bandpass transfer function, and exactly 90° phase difference at all frequencies. This output is obtained by using the same kind of Lerner grouping on the $V_{q,i}$ outputs of the prototype filter-bank. Hence,

$$H_{lq,i}(z) = -0.5 H^f_{bq,2i-1}(z) + H^f_{bq,2i}(z) - 0.5 H^f_{bq,2i+1}(z) \, , \tag{91}$$

and from (85),

$$H_{lq,i}(z) = b_1 z^{-1} + \cdots + b_{N-1} z^{-(N-1)} \, , \tag{92}$$

where $b_i = -0.5 \sin(\omega_{2i-1}) + \sin(\omega_{2i}) - 0.5 \sin(\omega_{2i+1}) \, .$

Now using the fact that

$$\sin\left((N-1)\omega_i\right) = -\sin(\omega_i) \, , \tag{93}$$

we have

$$b_i = -b_{N-i} \, . \tag{94}$$

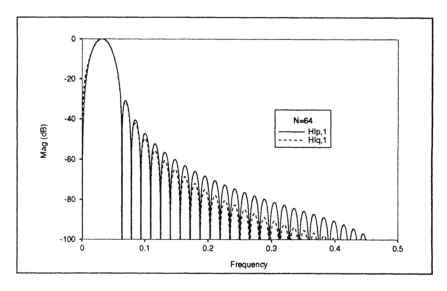

Figure 17: (a) Magnitude response of $H_{lp,i}$ and $H_{lq,i}$. (©1992 IEEE)

Hence, the impulse response corresponding to the transfer function $H_{lq,i}(z)$ has odd symmetry, and its length is exactly equal to that of $H_{lp,i}(z)$. This implies that $H_{lq,i}(z)$ are also linear phase filters, with the same group delay as $H_{lp,i}(z)$, but exactly 90° out of phase. The magnitude characteristic of $H_{lp,1}(z)$ and $H_{lq,1}(z)$ is shown in Fig. 17a, b for the case where three resonator outputs are grouped to form the Lerner outputs. It may be seen that the magnitude of the two transfer functions are very similar.

(d) The sum of the bandpass Lerner outputs is a pure delay (for even N). This fact has significance in applications such as analysis-synthesis systems for speech [59]. The proof of the allpass property is outlined below. From Fig. 15, the sum of the Lerner bandpass outputs is given by

$$\sum H_{lp,i}(z) = \sum_{i=0}^{\lfloor \frac{N}{2} \rfloor} e^{j\pi i} H_{fb,i}^{f}(z) \tag{95}$$

Now, assuming N to be even, and substituting for $H_{fb,i}(z)$ from (80), we get

$$\begin{aligned}
\sum H_{lp,i}(z) &= e^{j\pi 0} \sum_{k=1}^{N} z_0^k z^{-k} + \sum_{i=1}^{\frac{N-2}{2}} e^{j\pi i} \sum_{k=1}^{N} (z_i^k + z_i^{*k}) z^{-k} \\
&\quad + \sum_{k=1}^{N} z_{N/2}^k z^{-k} ,
\end{aligned} \tag{96}$$

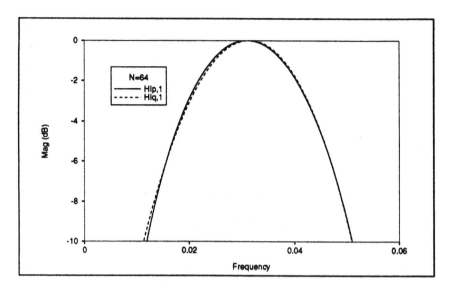

Figure 17: (b) Magnitude response of $H_{lp,i}$ and $H_{lq,i}$ (Passband).

$$= \sum_{k=1}^{N} e^{j\pi 0} z_0^k z^{-k} + \sum_{i=1}^{\frac{N}{2}-1} e^{j\pi i} \sum_{k=1}^{N} z_i^k z^{-k} \,,$$

$$+ \sum_{l=\frac{N}{2}+1}^{N-1} e^{j\pi N} e^{j\pi l} \sum_{k=1}^{N} z_l^k z^{-k} + \sum_{k=1}^{N} z_{N/2}^k z^{-k} \,, \qquad (97)$$

$$= \sum_{i=0}^{N-1} e^{j\pi i} \sum_{k=1}^{N} z_i^k z^{-k} \,, \qquad (98)$$

$$= \sum_{k=1}^{N} z^{-k} \sum_{i=0}^{N-1} e^{j\pi i} z_i^k = \sum_{k=1}^{N} z^{-k} \sum_{i=0}^{N-1} e^{j\frac{2\pi}{N}(\frac{N}{2}+k)i} \,. \qquad (99)$$

Using the identity

$$\sum_{i=0}^{N-1} e^{j2\pi k/N} = \left\{ \begin{array}{ll} 1 & \text{if} \quad k = lN \\ 0 & \text{else} \end{array} \right\} \,, \qquad (100)$$

we obtain

$$\sum H_{lp,i} = z^{-N/2} \,. \qquad (101)$$

Hence the sum of the Lerner bandpass outputs may be seen to be a pure delay.

2   Non-Uniform Filter-Bank

The filter-bank described earlier had all its bandpass channels spaced uniformly around the unit circle. However, there are several applications that

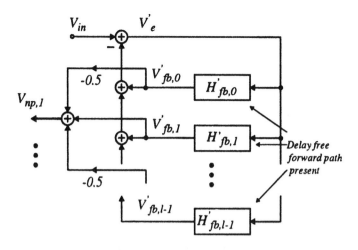

Figure 18: (a) Non-Uniform Filter-Bank with delay-free loop. (©1992 IEEE)

require the lower frequencies of the input signal to be resolved more accurately than the higher frequencies. For such applications, a filter-bank with non-uniformly spaced center frequencies is required. Rather than start the design of such a filter-bank from scratch, an easier alternative is to start with the uniform Lerner based filter-bank, which was shown to have good characteristics, and convert it to a non-uniform filter-bank by applying an allpass transformation [57, 53]. The merit of using such a transformation is that it retains the good magnitude characteristics of the bandpass channels. This procedure is similar to the one used in [57]; the main difference between them is the manner in which the allpass transformation is applied. The nature of the prototype filter-bank that we start with precludes a straight-forward application of the transformation, as it leads to a delay free loop.

To convert the uniform filter-bank to a non-uniform filter-bank, we use the allpass transformation proposed in [57], i.e.,

$$z \leftarrow \frac{z + \alpha}{1 + \alpha z} . \tag{102}$$

Applying the above transformation to each resonator of the prototype filter-bank results in the structure of Fig. 18a, where the primes are used to denote quantities after the transformation. Now, grouping the fedback outputs $V'_{fb,i}$ just as for the uniform filter-bank case, yields the outputs of the non-uniform filter-bank $V_{np,i}$ as shown in Fig. 18a. However, the filter structure of Fig. 18a is not implementable because the $H'_{fb,i}(z)$, turn out to have a delay free forward path. This problem is solved just as in Section II (see 3, 4), i.e., the transfer function $H'_{fb,i}(z)$ is expressed as the sum of a

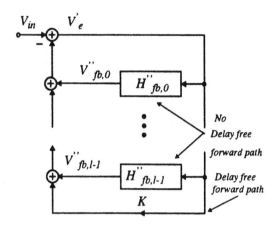

Figure 18: (b) Non-Uniform Filter-Bank with delay-free loop. (©1992 IEEE)

proper transfer function and a delay free forward path as

$$H'_{fb,i}(z) = H''_{fb,i}(z) + K_i \,, \tag{103}$$

and the delay-free forward paths of all resonators are grouped together as in Fig. 18b, where

$$H''_{fb,i} = r_i \frac{2a_i z^{-1} - 2z^{-2}}{1 - 2a_i z^{-1} + z^{-2}} \,, \tag{104}$$

$$r_i = \frac{1}{N} \frac{(1 - \alpha^2)}{(1 - \alpha z_i)(1 - \alpha z_i^*)} = \frac{1}{N} \frac{1 - \alpha^2}{1 + \alpha^2 - 2\alpha \cos(\frac{2\pi i}{N})} \,, \tag{105}$$

$$a_i = \frac{1}{2} \left[\frac{z_i - \alpha}{1 - \alpha z_i} + \frac{z_i^* - \alpha}{1 - \alpha z_i^*} \right] = \frac{(1 + \alpha^2) \cos(\frac{2\pi i}{N}) - 2\alpha}{1 + \alpha^2 - 2\alpha \cos(\frac{2\pi i}{N})} \,, \tag{106}$$

$$K_i = \frac{\alpha}{N} \left[\frac{z_i}{1 - \alpha z_i} + \frac{z_i^*}{1 - \alpha z_i^*} \right] = \frac{2\alpha}{N} \left[\frac{\cos(\frac{2\pi i}{N}) - \alpha}{1 + \alpha^2 - 2\alpha \cos(\frac{2\pi i}{N})} \right] \,, \tag{107}$$

$$K = \sum_{i=0}^{l-1} K_i = \frac{\alpha^N}{1 - \alpha^N} \,. \tag{108}$$

Consider now the error signal V'_e of Fig. 18b. It may be expressed as

$$V'_e = V_{in} - \sum_{i=0}^{l-1} H''_{fb,i} V'_e - K V'_e \,. \tag{109}$$

The equation (109) may be re-written as

$$V'_e = \frac{1}{1 + K} \left[V_{in} - \sum_{i=0}^{l-1} H''_{fb,i} V'_e \right] \,, \tag{110}$$

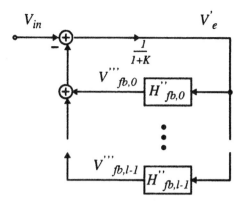

Figure 18: (c) Non-Uniform Filter-Bank without delay-free loop. (©1992 IEEE)

and the equation (110) may be implemented by a structure of the form shown in Fig. 18c.

The signal V_e' of Fig. 18c is exactly the same as that in Fig. 18a; also from Fig. 18a and (103),

$$V_{fb,i}' = H_{fb,i}'' V_e' + K_i V_e' . \qquad (111)$$

Hence, we may obtain $V_{fb,i}'$ from the structure of Fig. 18c as

$$V_{fb,i}' = V_{fb,i}''' + K_i V_e' , \qquad (112)$$

$$= V_{fb,i}''' + \frac{\alpha}{N} \frac{2\cos(\frac{2\pi i}{N}) - 2\alpha}{1 + \alpha^2 - 2\alpha\cos(\frac{2\pi i}{N})} V_e' . \qquad (113)$$

Subsequently, the $V_{fb,i}'$ may be grouped to form the Lerner bandpass outputs $V_{np,i}$. The magnitude response of some channels of the Lerner grouped filter-bank before and after the frequency transformation, are shown in Fig. 19. Five resonator outputs have been grouped to form the Lerner outputs, and α is chosen to be -0.7.

C   Hardware Complexity

The main feature of these filter-banks that makes them desirable is their low computational complexity and low sensitivity to coefficient inaccuracies. For the uniform case, the filter-bank requires $N/2 + 1$ real multipliers (complex conjugate resonators are grouped and implemented using a single undamped direct-form biquad), and yields $N/4$ bandpass channels; to obtain their quadrature components, an additional $N/4$ multipliers are

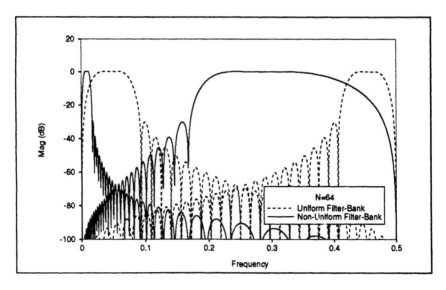

Figure 19: Transfer Function before and after allpass transformation $V_{lp,1}$ and $V_{np,1}$.

needed. For the nonuniform case, $N + N/4$ multipliers are required to generate $N/4$ bandpass outputs, and an additional $N/4$ multipliers are required to generate the quadrature outputs.

D   Applications

Earlier it was seen that the Lerner bandpass transfer function $H_{lp,i}$ and the quadrature Lerner bandpass transfer function $H_{lq,i}$ had an almost identical magnitude response, and were exactly 90° out of phase over the entire band. There are several applications in communication systems, where the need to obtain the quadrature component of a signal arises; the Lerner based filter-bank is expected to find wide application in such situations, as a Hilbert transformer. The situation shown in Fig. 17a corresponded to the case where three resonator outputs were grouped to form the bandpass and quadrature Lerner outputs, and gave rise to a narrowband Hilbert transformer. In contrast, if more than three resonator outputs are used, then a wideband Hilbert transformer can be synthesized, where the passband wider, and also flatter. This is shown in Fig. 20, where all the resonator outputs except the ones at 0 and $f_s/2$, have been grouped together.

In addition to these, the non-uniform filter-bank may be used for equalizing high quality audio signals or music, which require processing in non-uniformly spaced frequency bands. It is conjectured that it may also be used for subband coding, as the low stopband attenuation would reduce aliasing distortion to acceptable levels, though this has not been verified.

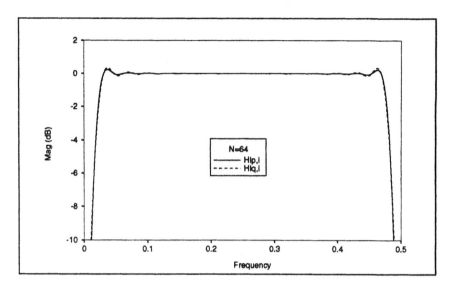

Figure 20: Wideband Hilbert Transformer. (©1992 IEEE)

VI   Summary

In summary, in this chapter

• The motivation for studying and developing low complexity filter-banks was given.

• A low complexity resonator-based filter structure was developed, starting from an analog prototype circuit, and some of its properties were examined. It must be metioned at this stage, that as the main emphasis in the chapter is the development of applications for the filter structure, only those properties that are necessary and relevant to these applications were described. However, for the interested reader, a more detailed analysis of these properties is available in [14, 15, 13].

• Some applications of the filter structure were then examined. These applications are somewhat different from conventional subband-coding application, that is typically associated with filter-banks.

• The first application examined was that of adaptive line enhancement; an ALE was designed, based on the filter structure, that avoided the convergence problems typically associated with IIR adaptive filters. It was also shown to have the merits of linear complexity, and the capability to provide enhanced sinusoidal outputs which were exactly in phase with their corresponding components in the input.

• The next application examined was that of transform computation. The filter-structure was used to compute trigonometric transforms such as the

DFT, DCT, DST, etc., efficiently, in a time-recursive manner. Its linear complexity, good finite precision behaviour, and pipelineability were shown to make it a good candidate for VLSI implementation.

• In the final application, the filter-structure was used to synthesize low complexity filter-banks with good magnitude and phase properties. The design methodology started out with a poor prototype filter-bank, and by appropriately grouping the outputs of this prototype, synthesized filter-banks with good magnitude and phase properties. Possible applications of these filter-banks include their use as Hilbert Transformers, for frequency domain filtering, etc.

References

[1] N. S. Jayant and P. Noll, *Digital Coding of waveforms.* Englewood Cliffs, New Jersey: Prentice-Hall Inc., 1984.

[2] D. F. Marshall, W. K. Jenkins, and J. J. Murphy, "The Use of Orthogonal Transforms for Improving Performance of Adaptive Filters," *"IEEE Transactions on Circuits and Systems"*, vol. 36, pp. 474–483, Apr 1989.

[3] A. Gilloire and M. Vetterli, "Adaptive Filtering in Sub-Bands," in *"International Conference on Acoustics Speech and Signal Processing"*.

[4] M. Padmanabhan and K. Martin, "Resonator-Based Filter-banks for Frequency-Domain Applications," *"IEEE Transactions on Circuits and Systems"*, vol. 38, pp. 1145–1159, Oct 1991.

[5] M. Padmanabhan and K. Martin, "Filter-Banks for Time-Recursive Implementation of Transforms," *"IEEE Transactions on Circuits and Systems"*, Jan 1993.

[6] H. S. Malvar, "Modulated QMF Filter Banks with Perfect Reconstruction," *"Electron Letters"*, pp. 906–907, June 1990.

[7] H. S. Malvar, "Lapped Transforms for Efficient Transform/Subband Coding," *"IEEE Transactions on Acoustics Speech and Signal Processing"*, pp. 969–978, June 1990.

[8] H. S. Malvar, "Extended Lapped Transforms: Fast Algorithms and Applications," in *"International Conference on Acoustics Speech and Signal Processing"*, pp. 1797–1800, 1991.

[9] P. P. Vaidyanathan, *Multirate Systems and Filter Banks.* New Jersey: Prentice-Hall, 1993.

[10] D. Koilpillai and P. P. Vaidyanathan, "Cosine-Modulated Filter-Banks Satisfying Perfect Reconstruction," *"IEEE Transactions on Acoustics Speech and Signal Processing"*, pp. 770–783, Apr 1992.

[11] J. Mau, "Perfect Reconstruction Modulated Filter Banks," in *"International Conference on Acoustics Speech and Signal Processing"*, pp. IV 273–276, 1992.

[12] M. Padmanabhan and K. Martin, "Some further results on modulated/extended lapped transforms," in *"International Conference on Acoustics Speech and Signal Processing"*, pp. IV 265–269, 1992.

[13] G. Peceli, "Resonator-based Digital Filters," *"IEEE Transactions on Circuits and Systems"*, vol. 36, pp. 156–159, jan 1989.

[14] M. Padmanabhan, *Feedback-Based Orthogonal Digital Filters, their Application in Signal Processing, and their VLSI Implementation.* University of California, Los Angeles: Ph.D Thesis, 1992.

[15] M. Padmanabhan and K. Martin, "Feedback-Based Orthogonal Digital Filters," *"IEEE Transactions on Circuits and Systems"*, to appear, 1993.

[16] B. Widrow *et al.*, "Fundamental Relations Between the LMS Algorithm and the DFT," *"IEEE Transactions on Circuits and Systems"*, vol. 34, pp. 814–820, Jul 1987.

[17] J. R. Glover Jr., "Adaptive Noise Cancelling applied to Sinusoidal Interferences," *"IEEE Transactions on Acoustics Speech and Signal Processing"*, vol. 25, pp. 484–491, Dec 1977.

[18] W. F. McGee, "Frequency Interpolation Filter Bank," in *"International Symposium on Circuits and Systems"*, pp. 1563–1566, 1989.

[19] W. F. McGee, "Fundamental Relations Between LMS Spectrum Analyzer and Recursive Least Squares Estimation," *"IEEE Transactions on Circuits and Systems"*, vol. 36, pp. 151–153, Jan 1989.

[20] G. Peceli, "A Common Structure for Recursive Discrete Transforms," *"IEEE Transactions on Circuits and Systems"*, vol. 33, pp. 1035–1036, Oct 1986.

[21] G. H. Hostetter, "Recursive Discrete Fourier Transformation," *"IEEE Transactions on Acoustics Speech and Signal Processing"*, vol. 28, pp. 184–190, Apr 1980.

[22] J. T. Lim, "Private conversation." Bell Northern Research, Ottawa, 1977.

[23] M. Vetterli, "Tree Structures for orthogonal transforms and applications to the Hadamard Transform," *Signal Processing*, vol. 5, pp. 473–484, Nov 1983.

[24] J. A. Stuller, "Generalized running discrete transforms," *"IEEE Transactions on Acoustics Speech and Signal Processing"*, vol. 30, pp. 60–68, Feb 1982.

[25] S. S. Narayan, A. M. Peterson, and M. J. Narasimha, "Transform Domain LMS Algorithm," *"IEEE Transactions on Acoustics Speech and Signal Processing"*, vol. 31, pp. 609–615, Jun 1983.

[26] A. Fettweis, "Wave Digital Filters: Theory and Practice," *"Proceedings of the IEEE"*, vol. 74, pp. 270–327, Feb 1986.

[27] N. G. Kingsbury, "Second-order recursive filter elements for poles near the unit circle and the real axis," *"Electron Letters"*, vol. 8, pp. 155–156, Mar 1972.

[28] C. T. Mullis and R. A. Roberts, "Synthesis of minimum roundoff noise fixed point digital filters," *"IEEE Transactions on Circuits and Systems"*, vol. 23, pp. 551–562, Sep 1976.

[29] C. V. K. Prabhakara Rao and P. Dewilde, "On Lossless Transfer Functions and Orthogonal Realizations," *"IEEE Transactions on Circuits and Systems"*, vol. 34, pp. 677–678, June 1987.

[30] A. Fettweis, "Passivity and Losslessness in Digital Filtering," *"Archiv fuer Electrotechnik"*, vol. 42, pp. 1–6, Jan 1988.

[31] H. J. Orchard, "Inductorless filters," *"Electron Letters"*, vol. 2, pp. 224–225, Sep 1966.

[32] K. K. Parhi and D. G. Messerschmitt, "Pipeline Interleaving and Parallelism in Recursive Digital Filters- Part i: Pipelining using Scattered Look-Ahead and Decomposition," *"IEEE Transactions on Acoustics Speech and Signal Processing"*, vol. 37, pp. 1099–1117, July 1989.

[33] K. K. Parhi and D. G. Messerschmitt, "Pipeline Interleaving and Parallelism in Recursive Digital Filters-part ii: Pipelined Incremental Block Filtering," *"IEEE Transactions on Acoustics Speech and Signal Processing"*, vol. 37, pp. 1118–1134, Jul 1989.

[34] B. Widrow and S. D. Stearns, *Adaptive Signal Processing*. Englewood Cliffs, New Jersey: Printice Hall Inc., 1975.

[35] D. Hush, N. Ahmed, R. David, and S. D. Stearns, "An adaptive IIR structure for sinusoidal enhancement, frequency estimation, and detection," *"IEEE Transactions on Acoustics Speech and Signal Processing"*, vol. 34, pp. 1380–1390, Dec 1986.

[36] D. V. B. Rao and S. Y. Kung, "Adaptive notch filtering for the retrieval of sinusoids in noise," "*IEEE Transactions on Acoustics Speech and Signal Processing*", vol. 32, pp. 791–802, Aug 1984.

[37] A. Nehorai, "A minimal parameter adaptive notch filter with constrained poles and zeros," "*IEEE Transactions on Acoustics Speech and Signal Processing*", vol. 33, pp. 983–996, Aug 1985.

[38] T. Kwan and K. Martin, "Adaptive Detection and Enhancement of Multiple Sinusoids using a Cascade IIR Filter," "*IEEE Transactions on Circuits and Systems*", vol. 36, pp. 937–946, Jul 1989.

[39] C. R. Johnson Jr., "Adaptive IIR Filtering: Current Results and Open Issues," , vol. 30, pp. 237–250, Mar 1984.

[40] A. Ogunfunmi and A. Peterson, "Adaptive Methods for Estimating Amplitudes and Frequencies of Narrowband Signals," in "*International Symposium on Circuits and Systems*", pp. 2124–2127, 1989.

[41] W. F. McGee and G. Zhang, "Logarithmic Filter Banks," in "*International Symposium on Circuits and Systems*", pp. 661–664, 1990.

[42] T. Kwan and K. Martin, "A Notch-Filter-Based Frequency-Difference Detector and its Applications," in "*International Symposium on Circuits and Systems*", pp. 1343–1346, 1990.

[43] N. Ahmed and K. R. Rao, *Orthogonal Transforms for Digital Signal Processing*. New York: Springer-Verlag, 1975.

[44] K. R. Rao and P. Yip, *Discrete Cosine Transform*. Academic Press, 1990.

[45] H. S. Hou, "A fast Recursive Algorithm For Computing the Discrete Cosine Transform," "*IEEE Transactions on Acoustics Speech and Signal Processing*", vol. 35, pp. 1455–1461, Oct 1987.

[46] P. Yip and K. R. Rao, "On the Shift Property of DCT's and DST's," "*IEEE Transactions on Acoustics Speech and Signal Processing*", vol. 35, pp. 404–406, Mar 1987.

[47] A. V. Oppenheim and R. W. Schafer, *Digital Signal Processing*. Englewood Cliffs, New Jersey: Prentice-Hall Inc., 1975.

[48] C. L. Gundel, "Filter Bank Interpretation of DFT and DCT," in "*EURASIP*", pp. 643–646, 1988.

[49] C. Loeffler, A. Ligtenberg, and G. S. Moschytz, "Practical fast 1-D DCT algorithms with 11 multiplications," in "*International Conference on Acoustics Speech and Signal Processing*", pp. 988–991, May 1989.

[50] R. R. Bitmead and B. D. O. Anderson, "Adaptive Frequency Sampling Filters," *"IEEE Transactions on Circuits and Systems"*, vol. 28, pp. 524–534, Jun 1981.

[51] D. Lewis, "Private communication,"   *University of Toronto*, 1993.

[52] K. Martin and M. Padmanabhan, "Resonator-In-A-Loop Filter-Banks Based on a Lerner Grouping of Outputs," in *"International Conference on Acoustics Speech and Signal Processing"*, pp. IV 329–332, 1992.

[53] K. Martin and M. Padmanabhan, "Lerner-Based Filter-Banks and Some of Their Applications," in *"International Symposium on Circuits and Systems"*, pp. 1003–1006, 1992.

[54] R. M. Lerner, "Band-Pass Filters with Linear Phase," *"Proceedings of the IEEE"*, Mar 1964.

[55] P. R. Drouilhet Jr. and L. M. Goodman, "Pole-Shared Linear-Phase Bandpass Filter-Bank," *"Proceedings of the IEEE"*, vol. 54, pp. 701–703, Apr 1966.

[56] J. S. Chang and Y. C. Tong, "A Pole-Sharing Technique for Linear-Phase Switched-Capacitor Filter-Banks," *"IEEE Transactions on Circuits and Systems"*, vol. 37, pp. 1465–1479, Dec 1990.

[57] G. Doblinger, "An Efficient Algorithm for Uniform and Nonuniform Digital Filter-Banks," in *"International Symposium on Circuits and Systems"*, 1991.

[58] G. Zhang and W. F. McGee, "Windowing Techniques in the design of resonator based frequency interpolation bank," in *"International Conference on Acoustics Speech and Signal Processing"*, pp. 1821–1824, 1991.

[59] L. R. Rabiner and R. W. Schafer, *Digital Processing of Speech Signals*. Englewood Cliffs, New Jersey: Prentice-Hall Inc., 1978.

A Discrete Time Nonrecursive Linear Phase Transport Processor Design Technique

Peter A. Stubberud
University of Nevada, Las Vegas

Cornelius T. Leondes
University of California, San Diego

Abstract

A discrete time transport processor is a discrete time system that is comprised only of delays, adds and subtracts. For example, a linear time invariant digital filter whose implementation uses only coefficients of +1, -1 and 0 is a discrete time transport processor. This chapter develops a technique for the design and implementation of frequency selective linear phase discrete time transport processors. This technique determines an optimal integer valued finite impulse response and determines a linear time invariant transport processor structure that can realize this impulse response.

I. Introduction

A discrete time transport processor is a discrete time system that is comprised only of delays, adds and subtracts. For example, a linear time invariant digital filter whose implementation uses only coefficients of +1, -1 and 0 is a discrete time transport processor. A transport processor is useful for implementing filters in technologies where multiplies are not economical, not feasible or are too slow. For example, the performance of a programmable digital signal processing (DSP) chip is generally limited by the performance of the chip's multiplier. For high speed real time DSP applications where the performance of a programmable DSP chip is not adequate, more expensive custom application specific integrated circuits (ASIC's) are used to implement these systems. If a programmable digital transport processor chip that consisted of only registers and adders was available, it could serve as a low cost alternative to ASIC's.

CONTROL AND DYNAMIC SYSTEMS, VOL. 68
Copyright © 1995 by Academic Press, Inc.
All rights of reproduction in any form reserved.

In this chapter, a technique for the design and implementation of frequency selective linear phase discrete time transport processors is developed. Because a transport processor is comprised only of delays, adds and subtracts, its impulse response is constrained to have integer values. The transport processor design technique in this chapter determines an optimal integer valued finite impulse response (FIR) and then develops a structure that can realize this impulse response. Several methods [1; 2; 3; 4; 5] have been proposed for determining optimal coefficients for the design of FIR filters with power of two coefficients. Several of these techniques use optimization methods to determine optimal or near optimal integer impulse responses before scaling them into powers of two. The integer impulse responses determined by these methods can be used to construct a transport processor. In this chapter, a different optimization technique for determining optimal integer finite impulse responses is developed. This optimization technique determines an optimal integer finite impulse response that minimizes a linear combination of the mean square error between the desired and actual frequency responses in the passband and stopband while bounding the impulse response to a user specified value.

II. A Technique for Determining Optimal Integer Valued Finite Impulse Responses

If we design a linear phase filter which minimizes the mean square error between a desired frequency response, $H_d(e^{j\omega})$, and the filter's frequency response, $H(e^{j\omega})$, then the design method would determine the function, $H(e^{j\omega})$, that minimizes the quantity,

$$\frac{1}{2\pi}\int_{-\pi}^{\pi}\left|H(e^{j\omega}) - H_d(e^{j\omega})\right|^2 d\omega.$$

A filter's frequency response is generally specified in terms of passband and stopband requirements, but the transition band usually has no requirements. For these types of filter specifications, this mean square error criterion is overly restrictive because it requires the mean square error to be minimized over the transition band. The transition band error is minimized at the cost of further improvement in the filter's passband and stopband performance. The design method developed in this section determines the optimal frequency response, $H(e^{j\omega})$, which minimizes a linear combination of the mean square error in the stopband and passband and does not minimize transition band errors.

The frequency response of a discrete time filter which has a finite impulse response, $h(n)$, of length N can be expressed as

$$H(e^{j\omega}) = \sum_{n=0}^{N-1} h(n)e^{-j\omega n}$$

If the filter has linear phase and a real impulse response, then $h(n) = h(N-1-n)$, and it can be shown[6] that $H(e^{j\omega})$ can be written as

$$H(e^{j\omega}) = e^{-j\frac{N-1}{2}\omega} H_r(\omega)$$

where $H_r(\omega)$ is a real function given by

$$H_r(\omega) = \begin{cases} h\left(\frac{N-1}{2}\right) + \displaystyle\sum_{p=1}^{\frac{N-1}{2}} 2h\left(\frac{N-1}{2} - p\right)\cos(\omega p) & N \text{ odd} \\[2em] \displaystyle\sum_{p=1}^{N/2} 2h\left(\frac{N}{2} - p\right)\cos\left[\omega(p - 1/2)\right] & N \text{ even} \end{cases}$$

$H_r(\omega)$ is often referred to as the amplitude or the zero phase frequency response of $H(e^{j\omega})$. If we define

$$\mathbf{x} = \begin{bmatrix} h\left(\frac{N-1}{2}\right) \\ 2h\left(\frac{N-1}{2} - 1\right) \\ \vdots \\ 2h(1) \\ 2h(0) \end{bmatrix}_{N \text{ odd}} , \quad \begin{bmatrix} 2h\left(\frac{N}{2} - 1\right) \\ 2h\left(\frac{N}{2} - 2\right) \\ \vdots \\ 2h(1) \\ 2h(0) \end{bmatrix}_{N \text{ even}}$$

and

$$\mathbf{s}(\omega) = \begin{bmatrix} 1 \\ \cos(\omega) \\ \cos(2\omega) \\ \vdots \\ \cos\left[\frac{N-1}{2}\omega\right] \end{bmatrix}_{N \text{ odd}} , \quad \begin{bmatrix} \cos(\omega/2) \\ \cos(3\omega/2) \\ \cos(5\omega/2) \\ \vdots \\ \cos\left[\frac{N-1}{2}\omega\right] \end{bmatrix}_{N \text{ even}}$$

then

$$H_r(\omega) = \mathbf{x}^T \mathbf{s}(\omega) = \mathbf{s}^T(\omega) \mathbf{x} \qquad (1)$$

where the superscript T denotes transpose and the appropriate expressions for $\mathbf{x}$ and $\mathbf{s}(\omega)$ are used depending upon whether N is odd or even.

Consider a linear phase FIR filter that has a frequency response, $H(e^{j\omega})$, that approximates a desired frequency response, $H_d(e^{j\omega})$. If we let J_{pb} represent the mean square error over the passband frequencies, then

$$J_{pb} = \frac{1}{m(\omega_{pb})} \int_{\omega \in \omega_{pb}} \left[\frac{H_r(\omega)}{\beta} - H_d(\omega)\right]^2 d\omega \qquad (2)$$

where ω_{pb} is the set of passband frequencies, $m(\omega_{pb})$ is the linear measure of the set ω_{pb}, $H_d(\omega)$ is the amplitude of the desired frequency response, and β is a variable which scales $H_r(\omega)$. Substituting Equation (1) into Equation (2), J_{pb} can be written as

$$J_{pb}(\mathbf{x}, \beta) = \frac{1}{m(\omega_{pb})} \Bigg[\beta^{-2} \int_{\omega \in \omega_{pb}} \mathbf{x}^T s(\omega) s^T(\omega) \mathbf{x} \, d\omega$$

$$-2\beta^{-1} \int_{\omega \in \omega_{pb}} H_d(\omega) \mathbf{x}^T s(\omega) \, d\omega + \int_{\omega \in \omega_{pb}} H_d^2(\omega) \, d\omega \Bigg]$$

$$= \frac{1}{m(\omega_{pb})} \Bigg[\beta^{-2} \mathbf{x}^T \int_{\omega \in \omega_{pb}} s(\omega) s^T(\omega) \, d\omega \ \mathbf{x}$$

$$-2\beta^{-1} \mathbf{x}^T \int_{\omega \in \omega_{pb}} H_d(\omega) s(\omega) \, d\omega + \int_{\omega \in \omega_{pb}} H_d^2(\omega) \, d\omega \Bigg] \quad (3)$$

If we define $\mathbf{W}(\omega)$ as the matrix,

$$\mathbf{W}(\omega) = s(\omega) \, s^T(\omega)$$

and let $W_{rc}(\omega)$ represent the element in the rth row and the cth column of the matrix, $\mathbf{W}(\omega)$, then

$$\mathbf{W}(\omega) = s(\omega) s^T(\omega) = \left[W_{rc}(\omega) \right] = \begin{cases} \left[\cos(r\omega) \cos(c\omega) \right] & N \text{ odd} \\[2mm] \left[\cos\left[\left(r - \tfrac{1}{2}\right)\omega\right] \cos\left[\left(c - \tfrac{1}{2}\right)\omega\right] \right] & N \text{ even} \end{cases}$$

where $r, c = 0, 1, \ldots , (N-1)/2$ when N is odd and $r, c = 1, 2, \ldots , N/2$ when N is even. Defining $\mathbf{Q}(\omega)$ as the matrix,

$$\mathbf{Q}(\omega) = \int \mathbf{W}(\omega) \, d\omega,$$

the element in the rth row and the cth column of the matrix, $\mathbf{Q}(\omega)$, can be written as

$$Q_{rc}(\omega) = \begin{cases} \omega & r = c = 0 \\[2mm] \dfrac{\omega}{2} + \dfrac{\sin(2r\omega)}{4r} & r = c \neq 0 \\[3mm] \dfrac{\sin(r+c)\omega}{2(r+c)} + \dfrac{\sin(r-c)\omega}{2(r-c)} & r \neq c \end{cases}$$

where $r, c = 0, 1, \ldots , (N-1)/2$ when N is odd and

$$Q_{rc}(\omega) = \begin{cases} \dfrac{\omega}{2} + \dfrac{\sin[2(r-1/2)\omega]}{4(r-1/2)} & r = c \\[2ex] \dfrac{\sin(r+c-1)\omega}{2(r+c-1)} + \dfrac{\sin(r-c)\omega}{2(r-c)} & r \neq c \end{cases}$$

where $r, c = 1, 2, \ldots, N/2$ when N is even. Defining

$$\mathbf{Q}_p = \int_{\omega \in \omega_{pb}} \mathbf{W}(\omega) \, d\omega$$

where $\mathbf{Q}_p$ is calculated by evaluating $\mathbf{Q}(\omega)$ at the appropriate limits, the term,

$$\mathbf{x}^T \int_{\omega \in \omega_{pb}} \mathbf{s}(\omega)\mathbf{s}^T(\omega) \, d\omega \, \mathbf{x},$$

in Equation (3) can be written as

$$\mathbf{x}^T \mathbf{Q}_p \mathbf{x}$$

If we also define the terms,

$$\mathbf{R}(\omega) = \int H_d(\omega)\mathbf{s}(\omega) \, d\omega \qquad\qquad \mathbf{R}_p = \int_{\omega \in \omega_{pb}} H_d(\omega)\mathbf{s}(\omega) \, d\omega$$

$$\gamma(\omega) = \int H_d^2(\omega) \, d\omega \qquad\qquad \gamma_p = \int_{\omega \in \omega_{pb}} H_d^2(\omega) \, d\omega$$

then Equation (3) can be written as

$$J_{pb}(\mathbf{x}, \beta) = \frac{1}{m(\omega_{pb})}\left[\beta^{-2}\mathbf{x}^T\mathbf{Q}_p\mathbf{x} - 2\beta^{-1}\mathbf{x}^T\mathbf{R}_p + \gamma_p\right] \qquad (4)$$

For filters which approximate constant values in their passbands, the expression for $\mathbf{R}(\omega)$ can be simplified. If $H_d(\omega)$ is equal to a constant in the passband, then without loss of generality, we can let $H_d(\omega) = 1$, and Equation (3) becomes

$$J_{pb}(\mathbf{x}, \beta) = \frac{1}{m(\omega_{pb})}\left[\beta^{-2}\mathbf{x}^T\mathbf{Q}_p\mathbf{x} - 2\beta^{-1}\mathbf{x}^T\mathbf{R}_p\right] + 1 \qquad (5)$$

where

$$\mathbf{R}(\omega) = \int \mathbf{s}(\omega) \, d\omega = \begin{bmatrix} \omega \\ \sin(\omega) \\ \frac{1}{2}\sin(2\omega) \\ \vdots \\ \frac{2}{N-1}\sin\left(\frac{N-1}{2}\omega\right) \end{bmatrix}_{N \text{ odd}} , \begin{bmatrix} 2\sin\left(\frac{\omega}{2}\right) \\ \frac{2}{3}\sin\left(\frac{3}{2}\omega\right) \\ \vdots \\ \frac{2}{N-1}\sin\left[\left(\frac{N-1}{2}\right)\omega\right] \end{bmatrix}_{N \text{ even}}$$

This expression for $\mathbf{R}(\omega)$ eliminates the need for integrating $H_d(\omega)\mathbf{s}(\omega)$ when calculating $\mathbf{R}_p$.

If we let J_{sb} represent the mean square error over the stopband frequencies, then

$$J_{sb} = \frac{1}{m(\omega_{sb})} \int_{\omega \in \omega_{sb}} \left[\frac{H_r(\omega)}{\beta} - H_d(\omega) \right]^2 d\omega \qquad (6)$$

where ω_{sb} is the set of stopband frequencies, $m(\omega_{sb})$ is the linear measure of the set ω_{sb}, $H_r(\omega)$ is the amplitude of the desired frequency response, and β is a variable which scales $H_r(\omega)$. Because $H_d(\omega)$ equals zero in the stopband, Equation (6) can be written as

$$J_{sb} = \frac{\beta^{-2}}{m(\omega_{sb})} \int_{\omega \in \omega_{sb}} H_r^2(\omega) \, d\omega \qquad (7)$$

Substituting Equation (1) into Equation (7),

$$J_{sb}(\mathbf{x}, \beta) = \frac{\beta^{-2}}{m(\omega_{sb})} \int_{\omega \in \omega_{sb}} \mathbf{x}^T \mathbf{s}(\omega) \mathbf{s}^T(\omega) \mathbf{x} \, d\omega$$

$$= \frac{\beta^{-2}}{m(\omega_{sb})} \mathbf{x}^T \int_{\omega \in \omega_{sb}} \mathbf{s}(\omega) \mathbf{s}^T(\omega) \, d\omega \, \mathbf{x}$$

Recall that earlier, we let $\mathbf{W}(\omega) = \mathbf{s}(\omega) \, \mathbf{s}^T(\omega)$ and

$$\mathbf{Q}(\omega) = \int \mathbf{W}(\omega) \, d\omega .$$

Thus, if we define

$$\mathbf{Q}_s = \int_{\omega \in \omega_{sb}} \mathbf{W}(\omega) \, d\omega$$

where $\mathbf{Q}_s$ is calculated by evaluating $\mathbf{Q}(\omega)$ at the appropriate limits, Equation (6) can be written as

$$J_{sb}(\mathbf{x}, \beta) = \frac{\beta^{-2}}{m(\omega_{sb})} \mathbf{x}^T \mathbf{Q}_s \, \mathbf{x} .$$

The design problem can now be stated as follows. Minimize the error function

$$J(\mathbf{x}, \beta) = \alpha J_{pb}(\mathbf{x}, \beta) + (1 - \alpha) J_{sb}(\mathbf{x}, \beta) \qquad (8)$$

where $1 \le \beta \le B$, B is a finite real number and $0 \le \alpha \le 1$. Because transport processors have integer valued impulse responses, the elements of the vector, $\mathbf{x}$, are constrained to be even integers except for the first element of the vector, $\mathbf{x}$, when N is odd. When N is odd, the first element of the vector, $\mathbf{x}$, corresponds to $h[(N-1)/2]$ and therefore can be either an even or odd integer. The variable, β, scales the amplitude of the filter's passband. By bounding its value, the passband amplitude is bounded as well as the values in the filter's impulse response. The scalar term, α, which can assume the values $0 \le \alpha \le 1$, weights the relative importance between the mean square error in the stopband and the mean square error in the passband. For example, if $\alpha = 1$, the mean square error is minimized over the passband and not the stopband, and if

$\alpha = 0$, the mean square error is minimized over the stopband and not the pass-band.

To solve this optimization problem, consider a similar problem where the vector, $\mathbf{x}$, and the variable, β, are unconstrained continuous variables. Substituting the appropriate expressions for J_{pb} and J_{sb} into Equation (8) yields

$$J(\mathbf{x},\beta) = \frac{\alpha}{m(\omega_{pb})}\left[\beta^{-2}\mathbf{x}^T\mathbf{Q}_p\mathbf{x} - 2\beta^{-1}\mathbf{x}^T\mathbf{R}_p + \gamma_p\right] + \frac{(1-\alpha)}{m(\omega_{sb})}\beta^{-2}\mathbf{x}^T\mathbf{Q}_s\,\mathbf{x}.$$

The necessary conditions for an optimal solution are

$$\frac{\partial J(\mathbf{x},\beta)}{\partial \mathbf{x}} = \mathbf{0} \tag{9}$$

and

$$\frac{\partial J(\mathbf{x},\beta)}{\partial \beta} = 0. \tag{10}$$

Because $\mathbf{Q}_p$ and $\mathbf{Q}_s$ are symmetric matrices, Equation (9) becomes

$$\frac{\partial J(\mathbf{x},\beta)}{\partial \mathbf{x}} = \frac{2\alpha}{m(\omega_{pb})}\left[\beta^{-2}\mathbf{Q}_p\mathbf{x} - \beta^{-1}\mathbf{R}_p\right] + \frac{2(1-\alpha)}{m(\omega_{sb})}\beta^{-2}\mathbf{Q}_s\,\mathbf{x} = 0. \tag{11}$$

Equation (10) implies

$$\frac{\alpha}{m(\omega_{pb})}\left[-2\beta^{-3}\mathbf{x}^T\mathbf{Q}_p\mathbf{x} + 2\beta^{-2}\mathbf{x}^T\mathbf{R}_p\right] - \frac{(1-\alpha)}{m(\omega_{sb})}2\beta^{-3}\mathbf{x}^T\mathbf{Q}_s\,\mathbf{x} = 0. \tag{12}$$

Rearranging Equation (11),

$$\left[\frac{\alpha}{m(\omega_{pb})}\mathbf{Q}_p + \frac{(1-\alpha)}{m(\omega_{sb})}\mathbf{Q}_s\right]\mathbf{x} = \beta\frac{\alpha}{m(\omega_{pb})}\mathbf{R}_p \tag{13}$$

which implies that

$$\mathbf{x} = \beta\mathbf{x}_o$$

where

$$\mathbf{x}_o = \left[\frac{\alpha}{m(\omega_{pb})}\mathbf{Q}_p + \frac{(1-\alpha)}{m(\omega_{sb})}\mathbf{Q}_s\right]^{-1}\frac{\alpha}{m(\omega_{pb})}\mathbf{R}_p.$$

Substituting $\mathbf{x} = \beta\mathbf{x}_o$ into Equation (12),

$$\frac{\alpha}{m(\omega_{pb})}\left[-2\beta^{-1}\mathbf{x}_o^T\mathbf{Q}_p\mathbf{x}_o + 2\beta^{-1}\mathbf{x}_o^T\mathbf{R}_p\right] - \frac{(1-\alpha)}{m(\omega_{sb})}2\beta^{-1}\mathbf{x}_o^T\mathbf{Q}_s\,\mathbf{x}_o = 0$$

which implies that

$$\frac{\alpha}{m(\omega_{pb})}\left[\mathbf{x}_o^T\mathbf{Q}_p\mathbf{x}_o - \mathbf{x}_o^T\mathbf{R}_p\right] + \frac{(1-\alpha)}{m(\omega_{sb})}\mathbf{x}_o^T\mathbf{Q}_s\,\mathbf{x}_o = 0. \tag{14}$$

Equation (14) shows that when the optimal vector, $\beta\mathbf{x}_o$, is substituted into Equation (12) that the equation is no longer a function of β. This implies that the optimal solution is $\mathbf{x} = \beta \mathbf{x}_o$ where β is any real number.

Consider the original problem where the elements of the vector, $\mathbf{x}$, are constrained to be even integers except when N is odd in which case the first element of the vector, $\mathbf{x}$, is constrained to be either an even or odd integer. From the above discussion, the cost function in Equation (8) is minimized when $\mathbf{x} = \beta \mathbf{x}_o$ for any real β. For a fixed value of β, the cost function in Equation (8) is quadratic and the constrained values of $\mathbf{x}$ that minimize a quadratic cost function can be determined by rounding to the nearest allowable value. Thus, for a fixed range of β's, such as $1 \leq \beta \leq B$, a set of optimal vectors can be determined as a function of β. The optimal vector, $\mathbf{x}$, for this problem can be determined by substituting the set of optimal vectors and their corresponding β's into the cost function in Equation (8) and choosing the vector, $\mathbf{x}$, which minimizes the cost function in Equation (8).

III. A Technique for the Implementation of Nonrecursive Discrete Time Transport Processors

A FIR discrete time system can be implemented using the direct convolution flowgraph shown in Figure 1. A FIR discrete time transport processor is a discrete time system that is comprised only of delays, adds and subtracts, and thus, the coefficients in its implementation are constrained to values of +1, -1 and 0. If a transport processor has an impulse response whose values are comprised only of +1, -1 and 0, it can be implemented using the direct convolution flowgraph shown in Figure 1. But, if a transport processor has an impulse response that contains integer values other than +1, -1 and 0, the transport processor cannot be implemented using the direct convolution structure shown in Figure 1. However, a combination of parallel and cascaded direct convolution structures that have only coefficients of +1, -1 and 0 can be used to implement FIR discrete time transport processors that have impulse responses which contain integer values other than +1, -1 and 0. In this section, a technique for implementing transport processors with integer valued

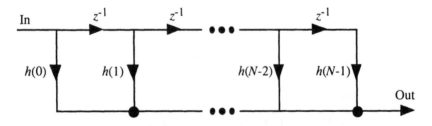

Figure 1. Flowgraph of a Direct Convolution Filter.

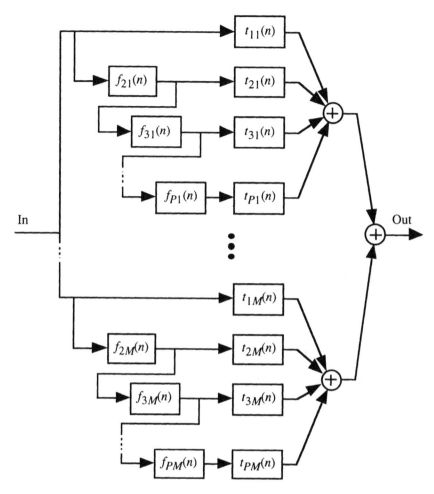

Figure 2. Interconnection of Direct Convolution Structures Used to Implement Nonrecursive Transport Processors.

impulse responses is developed that uses the structure in Figure 2 where the subfilters, $t_{11}(n), f_{21}(n), t_{21}(n), ..., t_{PM}(n)$, are implemented using the direct convolution flowgraph in Figure 1 with coefficients of +1, -1 and 0.

Consider the filter structure in Figure 2 where $t_{11}(n), f_{21}(n), t_{21}(n), ..., t_{PM}(n)$ are the impulse responses of the subfilters. The structure's impulse response, $h(n)$, is

$$h(n) = t_{11}(n) + f_{21}(n) * t_{21}(n) + \cdots + f_{21}(n) * f_{31}(n) * \cdots * f_{P1}(n) * t_{P1}(n)$$

$$+ \cdots + t_{1M}(n) + \cdots + f_{2M}(n) * f_{3M}(n) * \cdots * f_{PM}(n) * t_{PM}(n)$$

where * represents convolution. Because each of the subfilters are implemented using direct convolution with coefficients of +1, -1 and 0, each of the subfilter's impulse responses are constrained to have values of +1, -1 or 0.

To show that the filter structure in Figure 2 can realize any finite length integer valued impulse response, consider using only the $t_{1k}(n)$ subfilters to implement an integer valued impulse response, $h(n)$. If $h(n)$ is bounded such that

$$|h(n)| \le L,$$

then the system can be implemented by choosing values of +1, -1 or 0 for each of the $t_{1k}(n)$ subfilters such that

$$h(n) = \sum_{k=1}^{L} t_{1k}(n).$$

Although the choice of coefficients for the $t_{1k}(n)$ subfilters is not unique, such an implementation requires a minimum of L subfilters. By utilizing the other subfilters, fewer than L subfilters can be used to realize $h(n)$.

To assure that the transport processor has linear phase, each of the subfilters is designed so that the final implementation has linear phase. Consider a transport processor that has a finite length integer valued impulse response, $h(n)$. If the transport processor has linear phase,

$$h(n) = \begin{cases} h(N-1-n) & 0 \le n \le N-1 \\ 0 & \text{otherwise} \end{cases}$$

where N is the length of the impulse response. If we define

$$h_{11}(n) = t_{11}(n)$$
$$h_{21}(n) = f_{21}(n) * t_{21}(n)$$
$$h_{31}(n) = f_{21}(n) * f_{31}(n) * t_{31}(n)$$
$$\vdots$$
$$h_{PM}(n) = f_{2M}(n) * f_{3M}(n) * \cdots * f_{PM}(n) * t_{PM}(n),$$

then

$$h(n) = h_{11}(n) + h_{21}(n) + h_{31}(n) + \cdots + h_{PM}(n).$$

To ensure that

$$h(n) = \begin{cases} h(N-1-n) & 0 \le n \le N-1 \\ 0 & \text{otherwise} \end{cases}$$

each of the subfilters is designed so that

$$h_{rc}(n) = \begin{cases} h_{rc}(N-1-n) & 0 \le n \le N-1 \\ 0 & \text{otherwise} \end{cases}$$

for $r = 1, 2, ..., P$ and $c = 1, 2, ..., M$. A useful result for equating the lengths of $h_{11}(n)$, $h_{21}(n),..., h_{PM}(n)$ is that when two impulse responses of finite lengths, N_1 and N_2, are convolved, the result is a finite impulse response of length N_1+N_2-1.

Because the $t_{1k}(n)$ subfilters are in parallel with the other cascaded subfilters, each of these subfilters can change the impulse response created by the other subfilters by +1, -1 or 0. Although not effective for realizing large impulse response values, these subfilters can be used to make small adjustments to the impulse response created by the other subfilters. For example, the subfilter, $t_{11}(n)$, can adjust the impulse response created by the other subfilters by +1, -1 or 0. Thus, the $t_{1k}(n)$ subfilters are designed last to adjust the impulse response created by the other subfilters.

For many linear phase frequency selective filters, the center of the impulse response contains the impulse response's largest absolute value which would be $|h[(N-1)/2]|$ when N is odd, and $|h(N/2)|$ and $|h[(N/2)-1]|$ when N is even. The impulse response's smallest absolute values generally occur near the impulse response's ends which would be $h(0)$ and $h(N-1)$. If $M = M_0$ in the filter structure, then the subfilters, $t_{11}(n)$, $t_{12}(n)$, ..., $t_{1M_0}(n)$, can realize absolute values up to M_0, and the other subfilters do not need to realize those values in the impulse response. For example, if $M = 1$ and $h(n)$ is

n	0	1	2	3	4	5	6
$h(n)$	1	-2	5	9	5	-2	1

then $t_{11}(n)$ can be designed such that

$$t_{11}(0) = t_{11}(6) = 1,$$

and thus, the remaining subfilters only need to realize the impulse response, $h_{d21}(n)$, where

n	0	1	2	3	4	5	6
$h_{d21}(n)$	0	-2	5	9	5	-2	0

Similarly, if $M = 2$, $t_{11}(n)$ and $t_{12}(n)$ can be designed such that

$$t_{11}(0)+t_{12}(0) = t_{11}(6)+t_{12}(6) = 1$$

and

$$t_{11}(1)+t_{12}(1) = t_{11}(5)+t_{12}(5) = -2$$
$$\Rightarrow t_{11}(1) = t_{12}(1) = t_{11}(5) = t_{12}(5) = -1.$$

Thus, the remaining subfilters only need to realize the impulse response, $h_{d21}(n)$, where

n	0	1	2	3	4	5	6
$h_{d21}(n)$	0	0	5	9	5	0	0

After selecting a value for M and determining $h_{d21}(n)$, $f_{21}(n)$ is chosen so that it approximates the impulse response, $h_{d21}(n)$. The length of $f_{21}(n)$ is generally kept small compared to N since it is the first subfilter in cascade with $t_{21}(n)$, $f_{31}(n)$, ..., $t_{P1}(n)$. To approximate $h_{d21}(n)$, the values in $f_{21}(n)$ are chosen as the values nearest the values at the center of the impulse response, $h_{d21}(n)$. For example, if N is odd and the length of $f_{21}(n)$ is chosen to be three, then

$$f_{21}(0) = sgn\left[h_{d21}\left(\frac{N-3}{2}\right)\right]$$

$$f_{21}(1) = sgn\left[h_{d21}\left(\frac{N-1}{2}\right)\right]$$

$$f_{21}(2) = sgn\left[h_{d21}\left(\frac{N+3}{2}\right)\right]$$

where

$$sgn(x) = \begin{cases} 1 & x > 0 \\ 0 & x = 0 \\ -1 & x < 0 \end{cases}.$$

The next step in the implementation procedure is to determine $t_{21}(n)$ such that

$$h_{d21}(n) \approx f_{21}(n) * t_{21}(n) = \sum_{k=0}^{N_{f_{21}}-1} f_{21}(k)t_{21}(n-k)$$

where $N_{f_{21}}$ is the length of $f_{21}(n)$. For $f_{21}(n)*t_{21}(n)$ to have length N, $t_{21}(n)$ is designed to have a length, $N_{t_{21}}$, where

$$N = N_{f_{21}} + N_{t_{21}} - 1 \quad \Rightarrow \quad N_{t_{21}} = N - N_{f_{21}} + 1.$$

$t_{21}(n)$ is then determined recursively starting with $t_{21}(0)$. At $n = 0$, $t_{21}(0)$ should be chosen so that

$$h_{d21}(0) = \sum_{k=0}^{N_{f_{21}}-1} f_{21}(k)t_{21}(-k) = f_{21}(0)t_{21}(0)$$

which implies that

$$t_{21}(0) = \frac{h_{d21}(0)}{f_{21}(0)},$$

but because the values of $t_{21}(n)$ are constrained to +1, -1 and 0, $t_{21}(0)$ is chosen as

$$t_{21}(0) = sgn\left[\frac{h_{d21}(0)}{f_{21}(0)}\right]$$

Similarly, at $n = 1$, $t_{21}(1)$ should be chosen so that

$$h_{d21}(1) = \sum_{k=0}^{N_{f_{21}}-1} f_{21}(k)t_{21}(1-k) = f_{21}(0)t_{21}(1) + f_{21}(1)t_{21}(0)$$

which implies that

$$t_{21}(1) = \frac{h_{d21}(1) - f_{21}(1)t_{21}(0)}{f_{21}(0)},$$

but because the values of $t_{21}(n)$ are constrained to +1, -1 and 0, $t_{21}(1)$ is chosen as

$$t_{21}(1) = sgn\left[\frac{h_{d21}(1) - f_{21}(1)t_{21}(0)}{f_{21}(0)}\right].$$

This procedure continues until $n > (N_{t_{21}}-1)/2$. To ensure linear phase, the values of $t_{21}(n)$, for $n > (N_{t_{21}}-1)/2$, are chosen so that

$$h_{21}(n) = h_{21}(N - 1 - n).$$

After determining $f_{21}(n)$ and $t_{21}(n)$, $h_{d31}(n)$ is calculated by computing

$$h(n) - h_{21}(n) = h(n) - f_{21}(n) * t_{21}(n)$$

and then setting to zero this sequence's ends which are absolutely less than M as was done when determining $h_{d21}(n)$. $f_{31}(n)$ is then designed to minimize

$$\varepsilon = \frac{1}{2\pi}\int_{-\pi}^{\pi}|H_{d31}(\omega) - F_{21}(\omega)F_{31}(\omega)|^2 d\omega \tag{15}$$

where

$$H_{d31}(e^{j\omega}) = e^{-j\omega\frac{N-1}{2}}H_{d31}(\omega),$$

$$F_{21}(e^{j\omega})F_{31}(e^{j\omega}) = e^{-j\omega\frac{N_{f21}-1}{2}}F_{21}(\omega)\ e^{-j\omega\frac{N_{f31}-1}{2}}F_{31}(\omega)$$

and N_{f31} is the length of $f_{31}(n)$.

To determine ε, let

$$\mathbf{x}_{h_{d31}} = \begin{bmatrix} h_{d31}\left(\frac{N-1}{2}\right) \\ \vdots \\ 2h_{d31}(1) \\ 2h_{d31}(0) \end{bmatrix}_{N \text{ odd}} , \quad \begin{bmatrix} 2h_{d31}\left(\frac{N}{2}-1\right) \\ \vdots \\ 2h_{d31}(1) \\ 2h_{d31}(0) \end{bmatrix}_{N \text{ even}}$$

and

$$\mathbf{s}_N(\omega) = \begin{bmatrix} 1 \\ \cos(\omega) \\ \vdots \\ \cos\left[\frac{N-1}{2}\omega\right] \end{bmatrix}_{N \text{ odd}} , \quad \begin{bmatrix} \cos(\omega/2) \\ \cos(3\omega/2) \\ \vdots \\ \cos\left[\frac{N-1}{2}\omega\right] \end{bmatrix}_{N \text{ even}}$$

which implies that

$$H_{d31}(\omega) = \mathbf{x}_{h_{d31}}^T \mathbf{s}_N(\omega) = \mathbf{s}_N^T(\omega)\mathbf{x}_{h_{d31}}.$$

Define the sequence $c(n)$ as

$$c(n) = f_{21}(n)*f_{31}(n)$$

which has length $N_c = N_{f21} + N_{f31} - 1$, and let

$$
\mathbf{x}_c = \begin{bmatrix} c\left(\dfrac{N_c-1}{2}\right) \\ \vdots \\ 2c(1) \\ 2c(0) \end{bmatrix}_{N_c \text{ odd}}, \qquad \begin{bmatrix} 2c\left(\dfrac{N_c}{2}-1\right) \\ \vdots \\ 2c(1) \\ 2c(0) \end{bmatrix}_{N_c \text{ even}}
$$

then $C(\omega)$, the amplitude of the frequency response of $c(n)$, can be expressed as

$$
C(\omega) = F_{21}(\omega)F_{31}(\omega) = \mathbf{x}_c^T \mathbf{s}_{N_c}(\omega) = \mathbf{s}_{N_c}^T(\omega)\mathbf{x}_c.
$$

Thus, Equation (15) can be written as

$$
\varepsilon = \frac{1}{2\pi}\int_{-\pi}^{\pi}\left|\mathbf{x}_{h_{d31}}^T \mathbf{s}_N(\omega) - \mathbf{s}_{N_c}^T(\omega)\mathbf{x}_c\right|^2 d\omega
$$

$$
= \frac{1}{2\pi}\left[\mathbf{x}_{h_{d31}}^T \int_{-\pi}^{\pi} \mathbf{s}_N(\omega)\mathbf{s}_N^T(\omega)\,d\omega\,\mathbf{x}_{h_{d31}}\right.
$$

$$
\left. -2\mathbf{x}_{h_{d31}}^T \int_{-\pi}^{\pi} \mathbf{s}_N(\omega)\mathbf{s}_{N_c}^T(\omega)\,d\omega\,\mathbf{x}_c + \mathbf{x}_c^T \int_{-\pi}^{\pi} \mathbf{s}_{N_c}(\omega)\mathbf{s}_{N_c}^T(\omega)\,d\omega\,\mathbf{x}_c\right]
$$

$$
= \frac{1}{2\pi}\left[\mathbf{x}_{h_{d31}}^T \int_{-\pi}^{\pi} \mathbf{W}_N(\omega)\,d\omega\,\mathbf{x}_{h_{d31}}\right.
$$

$$
\left. -2\mathbf{x}_{h_{d31}}^T \int_{-\pi}^{\pi} \mathbf{s}_N(\omega)\mathbf{s}_{N_c}^T(\omega)\,d\omega\,\mathbf{x}_c + \mathbf{x}_c^T \int_{-\pi}^{\pi} \mathbf{W}_{N_c}\,d\omega\,\mathbf{x}_c\right]
$$

where $\mathbf{W}_N(\omega)$ is the matrix,

$$
\mathbf{W}_N(\omega) = \mathbf{s}_N(\omega)\mathbf{s}_N^T(\omega),
$$

which was defined in Section II. If we define

$$
\mathbf{Q}_N = \int_{-\pi}^{\pi} \mathbf{W}_N(\omega)\,d\omega = \begin{cases} \begin{bmatrix} 2\pi & 0 \\ \hline 0 & \pi\mathbf{I}_{(N-1)/2} \end{bmatrix} & N \text{ odd} \\[2ex] \pi\mathbf{I}_{N/2} & N \text{ even} \end{cases}
$$

where $\mathbf{I}_N$ is the $N{\times}N$ identity matrix, then

$$
\varepsilon = \frac{1}{2\pi}\left[\mathbf{x}_{h_{d31}}^T \mathbf{Q}_N\,\mathbf{x}_{h_{d31}}\right.
$$

$$
\left. -2\mathbf{x}_{h_{d31}}^T \int_{-\pi}^{\pi} \mathbf{s}_N(\omega)\mathbf{s}_{N_c}^T(\omega)\,d\omega\,\mathbf{x}_c + \mathbf{x}_c^T \mathbf{Q}_{N_c}\,\mathbf{x}_c\right]
$$

$$= \sum_{n=0}^{N-1} h_{d31}^2(n) - \frac{1}{\pi} \mathbf{x}_{h_{d31}}^T \int_{-\pi}^{\pi} \mathbf{s}_N(\omega) \mathbf{s}_{N_c}^T(\omega) \, d\omega \, \mathbf{x}_c + \sum_{n=0}^{N_c-1} c^2(n).$$

If N and N_c are both odd or both even, then

$$\int_{-\pi}^{\pi} \mathbf{s}_N(\omega) \mathbf{s}_{N_c}^T(\omega) \, d\omega = \begin{bmatrix} \mathbf{Q}_{N_c} \\ -\!-\!- \\ \mathbf{0} \end{bmatrix}$$

and

$$\varepsilon = \sum_{n=0}^{N-1} h_{d31}^2(n) - 2 \sum_{n=0}^{N_c-1} h_{d31}\left(n + \frac{N-1}{2} - \frac{N_c-1}{2}\right) c(n) + \sum_{n=0}^{N_c-1} c^2(n)$$

which can also be written as

$$\varepsilon = \sum_{n=0}^{N-1} \left| h_{d31}(n) - c\left(n + \frac{N_c-1}{2} - \frac{N-1}{2}\right) \right|^2. \tag{17}$$

If N is odd and N_c is even,

$$\varepsilon = \sum_{n=0}^{N-1} h_{d31}^2(n) - \frac{1}{\pi} \mathbf{x}_{h_{d31}}^T \mathbf{Q}_{N,N_c} \mathbf{x}_c + \sum_{n=0}^{N_c-1} c^2(n) \tag{18}$$

where

$$\mathbf{Q}_{N,N_c} = \begin{bmatrix} \dfrac{(-1)^{r+c+1}}{r+c-\frac{1}{2}} + \dfrac{(-1)^{r+c}}{r-c+\frac{1}{2}} \end{bmatrix}$$

for $r = 0, 1, \ldots, (N-1)/2$ and $c = 1, 2, \ldots, N_c/2$. And, if N is even and N_c is odd,

$$\varepsilon = \sum_{n=0}^{N-1} h_{d31}^2(n) - \frac{1}{\pi} \mathbf{x}_{h_{d31}}^T \mathbf{Q}_{N,N_c} \mathbf{x}_c + \sum_{n=0}^{N_c-1} c^2(n) \tag{19}$$

where

$$\mathbf{Q}_{N,N_c} = \begin{bmatrix} \dfrac{(-1)^{c+r+1}}{c+r-\frac{1}{2}} + \dfrac{(-1)^{c+r}}{c-r+\frac{1}{2}} \end{bmatrix}$$

for $r = 1, 2, \ldots, N/2$ and $c = 0, 1, \ldots, (N_c-1)/2$.

For example, if $f_{21}(n)$ was determined to be

n	0	1
$f_{21}(n)$	1	1

and $h_{d31}(n)$ is

n	0	1	2	3	4	5	6
$h_{d31}(n)$	0	−2	5	9	5	−2	0

then an optimal $f_{31}(n)$ can be determined by evaluating ε for the small number

Table 1. Results of ε's for determining $f_{31}(n)$.

	$f_{31}(n)$		$f_{21}(n)*f_{31}(n)$			ε
n	0	1	0	1	2	
	-1	-1	-1	-2	-1	201
	-1	1	-1	0	1	141
	1	-1	1	0	-1	141
	1	1	1	2	1	89

of various $f_{31}(n)$'s. The length, $N_{f_{31}}$, of $f_{31}(n)$ is generally kept small so that the length of $f_{21}(n)*f_{31}(n)$ does not exceed the number of the nonzero terms in $h_{d31}(n)$ and because it is in cascade with several other subfilters. For this example, let $N_{f_{31}} = 2$. The results are summarized in Table 1 where the ε's were determined using Equation (17). From the table, the $f_{31}(n)$ that minimizes ε is

n	0	1
$f_{31}(n)$	1	1

The design continues choosing the f and t subfilters in the same way $f_{31}(n)$ and $t_{21}(n)$ were chosen, respectively.

IV. Example

Consider a linear phase FIR discrete time transport processor that approximates the frequency response, $H_d(e^{j\omega}) = \exp[-(N-1)\omega/2] H_d(\omega)$, where

$$H_d(\omega) = \begin{cases} \beta & 0 \leq \omega \leq \omega_p \\ 0 & \omega_s \leq \omega \leq \pi \end{cases}$$

β is a constant greater than one, $\omega_p = 0.45\pi$ radians/sample, $\omega_s = 0.55\pi$ radians/sample and N is the length of the filter's impulse response. The transport processor should also approximate $H_d(e^{j\omega})$ in a minimum mean square error sense, satisfy the following requirements

$$-0.5 \text{ dB} \leq 20 \log\left|\frac{H(\omega)}{\beta}\right| \leq 0.5 \text{ dB} \qquad 0 \leq \omega \leq \omega_p$$

$$20 \log\left|\frac{H(\omega)}{\beta}\right| \leq 30 \text{ dB} \qquad \omega_s \leq \omega \leq \pi$$

and the transport processor's impulse response, $h(n)$, should be bounded such that

$$|h(n)| \leq L = 40.$$

The design begins by determining an integer valued impulse response using the technique described in Section II of this chapter. An impulse response, $h_o(n)$, that satisfies the filter specifications when $\beta = 1$ can be determined by solving Equation (13),

$$\left[\frac{\alpha}{m(\omega_{pb})} \mathbf{Q}_p + \frac{(1-\alpha)}{m(\omega_{sb})} \mathbf{Q}_s \right] \mathbf{x} = \beta \frac{\alpha}{m(\omega_{pb})} \mathbf{R}_p$$

where $\beta = 1$, $m(\omega_{pb}) = \omega_p - 0$, $m(\omega_{sb}) = \pi - \omega_s$, $\mathbf{Q}_p = \mathbf{Q}(\omega_p) - \mathbf{Q}(0)$, $\mathbf{Q}_s = \mathbf{Q}(\pi) - \mathbf{Q}(\omega_s)$ and $\mathbf{R}_p = \mathbf{R}(\omega_p) - \mathbf{R}(0)$. The values for N and α are chosen so that the passband and stopband specifications are satisfied. For example, if the values of N and α are chosen so that the passband specification is met, but the stopband specification is not, the value of α should be reduced. If both the passband and stopband specifications cannot be satisfied for a specific values N and α, the value of N should be increased. These values can be altered until the passband and stopband requirements are met. Because the impulse response, $h(n)$, of the transport processor will contain only integer values, the passband and stopband requirements only need to approximate the filter specifications. For this example, the values of N and α were determined to be 20 and 0.6, respectively. The resulting impulse response, $h_o(n)$, is listed in Table 2, and the magnitude of the frequency response is plotted in Figure 3A. Figure 3B shows the magnitude of the passband in detail.

To determine the optimal integer valued impulse response, $h(n)$, a set of integer valued impulse responses is generated by rounding the product of $h_o(n)$ and β for $1 \leq \beta \leq B$ where B is chosen so that $|h(n)| \leq L$. Because $h_o(10)$ has the largest absolute value in the sequence, $h_o(n)$, $B = (L+0.5)/h_o(10)$. Each of the integer valued impulse responses and their corresponding β's are substituted in the cost function, J, described in Equation (8). A plot of J vs.

Table 2. Impulse response, $h_o(n)$.

$$
\begin{aligned}
h_o(0) &= h_o(19) = 0.01125563141 \\
h_o(1) &= h_o(18) = 0.01640259639 \\
h_o(2) &= h_o(17) = -0.02035408668 \\
h_o(3) &= h_o(16) = -0.02640598838 \\
h_o(4) &= h_o(15) = 0.03365206042 \\
h_o(5) &= h_o(14) = 0.04406658762 \\
h_o(6) &= h_o(13) = -0.05958403968 \\
h_o(7) &= h_o(12) = -0.08662230534 \\
h_o(8) &= h_o(11) = 0.1480174823 \\
h_o(9) &= h_o(10) = 0.4494475632
\end{aligned}
$$

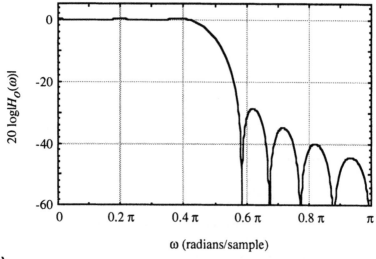

(A)

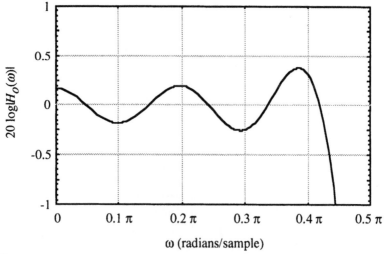

(B)

Figure 3. Magnitude of the frequency response for the filter that has the impulse response $h_o(n)$.

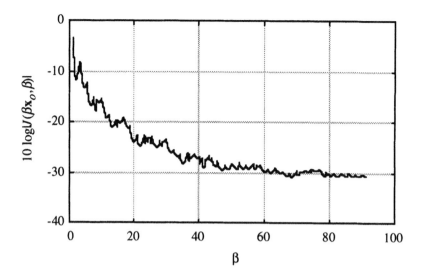

Figure 4. Plot of $J(\beta \mathbf{x}_o, \beta)$ vs. β.

β is shown in Figure 4. J is minimized when $\beta = 68.657$, and therefore the transport processor's optimal impulse response, $h(n)$, is

$$h(n) = \text{round}\left[\beta \, h_o(n)\right]\Big|_{\beta=68.657}$$

$h(n)$ is listed in Table 3, and the magnitude of its frequency response is plotted in Figure 5A. Figure 5B shows the passband in detail.

We are now ready to implement $h(n)$ using the transport processor structure shown in Figure 2. The implementation begins by choosing a value for M. For this example, we will let $M = 1$. If $h(n)$ cannot be implemented by the

Table 3. Impulse response, $h(n)$.

$h(0)$	=	$h(19)$	=	1
$h(1)$	=	$h(18)$	=	1
$h(2)$	=	$h(17)$	=	-1
$h(3)$	=	$h(16)$	=	-2
$h(4)$	=	$h(15)$	=	2
$h(5)$	=	$h(14)$	=	3
$h(6)$	=	$h(13)$	=	-4
$h(7)$	=	$h(12)$	=	-6
$h(8)$	=	$h(11)$	=	10
$h(9)$	=	$h(10)$	=	31

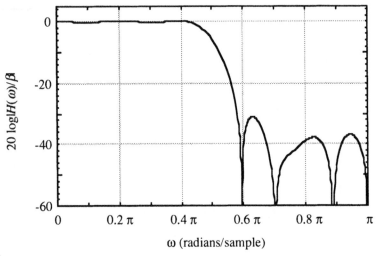

(A)

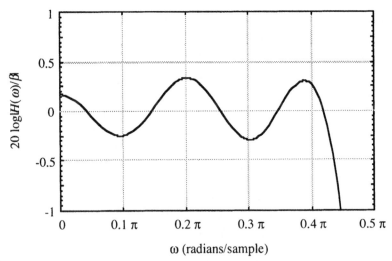

(B)

Figure 5. Magnitude of the frequency response for the filter that has
the impulse response $h(n)$.

structure in Figure 2 where $M = 1$, then let $M = 2$ and repeat the implementation procedure. If the structure in Figure 2 where $M = 2$ also cannot implement $h(n)$, then M is increased until the structure can implement $h(n)$. Letting $M = 1$, $h_{d21}(n)$ is

n	0	1	2	3	4	5	6	7	8	9	10	11	12	···
$h_{d21}(n)$	0	0	0	-2	2	3	-4	-6	10	31	31	10	-6	···

where $h_{d21}(n) = h_{d21}(19-n)$ for $0 \leq n \leq 19$.

The next step in the implementation procedure is to determine $f_{21}(n)$ so that it approximates the impulse response, $h_{d21}(n)$. For this example, the length, N_{f21}, of $f_{21}(n)$ is chosen to be two, and thus

$$f_{21}(0) = sgn\left[h_{d21}\left(\frac{N}{2} - 1\right)\right] = sgn\left[h_{d21}(9)\right] = 1$$

$$f_{21}(1) = sgn\left[h_{d21}\left(\frac{N}{2}\right)\right] = sgn\left[h_{d21}(10)\right] = 1.$$

The next step is to determine $t_{21}(n)$. For $f_{21}(n)*t_{21}(n)$ to have length $N = 20$, $t_{21}(n)$ is designed to have a length, N_{t21}, where

$$N_{t21} = N - N_{f21} + 1 = 20 - 2 + 1 = 19.$$

Starting at $n = 0$, $t_{21}(n)$ for $0 \leq n \leq 9$. is determined recursively using

$$t_{21}(n) = sgn\left[\frac{h_{d21}(n) - f_{21}(1)t_{21}(n-1)}{f_{21}(0)}\right]$$

For $10 \leq n \leq 19$,

$$t_{21}(n) = t_{21}(N_{t21} - 1 - n)$$

so that

$$h_{21}(n) = h_{21}(N - 1 - n).$$

Using this procedure $t_{21}(n)$ is

n	0	1	2	3	4	5	6	7	8	9	10	11	12	···
$t_{21}(n)$	0	0	0	-1	1	1	-1	-1	1	1	1	-1	-1	···

where $t_{21}(n) = t_{21}(18-n)$ for $0 \leq n \leq 18$.

Next, $h_{d31}(n)$ is calculated by computing

$$h(n) - h_{21}(n) = h(n) - f_{21}(n) * t_{21}(n)$$

which is

n	0	1	2	3	4	5	6	7	8	9	10	11	12	···
$h(n) - h_{21}(n)$	1	1	-1	-1	2	1	-4	-4	10	29	29	10	-4	···

and then setting to zero this sequence's ends which are absolutely less than M. This yields

n	0	1	2	3	4	5	6	7	8	9	10	11	12	$\cdots$
$h_{d31}(n)$	0	0	0	0	2	1	-4	-4	10	29	29	10	-4	$\cdots$

where $h_{d31}(n) = h_{d31}(19-n)$ for $0 \leq n \leq 19$.
   $f_{31}(n)$ is then designed to minimize

$$\varepsilon = \frac{1}{2\pi} \int_{-\pi}^{\pi} |H_{d31}(\omega) - F_{21}(\omega)F_{31}(\omega)|^2 \, d\omega$$

For this example, let $N_{f_{31}} = 2$. Using Equation (19), ε is calculated for the small number of various $f_{31}(n)$'s. The results are summarized in Table 4. From the table, the $f_{31}(n)$ that minimizes ε is

n	0	1
$f_{31}(n)$	1	1

.

   The design continues choosing the f and t subfilters in the same way $f_{31}(n)$ and $t_{21}(n)$ were chosen, respectively. The final filter implementation is summarized in Table 5. The only subfilter worth noting is $f_{61}(n)$ which is an impulse. Subfilter $f_{61}(n)$ was chosen as an impulse because $h_{d61}(n)$ was

n	0	1	2	3	4	5	6	7	8	9	10	11	12	$\cdots$
$h_{d61}(n)$	0	0	0	0	0	0	0	0	6	11	11	6	0	$\cdots$

which has 4 nonzero terms and the length of $f_{11}(n)*\cdots*f_{51}(n)$ was 5. Because $f_{61}(n)$ is an impulse, $t_{51}(n)$ and $t_{61}(n)$ are essentially connected in parallel from the output of $f_{51}(n)$. Thus, $f_{61}(n)$ can be omitted from the transport processors implementation.
   If $f_{11}(n)*\cdots*f_{61}(n)*t_{61}(n)$ had not been able to complete the implementation of the transport processor, then the implementation procedure would have been repeated for $M = 2$.

Table 4. Results of ε's for determining $f_{31}(n)$ in the Example.

	$f_{31}(n)$		$f_{21}(n)*f_{31}(n)$			ε
n	0	1	0	1	2	
	-1	-1	-1	-2	-1	2053.1
	-1	1	-1	0	1	2003.6
	1	-1	1	0	-1	1912.4
	1	1	1	2	1	1870.9

Table 5. Transport processor coefficients for the Example.

n	0	1	2	3	4	5	6	7	8	9	10	11	12	13	14	15	
$t_{11}(n)$	1	1	-1	-1	1	0	-1	0	1	1	1	1	0	-1	0	1	...
$f_{21}(n)$	1	1															
$t_{21}(n)$	0	0	0	-1	1	1	-1	-1	1	1	1	-1	-1	1	1	-1	...
$f_{31}(n)$	1	1															
$t_{31}(n)$	0	0	0	0	1	-1	-1	-1	1	1	-1	-1	-1	1	0	0	...
$f_{41}(n)$	1	1															
$t_{41}(n)$	0	0	0	0	0	0	-1	1	1	1	-1	0	0	0	0	0	0
$f_{51}(n)$	1	1															
$t_{51}(n)$	0	0	0	0	0	0	0	1	1	0	0	0	0	0	0	0	
$f_{61}(n)$	1																
$t_{61}(n)$	0	0	0	0	0	0	0	1	1	0	0	0	0	0	0	0	

If this filter had been implemented with multiplications, it would have required a minimum of 10 multiplies and 19 adds. This implementation of the filter requires 51 adds.

V. Summary

In this chapter, techniques were developed for the design and implementation of frequency selective linear phase discrete time transport processors. The design technique determines an optimal integer valued finite impulse response. References [1; 2; 3; 4; 5] describe other techniques for determining optimal integer valued finite impulse responses. The implementation technique developed in this chapter determines a nonrecursive transport processor structure that can realize integer valued finite impulse responses which could have been determined from the technique described in this chapter or from the techniques described in references [1; 2; 3; 4; 5]. As the example illustrates, these techniques are well suited for the design and implementation of frequency selective linear phase discrete time transport processors.

VI. References

[1]   D. M. Kodek, "Design of Optimal Finite Wordlength FIR Digital Filters Using Integer Programming Techniques", *IEEE Transactions on Acoustic, Speech, and Signal Processing*, Vol. ASSP-28, No. 3, pp. 304-308 (1980).

[2]   Y. C. Lim, S. R. Parker and A. G. Constantinides, "Finite Word Length FIR Filter Design Using Integer Programming Over a Discrete Coefficient Space", *IEEE Transactions on Acoustic, Speech, and Signal Processing*, Vol. ASSP-30, No. 4, pp. 661-664 (1982).

[3]   Y. C. Lim and S. R. Parker, "FIR Filter Design Over a Discrete Powers-of-Two Coefficient Space", *IEEE Transactions on Acoustic, Speech, and Signal Processing*, Vol. ASSP-31, No. 3, pp. 583-591 (1983).

[4]   Q. Zhao and Y. Tadokoro, "A Simple Design of FIR Filters with Powers-of-Two Coefficients", *IEEE Transactions on Circuits and Systems*, Vol. ASSP-35, No. 5, pp. 566-570 (1988).

[5]   H. Samueli, "An Improved Search Algorithm for the Design of Multiplierless FIR Filters with Powers-of-Two Coefficients", *IEEE Transactions on Circuits and Systems*, Vol. ASSP-36, No. 7, pp. 1044-1047 (1989).

[6]   A. V. Oppenheim and R. W. SCHAFER, *Discrete-Time Signal Processing*, Prentice Hall, Inc., Englewood Cliffs, New Jersey, (1989).

Blind Deconvolution
Channel Identification
and Equalization

D. Hatzinakos

Department of Electrical and Computer Engineering
University of Toronto
Toronto, Ontario, Canada M5S 1A4

I Introduction

There is a plethora of applications where the observed signal after sampling
can be written as

$$y(n) = f(n) * x(n) \;\; = \;\; \sum_k f(k)x(n-k), \tag{1}$$

$$n, k \in \mathcal{Z}$$

where $\mathcal{Z}$ is the set of integer numbers and $\{*\}$ denotes the linear convolution
operation. For example, this situation arises in communication links subject
to linear distortion or multipath propagation, [1-4], blurring of images, [5-8],
multiple signal reflections in seismology, [9,10], speech generation [11,12],
and restoration of old recordings, [13], among others, [14]. Without loss
of generality, the observed discrete time signal $\{y(n\}$ can be viewed as the
output of a stable discrete linear time invariant filter (channel) with impulse
response $\{f(n)\}$ which is driven by the discrete time signal $\{x(n)\}$.

When one of the signals $\{f(n)\}$ or $\{x(n)\}$ is known exactly the other
can be recovered from $\{y(n)\}$ via inverse filtering or deconvolution. This
is the classical problem of system/signal identification. The list of deconvolution methods proposed for this case is long. It includes a variety of well known techniques that are based on the zero forcing (ZF), the
minimum-mean-square error (MMSE), the least-squares (LS), the maximum likelihood (ML), and other optimality criteria; recursive algorithms
such as the least-mean square (LMS), the recursive least-squares (RLS),

Copyright © 1995 by Academic Press, Inc.
All rights of reproduction in any form reserved.

and the Kalman algorithms; linear and nonlinear equalization structures [15,16].

Blind deconvolution refers to the problem of separating the two convolved signals $\{f(n)\}$ and $\{x(n)\}$ when both signals are unknown or partially known. This is an important problem in seismic data analysis, transmission monitoring, deblurring of astronomical images, multipoint network communications, echo cancellation in wireless telephony, digital radio links over fading channels, and other applications when there is either limited knowledge of the signals due to practical constraints or there is sudden change in the properties of the signals [2,5,7,9,14]. Furthermore, in digital transmission applications, blind deconvolution eliminates the need for the transmission of training sequences which are required by the non-blind deconvolution approaches [1-4]. By nature the solution of this problem requires either partial knowledge or an intelligent guess of the properties of at least one of the convolved signals. For example, one of the earliest approaches was based on homomorphic filtering where differences in the length of the cepstrum of the two unknown signals were exploited in order to separate them [11]. In other applications, a prototype signal with characteristics similar to those of the signal to be recovered has been utilized for blind deconvolution [13]. In digital communications the statistical or geometrical properties of the transmitted signal $\{x(n)\}$ are usually known and are utilized by many algorithms to identify the unknown communication channel $\{f(n)\}$, and then to obtain $\{x(n)\}$ via inverse filtering [1,2,17].

A possible classification of the existing blind deconvolution methods is depicted in Figure 1. The objective of this chapter is to provide an overview of the basic principles and methodologies behind these methods and demonstrate their utilization in applications. Without loss of generality, emphasis will be given to the blind identification and equalization of digital communication channels; an area which has attracted a lot of interest lately.

We begin our discussion by stating the problem of blind deconvolution and discussing the constraints imposed on the input data $\{x(n)\}$ and the channel impulse response $\{f(n)\}$ necessary for a solution to exist. An analytic description of the algorithmic structure of the major blind deconvolution approaches follows. The convergence behaviour, the complexity and other implementation issues are discussed for each approach. An attempt is made to highlight the strengths and limitations of various approaches based on theoretical expectations and computer simulations.

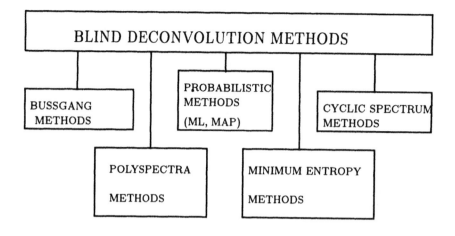

Figure 1: Classification of blind deconvolution methods

II   Problem Formulation

A   The General Blind Deconvolution Scenario

The discrete time blind deconvolution scenario that will be considered in this chapter is depicted in Figure 2 (a). In general the input data $\{x(n)\}$ is a zero-mean, independent identically distributed (i.i.d.) sequence with finite variance and higher-order statistics[1]. We assume that the input $\{x(n)\}$ is not known or accessible but is of known probability density function (p.d.f.)[2]. The discrete-time impulse response $\{f(n)\}$ of the linear time invariant (LTI) filter is completely unknown. Based on the above assumptions both the input and output sequences, $\{x(n)\}$ and $\{y(n)\}$, are stationary random processes. The objective is to recover first the characteristics of the $\{f(n)\}$ from the observed $\{y(n)\}$ and then obtain the $\{x(n)\}$ through

---

[1] The problem with dependent input sequences has not been investigated sufficiently when the nature of dependency is not known. Multichannel approaches towards this direction have been proposed recently [18]

[2] In many situations the p.d.f. is not known exactly but is substituted by appropriate models.

inverse filtering[3].

In theory, the discrete-time impulse responses $\{f(n)\}$ and $\{u(n)\}$ of the forward and inverse filters, respectively, are infinite length. However, in practice the length of the filters are chosen to be finite. Also, usually only a finite number of observed data samples $\{y(n)\}$ is available. Thus, in practice only a close approximation $\{\tilde{x}(n)\}$ of the input data is feasible.

The following cases can be considered for the impulse response $\{f(n)\}$ and the p.d.f. of the input signal $\{x(n)\}$:

1. The $\{f(n)\}$ is minimum phase, i.e., all the zeros and poles of the transfer function $F(z) = \sum_n f(n)z^{-n}$ lie inside the unit circle in the z-transform domain. In this case, one can utilize the second order statistics of the observed $\{y(n)\}$, exclusively, to recover $\{f(n)\}$ or $\{u(n)\}$. The classical linear prediction method that is based only on the autocorrelation function of $\{y(n)\}$ is an example of a second-order statistics based blind deconvolution method [16, 19].

2. The $\{f(n)\}$ is non-minimum phase, i.e., the transfer function $F(z) = \sum_n f(n)z^{-n}$ has zeros that lie inside and outside the the unit circle in the z-transform domain[4]. In this case, methods that are based exclusively on second-order statistics fail to identify the channel correctly. Actually, the autocorrelation based methods would identify a minimum phase filter that is power spectrally equivalent to $\{f(n)\}$ rather than $\{f(n)\}$ [11, 20]. Similar comments apply to the identification of $\{u(n)\}$. In other words, the second order statistics of a stationary sequence do not preserve non-minimum phase information. Therefore, one must employ the higher-order statistics of order greater than two of $\{y(n)\}$ to correctly identify the non-minimum phase characteristics of $\{f(n)\}$ or $\{u(n)\}$ [20].

3. The input sequence $\{x(n)\}$ is Gaussian distributed. Therefore, the observed sequence $\{y(n)\}$ is Gaussian distributed as well [14]. It is well known that the higher-order statistics of a Gaussian process provide no more information than that found in the second order statistics of the process. In this case solution to the blind deconvolution problem exists only if the impulse response $\{f(n)\}$ is minimum phase.

In most practical situations, the unknown impulse response $\{f(n)\}$ is non-minimum phase. Thus, most of the existing blind deconvolution meth-

---

[3] In many cases the characteristics of the inverse filter $\{u(n)\}$ are recovered directly without prior identification of $\{f(n)\}$.

[4] In theory, we can allow poles to lie outside the unit circle. Nevertheless, in practice all poles must lie inside the unit circle to guarantee stability of $\{f(n)\}$.

ods utilize directly or indirectly (through nonlinear transformations) the higher-order statistics (H.O.S.) of the observed sequence $\{y(n)\}$.

B   Blind Deconvolution of Communication Channels

Let us consider the baseband digital communication system of figure 2 (b). The observed signal, after being demodulated, low-pass filtered and synchronously sampled (i.e., at the symbol rate $\frac{1}{T}$), takes the form of equation 1. Therefore, the problem of blind deconvolution in this case closely follows the scenario of the previous sections. The following are some unique characteristics that are found in digital communications applications.

1. The input signal $\{x(n)\}$ is a complex, non-Gaussian, i.i.d., discrete random process with a discrete and symmetric probability density function. For example, $\{x(n)\}$ may be the complex baseband equivalent sequence of a Quadrature Amplitude Modulated signal $(L^2 - QAM)$, where $x(n) = x_R(n) + jx_I(n)$ where $x_R(n)$ and $x_I(n)$ are two i.i.d. random sequences independent from each other and each taking the values $\pm 1, \pm 3, \pm(L-1)$ with equal probability. Thus, the odd-order statistics (moments and cumulants) of $\{x(n)\}$ are all equal to zero.

2. The unknown impulse response $\{f(n)\}$ is nonminimum phase and accounts for the total linear distortion introduced in the communication channel (known as Intersymbol Interference (ISI)). In practice, $\{f(n)\}$ is time varying. However, we make the assumption that it changes slowly with time and therefore is practically constant over a large number of observed data. We will assume that the transfer function $F(z) = \sum_n f(n)z^{-n}$ can be factored as follows [11,21]:

$$F(z) = G \cdot z^{-d} \cdot I(z^{-1}) \cdot O(z) \tag{2}$$

where, the factor:

$$I(z^{-1}) = \frac{\prod_{k=1}^{L_a}(1 - a_k z^{-1})}{\prod_{k=1}^{L_c}(1 - c_k z^{-1})}, \qquad |a_k| < 1, \quad |c_k| < 1$$

is minimum phase, i.e., with zeros and poles inside the unit circle, and the factor:

$$O(z) = \prod_{k=1}^{L_b}(1 - b_k z), \qquad |b_k| < 1$$

is maximum phase , i.e., with zeros outside the unit circle. Finally, the parameters G and z^{-d} represent a constant complex factor and a delay factor, respectively.

3. It is sufficient that the channel $\{f(n)\}$ and therefore the reconstructed input data to be identified up to a constant phase shift θ and an unknown delay d [3, 14, 22]. In other words, the objective is to obtain:

$$\tilde{x}(n) = x(n-d)e^{j\theta} \tag{3}$$

or

$$x(n) * f(n) * u(n) = x(n-d)e^{j\theta} \tag{4}$$

Thus, the impulse response $\{u(n)\}$ of a blind equalizer which plays the role of inverse filtering must satisfy:

$$f(n) * u(n) = \delta(n-d) \cdot e^{j\theta} \tag{5}$$

where, $\delta(n)$ is the discrete delta function. Note that in digital communications the constant delay d does not affect the recovery of the input data sequence. Also, in general, the constant phase θ can be removed after equalization by a decision device[5]. Thus, in the derivation of blind equalization algorithms we may assume that $d = 0$ and $\theta = 0$. Taking into account these assumptions, the "zero forcing" (ZF) equalization constraint is expressed as, [15, 21]:

$$f(n) * u(n) = \delta(n), \quad or \quad F(z) \cdot U(z) = 1 \tag{6}$$

where, $U(z)$ is the transfer function of the equalizer.

4. Because the input data sequence $\{x(n)\}$ takes values from a discrete set of numbers, it is sufficient to identify the $\{f(n)\}$ or $\{u(n)\}$ to such a degree so that residual linear distortion at the output of the equalizer can be removed by means of a decision device (threshold decoder), as indicated in figure 2.

5. In addition to the linear distortion the observed signal $\{y(n)\}$ is subject to additive Gaussian noise, carrier frequency offset due to imperfect demodulation and other types of distortion that affect the performance of blind deconvolution methods. Thus, in conjunction to blind equalization, blind phase tracking/cancellation and noise cancellation algorithms should be utilized [2, 23, 24].

---

[5] In $L^2 - QAM$ signaling, the signals $x(n)$, $x(n)e^{\pm j90°}$, $x(n)e^{\pm j180°}$ are statistically identical and therefore may not be distinguishable by blind deconvolution algorithms. Similar phase restrictions may apply to other signal constellations as well.

In the next sections, we describe in detail the principles and algorithmic structure of the major blind deconvolution methods that have appeared in the literature.

a)

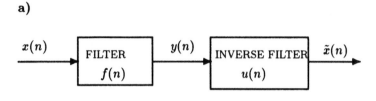

b)

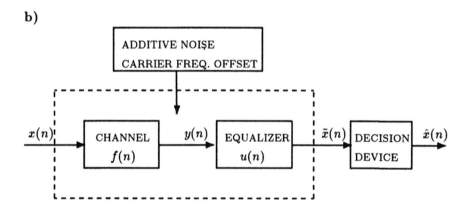

Figure 2: a) Basic deconvolution set up, b) Baseband digital communication system subject to linear distortion.

III   Bussgang Approaches to Blind Deconvolution

A   Bussgang Iterative Deconvolution Method

This method which was first introduced by Godfrey and Rocca [25], under the name "zero-memory nonlinear deconvolution", proceeds as follows. An initial guess for the impulse response of the equalizer (inverse filter) is made and is denoted as $u^{(0)}(n)$, $n = 1, 2, \ldots, N$. Then, the output of the equalizer is written as:

$$\tilde{x}^{(0)}(n) = y(n) * u^{(0)}(n) = x(n) * \left[f(n) * u^{(0)}(n) \right]. \tag{7}$$

Comparing relations (5) and (7), and assuming $d = 0$, $\theta = 0$, we conclude that :

$$f(n) * u^{(0)}(n) = \delta(n) + e^{(0)}(n) \tag{8}$$

where $e^{(0)}(n)$ is the error resulting from the difference between $\{u(n)\}$ and $\{u^{(0)}(n)\}$. By substituting (8) into (7), we find that:

$$\tilde{x}^{(0)}(n) = x(n) + \left[x(n) * e^{(0)}(n) \right] = x(n) + w^{(0)}(n). \tag{9}$$

The $\{w^{(0)}(n)\}$ is the residual linear distortion (residual ISI) at the output of the equalizer with coefficients $\{u^{(0)}(n)\}$. It is also known as "convolutional noise".

Given $\{\tilde{x}^{(0)}(n)\}$, the problem then becomes to find an estimate of the input data $\{x(n)\}$, namely $\{d^{(0)}(n)\}$. In general , $\{d^{(0)}(n)\}$ will be a nonlinear function of a subset of $\{\tilde{x}^{(0)}(n)\}$. In practice, it is simpler to choose a memoryless nonlinear estimator where $d^{(0)}(n)$ at instant n depends only on $\tilde{x}^{(0)}(n)$ at the same instant; that is $d^{(0)}(n) = g^{(0)} \left[\tilde{x}^{(0)}(n) \right]$. We may obtain an optimum memoryless nonlinearity by minimizing the mean-square-error (MSE) criterion:

$$E \left\{ \left| g^{(0)} \left[x^{(0)}(n) \right] - x(n) \right|^2 \right\} \tag{10}$$

where $E\{.\}$ denotes statistical expectation. The solution is the well known conditional expectation [14, 26]:

$$\begin{aligned} d^{(0)}(n) = g^{(0)} \left[x^{(0)}(n) \right] &= E \left\{ x(n) / \tilde{x}^{(0)}(n) \right\} \\ &= \frac{P_{\tilde{x}/x}(\tilde{x}/x) \cdot P_x(x)}{P_{\tilde{x}}(\tilde{x})}. \end{aligned} \tag{11}$$

The p.d.f. $P_x(x)$ of the input data is assumed known, the $P_{\tilde{x}/x}(\tilde{x}/x)$ is assumed to be Gaussian[6] with mean $x(n)$ and variance $\sigma^2_{w^{(0)}}$, and the $P_{\tilde{x}}(\tilde{x})$ plays the role of a normalization constant.

The next step is to utilize the estimated sequence $\{d^{(0)}(n)\}$ to obtain a better estimate $\{u^{(1)}(n)\}$, $n = 1, 2, \ldots, N$ of the equalizer coefficients. To accomplish this we write:

$$d^{(0)}(n) = y(n) * u^{(1)}(n) = \sum_{k=1}^{N} u^{(1)}(k) \cdot y(n - k) \qquad (12)$$

and then repeat (12) for N_{OV} different values of n to form an overdetermined system of equations:

$$\mathbf{Y} \cdot \mathbf{u}^{(1)} = \mathbf{d}^{(0)} \qquad (13)$$

where, $\mathbf{Y}$ is an $N_{OV} \times N$, $(N_{OV} > N)$, matrix with elements from the observed data samples $\{y(n)\}$, $\mathbf{d}^{(0)}$ is an $N_{OV} \times 1$ matrix with elements from $\{d^{(0)}(n)\}$ and $\mathbf{u}^{(1)} = [u^{(1)}(1), \ldots, u^{(1)}(N)]^T$. Finally, we obtain the least squares solution

$$\mathbf{u}^{(1)} = (\mathbf{Y}^H \mathbf{Y})^{-1} \cdot (\mathbf{Y}^H \mathbf{d}^{(0)}) = \hat{\mathbf{R}}_y^{-1} \cdot \hat{\mathbf{r}}_{yd}^{(0)} \qquad (14)$$

where, $\hat{\mathbf{R}}_y = \mathbf{Y}^H \mathbf{Y}$ is the $N \times N$ deterministic autocorrelation matrix of the observed data $\{y(n)\}$, the $\hat{\mathbf{r}}_{yd}^{(0)} = \mathbf{Y}^H \mathbf{d}^{(0)}$ is the $N \times 1$ crosscorrelation vector between $\{y(n)\}$ and $\{d^{(0)}(n)\}$, and the H denotes transpose conjugate.

Given $u^{(1)}(n)$, $n = 1, 2, \ldots, N$, the output of the equalizer is written as:

$$\tilde{x}^{(1)}(n) = y(n) * u^{(1)}(n) = x(n) + w^{(1)}(n) \qquad (15)$$

an estimate of the input data is obtained as follows:

$$d^{(1)}(n) = g^{(1)}\left[\tilde{x}^{(1)}(n)\right] = E\left\{x(n)/\tilde{x}^{(1)}(n)\right\} \qquad (16)$$

an update of the equalizer coefficients is found as follows:

$$\mathbf{u}^{(2)} = \hat{\mathbf{R}}_y^{-1} \cdot \hat{\mathbf{r}}_{yd}^{(1)} \qquad (17)$$

and so on for iterations $3, 4, 5, \ldots$. Assuming that the process converges, the equalizer coefficients estimates $\mathbf{u}^{(i)} = [u^{(i)}(1), \ldots, u^{(i)}(N)]^T$ get closer to their ideal value and the variance of the convolutional noise $\sigma^2_{w^{(i)}}$ at the output of the equalizer decreases, as i increases.

---

[6] In (9), the $w^{(0)}(n) = x(n) * e^{(0)}(n)$ can be assumed to be approximately zero-mean Gaussian by means of the Central Limit Theorem (C.L.T). Furthermore, the variance of $w^{(0)}(n)$ is large compared to the magnitude of the crosscorrelation between $x(n)$ and $w^{(0)}(n)$ and thus, the $\{w^{(0)}(n)\}$ may be considered to be statistically independent of $\{x(n)\}$ [14,16,27].

B   Adaptive Bussgang Algorithm

At iteration n the output of the equalizer is written as:

$$\tilde{x}(n) = \mathbf{y}^T(n)\mathbf{u}(n) \tag{18}$$

where, $\mathbf{u}(n) = [u_1(n), \ldots, u_N(n)]^T$ is the equalizer coefficient vector and $\mathbf{y}(n) = [y(n), y(n-1), \ldots, y(n-N+1)]^T$ is the input data vector to the equalizer.

To estimate $\mathbf{u}(n)$, we minimize the cost function:

$$J(n) = E\{G[\tilde{x}(n)]\} \tag{19}$$

where, $G[\tilde{x}(n)]$ is a memoryless nonlinear function of the equalizer output $\tilde{x}(n)$. Then, by choosing a steepest descent adaptation procedure we obtain the following adaptation rule for the equalizer coefficients:

$$\mathbf{u}(n+1) = \mathbf{u}(n) - \gamma(n) \cdot \nabla_{\mathbf{u}} J(n) \tag{20}$$

where, $\gamma(n) \neq 0$ is a step size and $\nabla_{\mathbf{u}} J(n)$ the gradient of J with respect to the equalizer coefficient vector at iteration n. Furthermore, it can be shown that [2, 16]

$$\nabla_{\mathbf{u}} J(n) = -E\{e(n) \cdot \mathbf{y}^*(n)\} \tag{21}$$

where, $e(n)$ is an error quantity defined as $e(n) = \nabla_{\tilde{x}} G[\tilde{x}(n)]$ and $*$ denotes conjugation operation. Therefore, the coefficient adaptation rule becomes:

$$\mathbf{u}(n+1) = \mathbf{u}(n) + \gamma(n) \cdot E\{e(n) \cdot \mathbf{y}^*(n)\} \tag{22}$$

Notice, that according to the last equation optimality is achieved when the error $e(n)$ is orthogonal to $\mathbf{y}^*(n)$. Depending on the choice for the memoryless function $G[\cdot]$ (or equivalently the corresponding $e(n)$), different types of blind equalization algorithms can be derived.

In the Bussgang adaptive equalizers the error at iteration n is defined as:

$$e(n) = g^{(n)}[\tilde{x}(n)] - \tilde{x}(n) \tag{23}$$

where $g^{(n)}[\tilde{x}(n)] = E[x(n)/\tilde{x}(n)]$ is a memoryless nonlinear function. Thus, at the point of optimality, where $E\{e(n) \cdot \mathbf{y}^*(n)\} = 0$, we find after some calculations that:

$$\underbrace{E\{\mathbf{y}^*(n) \cdot g^{(n)}[\tilde{x}(n)]\}}_{\mathbf{r}_{yd}} = \underbrace{E\{\mathbf{y}^*(n) \cdot \mathbf{y}^T(n)\}}_{\mathbf{R}_y} \cdot \mathbf{u}_{opt.} \tag{24}$$

By comparing relations (24) and (17) we observe the similarity between the solutions for the equalizer coefficients obtained from the adaptive Bussgang algorithm and the iterative Bussgang method.

In practice, the calculation of the optimum nonlinearity $g^{(n)}[\tilde{x}(n)]$ at iteration n is tedious and in general suboptimum because of the approximations made in modeling the convolutional noise at the output of the equalizer (see footnote 6). Thus, for simplicity a nonlinear estimator $g[\tilde{x}(n)]$ which is independent of n is chosen. Also, in the adaptation rule the expectation operation $E\{e(n) \cdot \mathbf{y}^*(n)\}$ is replaced by its instantaneous estimate $e(n) \cdot \mathbf{y}^*(n)$. Finally, a constant step size γ is often utilize because it is simple and provides a constant convergence rate. With these modifications **the Bussgang adaptive algorithm can be summarized as follows,** [14]:

$$\mathbf{u}(n) = [u_1(n), \ldots, u_N(n)]^T$$
$$\mathbf{y}(n) = [y(n), \ldots, y(n - N + 1)]^T$$
$$\gamma: \quad step \quad size$$

Initialization:[7]

$$\mathbf{u}(0) = [0, \ldots, 0, 1, 0, \ldots, 0]^T$$

At iteration n $= 0, 1, 2, \ldots$

$$\tilde{x}(n) = \mathbf{y}^T(n) \cdot \mathbf{u}(n)$$
$$e(n) = g[\tilde{x}(n)] - \tilde{x}(n)$$
$$\mathbf{u}(n + 1) = \mathbf{u}(n) + \gamma \cdot e(n) \cdot \mathbf{y}^*(n) \tag{25}$$
$$\hat{x}(n) = Q[\tilde{x}(n)], \qquad Q[\cdot]: threshold \quad decoding$$

A large number of adaptive Bussgang algorithms have been proposed in the literature [1-3, 28-46]. The error quantities utilized by few widely known Bussgang algorithms are given in Table I. the block diagram of a linear Bussgang equalizer is depicted in figure 3(a).

---

[7] This form of initialization has been found in practice to work well with Bussgang equalizers [3,14].

Table I: Error quantities used in the coefficient adaptation rule of several Bussgang blind equalizers

$e(n) = g[\tilde{x}(n)] - \tilde{x}(n),$ $\qquad$ $g[\tilde{x}(n)]$: memoryless nonlinear function

Algorithm	Error: $e(n)$	Comments
LMS training mode, [15, 16]	$e(n) = x(n) - \tilde{x}(n)$	$x(n)$ is known
Bussgang Optimum ,[25]	$e_{opt}(n) = E\left[x(n)/\tilde{x}(n)\right] - \tilde{x}(n)$	
Decision Directed mode, [15, 16]	$e_D(n) = \hat{x}(n) - \tilde{x}(n)$	$\hat{x}(n) = Q[\tilde{x}(n)]$ $Q[.]$: Threshold Decoder
Sato, [1]	$e_S(n) = \alpha \cdot csgn[\tilde{x}(n)] - \tilde{x}(n)$	$\alpha = \frac{E\{(Re[x(n)])^2}{E\{|Re[x(n)]|\}}$
GSSA, [28]	$e_{GS}(n) = csgn[\gamma \cdot csgn[\tilde{x}(n)] - \tilde{x}(n)]$	$\gamma = \frac{maxRe[x(n)]}{\sqrt{2}}$
Godard, [2]	$e_{Gp}(n) = \tilde{x}(n)|\tilde{x}(n)|^{p-2}(|\tilde{x}(n)|^p - R_p)$	$R_p = \frac{E|x(n)|^{2p}}{E|x(n)|^p}$
CMA (p=2), -[30]-	$e_{G2}(n) = \tilde{x}(n)(R_2 - |\tilde{x}(n)|^2)$	$R_2 = \frac{E|x(n)|^4}{E|x(n)|^2}$
Benveniste-Goursat, [32]	$e_{BG}(n) = k_1 \cdot e_D(n) + k_2 \cdot e_S(n) \cdot |e_D(n)|$	$k_1, k_2 > 0$
Stop-and-Go, -[33]-	$e_{SG}(n) = \frac{1}{2} \cdot [A \cdot e_D(n) + B \cdot e_D^*(n)]$ $A = I_R(n) + I_I(n),\ B = I_R(n) - I_I(n)$ $I_K(n) = \frac{1 + sgn(e_D^K(n)) \cdot sgn(e_S^K(n))}{2}$	R: real I: Imag.
Stop-and-Go sign, [35]	$e_{SG}(n) = \frac{1}{2} \cdot [A \cdot csgn[e_D(n)] + B \cdot csgn[e_D^*(n)]]$	csgn: complex sign function

C   Convergence Behavior of Bussgang Algorithms

Observing (25), we conclude that the coefficient adaptation rule for adaptive Bussgang equalizers is similar to the classical Least-Mean-Square (LMS) adaptation that is used in non-blind equalization approaches. As such, the Bussgang equalizers are simple and easy to implement. The difference between classical and Bussgang methods lies in the definition of the error quantity $e(n)$. In the classical LMS approach $e(n) = x(n) - \tilde{x}(n)$ where $x(n)$ is the true or "desired" value supplied by training sequences. In Bussgang adaptive algorithms the "desired" value, or in other words our "guess" for the true value, is provided by the nonlinear function $g[\tilde{x}(n)]$ (or $g^{(n)}[\tilde{x}(n)]$). An example that shows the true signal constellation and the assumed constellation by two Bussgang algorithms is given in figure 3(b).

The convergence behaviour of the adaptive Bussgang algorithms resembles the convergence behaviour of the LMS algorithm in the sense that it depends on the value of the step size γ as well as the eigenvalue spread of the autocorrelation matrix $\mathbf{R}_y$ of the observed data samples at the input of the equalizer [21, 47]. In general, the convergence analysis of these algorithms is difficult due to the presence of the nonlinearity [48-52]. Assuming convergence is achieved, the following relations can be written at equilibrium:

$$E\{e(n) \cdot \mathbf{y}^*(n)\} = 0 \qquad (26)$$

or after multiplying with $\mathbf{u}^H(n)$:

$$E\{e(n) \cdot \mathbf{u}^H(n) \cdot \mathbf{y}^*(n)\} = E\{e(n) \cdot \tilde{x}^*(n)\} = 0. \qquad (27)$$

and after simple calculations:

$$E\{g[\tilde{x}(n)] \cdot \tilde{x}(n)\} = E\{\tilde{x}(n) \cdot \tilde{x}^*(n)\}. \qquad (28)$$

Note that a process $\tilde{x}(n)$ that satisfies (28) with a memoryless nonlinearity $g[\cdot]$ is called a "Bussgang process" [53]. Thus, at the equilibrium the output of a Bussgang adaptive equalizer will be a Bussgang process, hence the name of this class of algorithms.

One problem with the Bussgang equalizers is that the underlying cost function $G[\tilde{x}(n)]$ being minimized is non-convex with respect to the equalizer coefficients. Thus, the $G[\tilde{x}(n)]$ is in general multimodal. As a result the utilization of a gradient LMS algorithm may provide an undesired solution that corresponds to a local equilibrium of the cost function [54-59]. Thus, research has been directed towards deriving new desirable cost functions, or applying non-gradient adaptation procedures capable of escaping from local equilibria [60-63].

a)

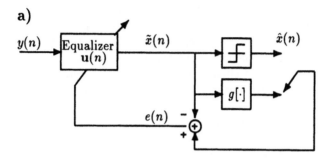

b)

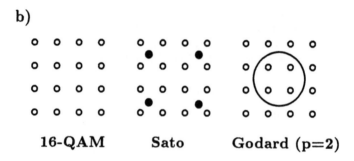

16-QAM   Sato   Godard (p=2)

Figure 3: a) Typical Bussgang adaptive equalizer, b) True (16-QAM) and assumed ($g[\tilde{x}(n)]$) constellation by the Sato (4 solid dots) and the Godard (circle of radius R_p) equalizers.

IV Polyspectra Approaches to Blind Deconvolution

A The General Approach

Polyspectra blind deconvolution approaches utilize the higher-order statistics (moments and cumulants of order greater than two) of the observed signal $\{y(n)\}$ to identify the characteristics of the channel impulse response $\{f(k)\}$ or the equalizer $\{u(k)\}$. Let us consider the linear filtering model with additive noise, that is:

$$y(n) = \sum_k f(k)x(n-k) + w(n) \qquad (29)$$

where, $\{w(n)\}$ is a stationary zero-mean additive Gaussian noise sequence independent from the data $\{x(n)\}$. The rest of the assumptions are as in section II.B. Then, using the properties of higher-order cumulants, [20,64,65], the following relations are written:

$$C_{y,1} = 0 \qquad (30)$$

$$C_{y,2}(\tau_1) = \gamma_{x,2} \cdot \sum_k f(k)f(k+\tau_1) + C_{w,2}(\tau_1) \qquad (31)$$

$$C_{y,3}(\tau_1, \tau_2) = \gamma_{x,3} \cdot \sum_k f(k)f(k+\tau_1)f(k+\tau_2) \qquad (32)$$

$$C_{y,4}(\tau_1, \tau_2, \tau_3) = \gamma_{x,4} \cdot \sum_k f(k)f(k+\tau_1)f(k+\tau_2)f(k+\tau_3) \qquad (33)$$

$$\tau_1, \tau_2, \tau_3 \in \mathcal{Z}$$

where, $C_{y,n}(\tau_1, \ldots, \tau_{n-1})$ is the n-th order cumulant sequence of $\{y(n)\}$. For the zero-mean sequence $\{y(n)\}$, the second, third, and fourth-order cumulants are defined as follows:

$$
\begin{aligned}
C_{y,2}(\tau_1) &= E\{y(n)y(n+\tau_1)\} & (34)\\
C_{y,3}(\tau_1, \tau_2) &= E\{y(n)y(n+\tau_1)y(n+\tau_2)\} & (35)\\
C_{y,4}(\tau_1, \tau_2, \tau_3) &= E\{y(n)y(n+\tau_1)y(n+\tau_2)y(n+\tau_3) & (36)\\
&\quad - C_{y,2}(\tau_1) \cdot C_{y,2}(\tau_2 - \tau_3) - C_{y,2}(\tau_2) \cdot C_{y,2}(\tau_3 - \tau_1)\\
&\quad\quad - C_{y,2}(\tau_3) \cdot C_{y,2}(\tau_1 - \tau_2)
\end{aligned}
$$

The constants $\gamma_{x,2} = C_{x,2}(0)$, $\gamma_{x,3} = C_{x,3}(0,0)$ and $\gamma_{x,4} = C_{x,4}(0,0,0)$ are the variance, skewness, and kurtosis of the i.i.d. sequence $\{x(n)\}$, respectively.

The polyspectra are defined as multidimensional (m-d) Fourier transforms of cumulants. For example, the power spectrum, the bispectrum and the trispectrum are the 1-d, 2-d and 3-d Fourier transforms of the second-, third-, and fourth-order cumulants, respectively.

In relations (30) to (33), we observe that the second-order cumulants of $\{y(n)\}$ are corrupted by the Gaussian noise. On the other hand, the higher order cumulants of $\{y(n)\}$ are Gaussian noise free and proportional to the corresponding order correlations of the channel impulse response. This is because the higher-order cumulants (of order greater than two) of a Gaussian process are in theory equal to zero. Furthermore, as it has been already mentioned in section II, the second-order cumulants of $\{y(n)\}$ do not preserve the true phase character of nonminimum phase channels while the cumulants of order greater than two preserve the true phase characteristics of minimum as well nonminimum phase channels. For these reasons, higher-order cumulants and their polyspectra are useful in blind deconvolution.

In digital communication applications the third-order cumulants of $\{y(n)\}$ are equal to zero because both the $\{x(n)\}$ and the $\{w(n)\}$ are symmetrically distributed around zero. Therefore, the fourth-order cumulants are utilized for blind identification/deconvolution of the communication channel. The basic problem is to find efficient ways to extract the channel impulse response from its fourth-order correlation function.

Three types of polyspectra blind equalization schemes have been proposed in the literature: the parametric methods, the nonlinear least-squares optimization approaches, and the polycepstra-based techniques [66]. The parametric methods [67-72] are sensitive to the estimation of the order of the underlying parametric channel model especially in adaptive implementations. The nonlinear optimization approaches, [73-77], are based on the minimization of a non-linear cost function that is formed from the higher-order cumulants or moments of the channel output. This cost function is in general multimodal. Thus, in adaptive realizations these methods require good initialization procedures to avoid convergence to local solutions. Finally, the polycepstra-based techniques utilize the complex cepstrum of the higher-order cumulants of the channel output to estimate directly either the channel characteristics or the coefficients of appropriate linear and decision feedback equalization filters [6,17,18,21,24], [78-82]. These schemes do not suffer from the above mentioned limitations of other polyspectra equalization approaches and they have been proven to be very successful blind deconvolution techniques. One of the well known polycepstra algorithms is described next.

B The Tricepstrum Equalization Algorithm (TEA)

This algorithm exploits the relations between the differential cepstrum parameters of the channel impulse response $\{f(k)\}$ and the "tricepstrum" of the channel output $\{y(n)\}$ to identify the channel. The differential cepstrum parameters of $\{f(k)\}$ are defined as follows [21, 78]:

$$A(k) = \sum_i a_i^k - \sum_i c_i^k, \qquad B(k) = \sum_i b_i^k, \qquad k = 1, 2, 3, \ldots \quad (37)$$

The $\{A(k)\}$ are called minimum phase differential cepstrum parameters because they are functions of the minimum phase zeros and poles of the channel transfer function $F(z)$ (see Section II.B(2)). The $\{B(k)\}$ are called maximum phase differential cepstrum parameters because they are functions of the maximum phase zeros of $F(z)$. Because we have assumed that $|a_i| < 1$, $|b_i| < 1$ and $|c_i| < 1$ for all i, the differential cepstrum parameters decay exponentially with k. Thus, in practice we can place $A(k) \simeq 0$, $k > p$ and $B(k) \simeq 0$, $k > q$, where the truncating parameters p and q are chosen accordingly[8].

The relation between these parameters and the channel impulse response becomes clear from the following recursive relations [11, 78]. Let $i(k) = Z^{-1}[I(z^{-1})]$ and $o(k) = Z^{-1}[O(z)]$, where $Z^{-1}[\cdot]$ denotes inverse Z-transform. Then,

$$i(k) = -\frac{1}{k} \sum_{i=2}^{k+1} A(i-1) \cdot i(k-i+1), \quad k = 1, 2, \ldots \quad (38)$$

$$o(k) = \frac{1}{k} \sum_{i=k+1}^{0} B(1-i) \cdot o(k-i+1), \quad k = -1, -2, \ldots \quad (39)$$

Then,

$$f_{norm}(k) = Z^{-1}[I(z^{-1})O(z)] = i(k) * o(k) \quad (40)$$

In other words, the minimum and maximum phase components of the channel impulse response can be obtained easily from the minimum and maximum phase differential cepstrum parameters of the channel, respectively. Furthermore, given the $\{A(k)\}$ and $\{B(k)\}$, one can easily obtain the coefficients of a zero-forcing linear equalizer $\{u_{norm}(k)\}$ (see figure 4(a)) or

---

[8]Modifications of the TEA that allow the channel to have zeros on the unit circle have been proposed in [81]

the coefficients of a zero-forcing decision feedback equalizer $\{u_{AP}(k)\}$ and $\{u_{FD}(k)\}$ (see figure 4(b))[9] as follows [21]: Let $i_{inv}(k) = Z^{-1}[\frac{1}{I(z^{-1})}]$ and $o_{inv}(k) = Z^{-1}[\frac{1}{O(z)}]$. Then,

$$i_{inv}(k) = -\frac{1}{k}\sum_{i=2}^{k+1}[-A(i-1)] \cdot i_{inv}(k-i+1), \quad k = 1, 2, \ldots \quad (41)$$

$$o_{inv}(k) = \frac{1}{k}\sum_{i=k+1}^{0}[-B(1-i)] \cdot o_{inv}(k-i+1), \quad k = -1, -2, \ldots \quad (42)$$

Then,

$$u_{norm}(k) = Z^{-1}[\frac{1}{I(z^{-1})O(z)}] = i_{inv}(k) * o_{inv}(k) \quad (43)$$

Also, let $io(k) = Z^{-1}[I(z^{-1})O^*(z^{-1})]$ and $oo(k) = Z^{-1}[O^*(z^{-1})]$. Then,

$$io(k) = -\frac{1}{k}\sum_{i=2}^{k+1}[A(i-1)+B^*(i-1)] \cdot i(k-i+1), \quad k = 1, 2, \ldots \quad (44)$$

$$oo(k) = -\frac{1}{k}\sum_{i=2}^{k+1}B^*(i-1) \cdot i(k-i+1), \quad k = 1, 2, \ldots \quad (45)$$

Then,

$$u_{AP}(k) = Z^{-1}[\frac{O^*(z^{-1})}{O(z)}] = o(k) * oo(k) \quad (46)$$

$$u_{AP}(k) = Z^{-1}[I(z^{-1})O^*(z^{-1}) - 1] = io(k) - \delta(k) \quad (47)$$

The tricepstrum of $\{y(n)\}$ is defined as follows [21, 78]:

$$c_{y,4}(\tau_1, \tau_2, \tau_3) = Z_{(3)}^{-1}\left[ln\left[Z_{(3)}\left[C_{y,4}(\tau_1, \tau_2, \tau_3)\right]\right]\right] \quad (48)$$

where, $Z_{(3)}$ and $Z_{(3)}^{-1}$ denote forward and inverse 3-d Z-transform and $ln[\cdot]$ denotes logarithmic operation.

---

[9] The decision feedback equalizer in general exhibits less noise enhancement than a linear equalizer [15, 21]

The tricepstrum takes the form, [78],

$$
c_{y,4}(\tau_1, \tau_2, \tau_3) = \begin{cases}
\ln\left(\gamma_{x,4} \cdot G^4\right), & \tau_1 = \tau_2 = \tau_3 = 0 \\
-\frac{1}{\tau_1} A(\tau_1), & \tau_2 = \tau_3 = 0, \ \tau_1 > 0 \\
-\frac{1}{\tau_2} A(\tau_2), & \tau_1 = \tau_3 = 0, \ \tau_2 > 0 \\
-\frac{1}{\tau_3} A(\tau_3), & \tau_1 = \tau_2 = 0, \ \tau_3 > 0 \\
\frac{1}{\tau_1} B(-\tau_1), & \tau_2 = \tau_3 = 0, \ \tau_1 < 0 \\
\frac{1}{\tau_2} B(-\tau_2), & \tau_1 = \tau_3 = 0, \ \tau_2 < 0 \\
\frac{1}{\tau_3} B(-\tau_3), & \tau_1 = \tau_2 = 0, \ \tau_3 < 0 \\
-\frac{1}{\tau_2} B(\tau_2), & \tau_1 = \tau_2 = \tau_3 > 0 \\
\frac{1}{\tau_2} A(-\tau_2), & \tau_1 = \tau_2 = \tau_3 < 0 \\
0, & \text{otherwise}
\end{cases}
$$

(49)

We note that the differential cepstrum parameters of the channel appear repeatedly in four lines in the 3-d space of the tricepstrum. By properly combining relations (48) and (49), the following relation can be shown [78]

$$
\sum_{K=1}^{p} A(K)[C_{y,4}(\tau_1, \tau_2 - K, \tau_1) - C_{y,4}(\tau_1 + K, \tau_2 + K, \tau_1 + K)] +
$$

$$
\sum_{J=1}^{q} B(J)[C_{y,4}(\tau_1 - J, \tau_2 - J, \tau_1 - J) - C_{y,4}(\tau_1, \tau_2 + J, \tau_1)] \simeq
$$

$$
-\tau_2 \cdot C_{y,4}(\tau_1, \tau_2, \tau_1) \tag{50}
$$

The approximation in (50) is due to the truncation of the differential cepstrum parameters. Then, we can repeat (50) N_p times with different values of τ_1 and τ_2 to form a linear overdetermined system of equations:

$$
\mathbf{P} \cdot \mathbf{a} = \mathbf{p} \tag{51}
$$

where $\mathbf{P}$ is a $N_p \times (p+q)$ matrix ($N_p > (p+q)$) with entries of the form $\{C_{y,4}(\lambda, \rho, \lambda) - C_{y,4}(\sigma, \tau, \sigma)\}$; $\mathbf{p}$ is a $N_p \times 1$ vector with entries of the form $\{-\tau_2 \cdot C_{y,4}(\tau_1, \tau_2, \tau_1)\}$; $\mathbf{a} = [A(1), \ldots, A(p), B(1), \ldots, B(q)]^T$ is the $(p \times q) \times 1$ vector of unknown cepstrum parameters. We can obtain the least squares solution of (51) as follows [21, 78]:

$$
\mathbf{a} = [\mathbf{P}^H \mathbf{P}]^{-1} \mathbf{P}^H \mathbf{p} \tag{52}
$$

and then utilize the elements of $\mathbf{a}$ to obtain the coefficients of either the channel impulse response or the equalization filters. The general procedure to follow can be summarized in the following steps:

1. Estimate the fourth-order cumulants of $\{y(n)\}$

2. Form the matrices $\mathbf{P}$ and $\mathbf{p}$

3. Obtain the least squares solution for $\mathbf{a}$

4. Calculate $\{f_{norm}(k)\}$ or $\{u_{norm}(k)\}$ or $\{u_{AL}(k)\}$ and $\{u_{FD}(k)\}$

5. Calculate the output of the equalizer and make a decision.

However, in practice where the channel characteristics may change slowly, an adaptive procedure is more appropriate. The TEA is an adaptive algorithm that follows the next steps [21]:

At iteration $n = 0, 1, 2, \ldots$

1. Obtain recursively from the samples $\{y(n)\}$ the time estimates

$$M_{y,4}^{(n+1)}(\tau_1, \tau_2, \tau_1) = (1 - \eta(n+1))M_{y,4}^{(n)}(\tau_1, \tau_2, \tau_1)$$
$$+\eta(n+1)y(S_2^{(n+1)})y^2(S_2^{(n+1)} + \tau_1)y(S_2^{(n+1)} + \tau_2)$$
$$C_{y,2}^{(n+1)}(\tau) = (1 - \eta(n+1))C_{y,2}^{(n)}(\tau) + \eta(n+1)y(S_2^{(n+1)})y(S_1^{(n+1)} + \tau)$$
$$C_{y,4}^{(n+1)}(\tau_1, \tau_2, \tau_1) = M_{y,4}^{(n+1)}(\tau_1, \tau_2, \tau_1) - 2C_{y,2}^{(n+1)}(\tau_1)C_{y,2}^{(n+1)}(\tau_2 - \tau_1)$$
$$-C_{y,2}^{(n+1)}(0)C_{y,2}^{(n+1)}(\tau_2) \tag{53}$$
$$-M \leq \tau_1, \tau_2 \leq M$$

where, $\eta(n) = \frac{1}{n}$, $S_1^{(n)} = min(n, n - \tau)$ and $S_2^{(n)} = min(n, n - \tau_1, n - \tau_2)$, $M_{y,4}^{(0)}(\tau_1, \tau_2, \tau_1) = 0$ and $C_{y,2}^{(0)}(\tau) = 0$.

2. Form the matrix $\mathbf{P}(n)$ and vector $\mathbf{p}(n)$ as $\mathbf{P}$ and $\mathbf{p}$ in (51), but with $C_{y,4}(\tau_1, \tau_2, \tau_1)$ being replaced by its <u>time estimate</u> $C_{y,4}^{(n)}(\tau_1, \tau_2, \tau_1)$.

3. Update the parameters $\{A^{(n)}(k)\}$, $\{B^{(n)}(k)\}$ using a gradient LMS type algorithm, i.e.,

$$\mathbf{a}(n+1) = \mathbf{a}(n) + \delta(n) \cdot \mathbf{P}^H(n) \cdot \mathbf{e}(n) \tag{54}$$
$$\mathbf{e}(n) = \mathbf{p}(n) - \mathbf{P}(n) \cdot \mathbf{a}(n)$$

where the step size $\delta(n)$ is such that $0 < \delta(n) < \frac{2}{trace[\mathbf{P}^H(n)\mathbf{P}(n)]}$.

4. Estimate either one of:

   i. The normalized channel impulse response $\{f_{norm}^{(n)}(k)\}$ using relations (38), (39) and (40).

   ii. The normalized coefficients of a linear equalizer $\{u_{norm}^{(n)}(k)\}$ using relations (41), (42) and (43).

   iii. The coefficients $\{u_{AP}^{(n)}(k)\}$ and $\{u_{FD}^{(n)}(k)\}$ using relations (44), (45), (46) and (47).

5. Calculate the input to the threshold decoder:

   i) For a linear equalizer:

$$\tilde{x}(n) = G(n) \cdot \sum_k u_{norm}^{(n)}(k)y(n-k)$$

   ii) For a decision feedback equalizer (DFE):

$$\tilde{x}(n) = G(n) \cdot \sum_k u_{AP}^{(n)}(k)y(n-k) - \sum_k u_{FB}^{(n)}(k)\hat{x}(n-k+1)$$

6. Make a decision $\hat{x}(n) = Q[\tilde{x}(n)]$ based on threshold decoding.

7. Recover the necessary gain and phase $G(n) = |G(n)|e^{j\phi(n)}$ by using blind gain and phase tracking algorithms. For example using the recursive algorithms

$$|G(n+1)| = \sqrt{\frac{PO}{Q(n+1)}}, \qquad Q(n+1) = (1-\rho)Q(n) + \rho\left|\frac{\tilde{x}(n)}{G(n)}\right|^2$$

$$\phi(n+1) = \phi(n) + \delta_\phi Imag[\tilde{x}^*(n)[\hat{x}(n) - \tilde{x}(n)]]$$

$$Q(0) = 0, \qquad \phi(0) = 0, \qquad PO = E\{|x(n)|^2\} \qquad (55)$$

where, $0 < \rho < 1$ is a small number (e.g. $\rho = 0.005$) and δ_ϕ is an appropriate step size.

C Properties of the TEA and other Polycepstra Algorithms

The least squares solution given by relation (52) exists and is unique provided that the matrix $\mathbf{P}$ has linearly independent columns. Note that this requirement is usually satisfied by taking a sufficiently large number of equations, i.e., $N_p \gg (p+q)$.

On the other hand, in the adaptive solution of (51), as described in step 3 of the TEA, the underlying cost function being minimized with respect to the vector $\mathbf{a}(n)$ is the $J = \mathbf{e}^H(n)\mathbf{e}(n)$, where $\mathbf{e}(n) = \mathbf{p}(n) - \mathbf{P}(n)\mathbf{a}(n)$. This cost function is quadratic with respect to $\mathbf{a}(n)$ and thus utilization of the LMS algorithm guarantees convergence to a global solution [16, 21]. This is an advantage of the TEA compared to the Bussgang family of algorithms.

Since the LMS approach is used for adaptation, the convergence rate of the TEA depends on the size of the step size $\delta(n)$ as well as the eigenvalue spread of the deterministic autocorrelation matrix $\mathbf{P}^H\mathbf{P}$. These are well known properties in the literature [16].

The complexity of the TEA depends on the estimation of the fourth-order cumulants and the size of the matrix $\mathbf{P}$ in (51). It has been found, that a value of $N_p \geq 3(p+q)$ is usually sufficient in practice. For large p or q the computational complexity is of the order $\mathcal{O}(p+q)$ [21]. Note that the values of the differential cepstrum truncating parameters p, and q increase as the zeros and poles of the channel get closer to the unit circle and the complexity increases accordingly.

The above properties of the TEA are shared by all polycepstra-based polyspectra blind equalizers [18,21,24,78-82].

a)

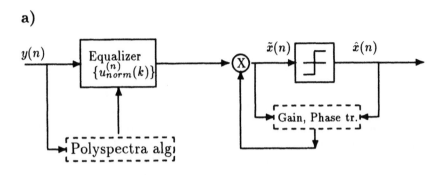

b)

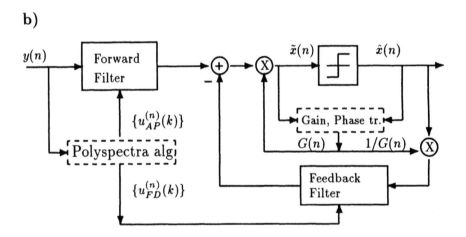

Figure 4: a) Linear polyspectra equalizer, b) Decision feedback polyspectra equalizer

V   Minimum Entropy Approaches to Blind Deconvolution

Let us consider once again the deconvolution scenario depicted in figure 2 where,

$$\tilde{x}(n) = u(n) * y(n) = u(n) * [f(n) * x(n)] = [u(n) * f(n)] * x(n) \qquad (56)$$

If the input random process $x(n)$ to the linear time invariant channel $f(n)$ is Gaussian distributed then the output of the channel $y(n)$ is also Gaussian distributed. On the other hand, if $x(n)$ is non-Gaussian then $y(n)$ is in general non-Gaussian, however, the "Gaussianity" of $y(n)$ is higher than $x(n)$. This is a result of the Central Limit Theorem that states that the linear superposition of similar random variables tends asymptotically in distribution to a Gaussian random variable. Note that the "Gaussianity" of a random process can be expressed by an appropriate statistical measure. Traditionally the kurtosis (i.e., the all zero lag of the fourth-order cumulant, see section IV.A) of a process has been used to indicate deviations form Gaussianity because the kurtosis of a Gaussian process is zero [20]. For example, sub-Gaussian distributed processes such as the uniformly distributed process have a negative kurtosis while super-Gaussian distributed processes have a positive kurtosis.

Since the effect of linear filtering (i.e., convolution with $f(n)$) increases the Gaussianity of a random process, then, inverse filtering (i.e., deconvolution with $u(n)$) must decrease the Gaussianity of the process. Based on this idea blind deconvolution can rely on maximizing or perhaps minimizing an appropriate measure of Gaussianity [9, 83]. The constraint or unconstraint maximization of the absolute value of the kurtosis has been proposed and utilized by various blind deconvolution schemes in the literature [9, 76, 83-86]. For example, in [84] it was proposed to:

$$Maximize \quad |C_{\tilde{x},4}(0,0,0)| \qquad (57)$$

$$Subject \quad to: \quad E\{|\tilde{x}(n)|^2\} = E\{|x(n)|^2\}$$

On the other hand, the entropy $EN(Y)$ of a random vector Y with probability density function $p_Y(y)$ is defined as [15]:

$$EN(Y) = - \int p_Y(y) log[p_Y(y)] dy. \qquad (58)$$

For a random process the entropy $EN(Y)$ can be used as a measure of deviation from Gaussianity as well. Among different distributions of a process the entropy is maximum for the Gaussian distribution. Thus, decreasing the Gaussianity of a process agrees with the minimization of the entropy [9, 83]. This is why the name "minimum entropy blind deconvolution" has been assigned to such deconvolution approaches.

VI   Probabilistic Approaches to Blind Deconvolution

Two general type of probabilistic approaches have appeared in the literature for blind deconvolution of communication channels. Those that employ the maximum likelihood criterion (ML) for jointly estimating the channel and detecting the data , [87-95], and those that employ Bayesian maximum a-posteriori (MAP) estimation principles [96-98].

To understand the basic principles behind the probabilistic approaches let us consider a block of N samples obtained from the linear filtering model with additive Gaussian noise, that is

$$y(n) = \sum_{k=0}^{L} f(k)x(n - k) + w(n), \qquad n = 1, 2, \ldots, N \tag{59}$$

where, the $\{x(n)\}$ is the communication data sequence, $\{f(n)\}$ is the communication channel and $\{w(n)\}$ is zero mean white Gaussian noise of variance σ^2. Then by defining the vectors $\mathbf{y} = [y(1), \ldots, y(n)]^T$, $\mathbf{f} = [f(0), \ldots, f(L)]^T$, and $\mathbf{x} = [x(1), \ldots, x(J)]^T$, we find that the joint conditional probability density function is [87,89]:

$$P(\mathbf{y}|\mathbf{f}, \mathbf{x}) = \frac{1}{(2\pi\sigma^2)^N} e^{-\frac{1}{2\sigma^2} \sum_{n=1}^{N} |y(n) - \sum_{k=0}^{L} f(k)x(n-k)|^2} \tag{60}$$

To obtain the ML estimate of the channel and the data we must maximize the $P(\mathbf{y}|\mathbf{f}, \mathbf{x})$ with respect to $\mathbf{f}$ and $\mathbf{x}$. This is equivalent to minimizing the term in the exponent of $P(\mathbf{y}|\mathbf{f}, \mathbf{x})$, that is

$$(\mathbf{x}_{ML}, \mathbf{f}_{ML}) = \min_{\mathbf{x}_{ML}, \mathbf{f}_{ML}} \sum_{n=1}^{N} |y(n) - \sum_{k=0}^{L} f(k)x(n - k)|^2 \tag{61}$$

Assuming that the true data vector $\mathbf{x}$ is known a-priori, then, the $\mathbf{f}_{ML}$ can be easily obtained from the least squares solution of (61), [87]. On the other hand, assuming that the channel vector $\mathbf{f}$ is known, then, the $\mathbf{x}_{ML}$ is

obtained by carrying out a trellis search by means of the Viterbi algorithm
(VA) [87,89]. The VA exhibits a complexity that is linear with respect to
the length (J) of the data vector $\mathbf{x}$ and exponential with respect to the
length $(L + 1)$ of the channel vector $\mathbf{f}$.

In the case of blind equalization the minimization of (61) must be car-
ried out jointly with respect to $\mathbf{x}$ and $\mathbf{f}$. An exhaustive search optimization
can be devised by taking into account the discrete nature of digital communication signals. Assuming that the communication data $\{x(n)\}$ take
M possible discrete values then, the vector $\mathbf{x}$ of length J will take one of
M^J possible solutions. Thus, we can calculate all possible data vectors
$\mathbf{x}^{(i)}$, $i = 1, 2, \ldots, M^J$, for each $\mathbf{x}^{(i)}$ obtain the corresponding ML channel
estimate $\mathbf{f}^{(i)}$ from the least squares solution of (61), and then choose, [90]:

$$(\mathbf{x}_{ML}, \mathbf{f}_{ML}) = \min_{(i)} P(\mathbf{y}|\mathbf{f}^{(i)}, \mathbf{x}^{(i)}) \tag{62}$$

Obviously, the exhaustive search approach exhibits high complexity that
increases fast with J and L. This has motivated research towards the
development of other approaches for joint data and channel estimation with
less computational complexity.

In [89,90], a generalized Viterbi algorithm was proposed where the underlying trellis search retains more than one in general "best" estimates of
the data sequence at each stage of the algorithm (in contrast to the classical VA that retains a single most probable estimate of the data sequence at
each stage). In addition, the corresponding channel estimates are updated
recursively by means of an adaptive LMS algorithm. This algorithm has
been shown to work well by retaining as few as four sequence estimates at
each stage at moderate signal to noise ratio environments. Its complexity
is lower than that of the exhaustive method, however, it is still much higher
that the complexity of the classical VA.

In [91], a recursive algorithm that recursively alternates between the estimation of the channel and the estimation of the data sequence has been
proposed. First an initial estimate of the channel is chosen. Given this
channel estimate the VA is applied to obtain the optimum data sequence for
this channel. Then, using this data sequence a new estimate of the channel
is obtained based on the least square solution. This procedure is repeated
until the algorithm converges. A similar and somehow improved procedure named "quantized-channel algorithm" has been proposed in [92,93].
At every step of the algorithm the channel is selected from a number of
candidate channels on the basis of selecting the channel with the smallest
accumulated error energy. Then, the classical VA is used to produce the
optimum data sequence. The algorithm needs to know the order of the
channel and its energy. However, these quantities are in general unknown

and in practice can only be approximated. An advantage of the algorithm is its inherent parallel structure that may compensate for its high computational complexity.

The above algorithms for joint data recovery and channel estimation based on the ML criterion require relatively few received data samples to estimate the communication channel. These algorithms have yet to be investigated thoroughly. However, they may become attractive methods for blind equalization of signal constellations that are approximately Gaussian distributed, as in the case of shaped signal constellations [94] , since they are not subject to the non-Gaussian signal restrictions required by the Bussgang, Polyspectra and other blind deconvolution approaches. The main drawbacks of the ML blind methods are their high complexity and the large decoding delay introduced by the utilized Viterbi decoder [87,89].

The other class of probabilistic methods for blind equalization, the MAP methods, in general have similar algorithmic structures to the ML approaches. The main difference is found in the utilization of a MAP data sequence estimation instead of the Viterbi algorithm, in conjunction with least squares procedures for the estimation of the channel. A few such methods have been proposed in the literature, [96-98], however, without demonstrating any significant advantages or disadvantages compared to the ML approaches.

VII   Cyclic Spectrum Approaches to Blind Deconvolution

Let us consider the the situation shown in fig. 5 where the discrete time signal $y(n)$, $n \in \mathcal{Z}$ is obtained by sampling the continuous time signal $\bar{y}(t)$, $t \in \mathcal{R}$ every T_s sec. The sequence $\{a_k\}$, $k \in \mathcal{Z}$ is a zero mean, i.i.d. data sequence, the $\bar{\delta}(t)$ and $\delta(n)$ are the continuous and discrete time delta functions respectively, the T is the symbol duration, the $\bar{f}(t)$ and $f(n)$ are the continuous time channel and the equivalent discrete time channel corresponding to the sampling instants respectively, and so on. This modeling is appropriate for digital communication links, digital recording systems and other digital transmission applications [15].

For any finite value of $T \neq 0$, the signal $\bar{x}(t)$ is cyclostationary, i.e., all statistics of $\bar{x}(t)$ are periodic in time t with period T sec [15,99]. Then, since the channel $\bar{f}(t)$ is linear time invariant, the $\bar{y}(t)$ is cyclostationary with period T sec. as well. On the other hand, assuming that $L \in \mathcal{Z}$, $L \neq 0$, then, we must consider two different cases for the discrete signal $x(n)$: i) $L > 1$, in which case the $x(n)$ is a cyclostationary signal with period L

samples, i.e., the statistics of $x(n)$ are periodic with period L samples, or ii) $L = 1$, in which case the $x(n)$ is a stationary signal, i.e., the statistics of $x(n)$ are time invariant. Since the $f(n)$ is linear time invariant, the $y(n)$ is also cyclostationary with period L samples or stationary, respectively [99].

a)

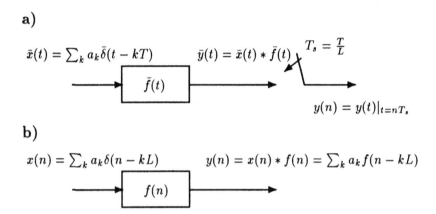

b)

Figure 5: a) Continuous time model, b) equivalent discrete time model

In the applications of blind channel deconvolution in digital communications that we have examined thus far the observed cyclostationary signal has been treated as a stationary random process. This has been the result of synchronously sampling (i.e., $L = 1$ or $T_s = T$) the observed output of the communication channel (see eq. (1). However, synchronous sampling does not preserve the inherent cyclostationary properties of the observed signal since in this case $y(n)$ is stationary as it has been argued above. Thus, another approach to the blind deconvolution of digital communication signals would be to preserve and utilize the cyclostationary properties of these signals (i.e., periodicity of statistics) by oversampling ($L > 1$). Actually, in few related papers that have appeared in the literature [100-109], it has been shown that the cyclic second order statistics (cyclic autocorrelation and cyclic spectrum) do not share the limitations of the second order statistics of stationary signals (autocorrelation and power spectrum) and can be utilized for the blind identification of a large class of nonminimum phase channels. Among the channels that cannot be identified by blind cyclic second order statistics methods are those with all their zeros and poles evenly spaced radially with angle $\frac{2\pi}{L}$ [101,104, 106]. However, this problem is rare and will occur for specific values of L only. Finally, we should mention that cyclic second-order statistics cannot be utilized for the identification of nonminimum phase discrete channels that are strictly bandlimited to a total bandwidth less than $\frac{2\pi}{L}$ rads/sample (or $\frac{2\pi}{T}$ rads/sec). In this case the cyclic spectrum reduces to the classical power spectrum [100, 103].

Compared to the blind deconvolution approaches that utilize the higher-order statistics of the observed signal, under stationary signal assumptions, the methods that utilize cyclic second order statistics possess the following attractive characteristics, [106]: i) fewer data samples are required and less complexity is encountered in the estimation the cyclic second order statistics, ii) no restrictions on the distribution of the input data are imposed, iii) they are insensitive (in theory) to any stationary additive noise, and iv) fractionally-spaced sampling is less sensitive than synchronous sampling to timing errors. For these reasons, it is expected that the cyclostationarity-based methods could be the appropriate choice in many applications. Lately, there is an increased interest towards this direction.

VIII   Simulation Examples

In this section we discuss the behaviour of different blind equalization algorithms by means of computer simulated results. The simulation set-up is based on a typical digital communications scenario with the assumptions of section II. For L-PAM signaling, $\{x(n)\}$ is an i.i.d sequence taking the equally probable values $(-L + 1, \ldots, -1, 1, \ldots, L - 1)$. For L^2-QAM signaling, $\{x(n) = x_R(n) + jx_I(n)\}$ where, the $\{x_R(n)\}$ and $\{x_I(n)\}$ are two independent and identically distributed L-PAM sequences. The channel is modeled as a nonminimum-phase finite impulse response (FIR) filter like the complex channels depicted in figure 6. White and zero-mean Gaussian noise was added at the output of the channel.

A number of performance metrics demonstrate the behaviour of the algorithms at each iteration:

1. The mean square error between the output of the equalizer and the corresponding desired value, that is $MSE = E\{|x(n - d) - \tilde{x}(n)|^2\}$.

2. The symbol error rate (SER) which measures the percentage of wrong decisions in intervals of 500 samples.

3. The "discrete eye patterns" consisting of the signal values at the output of the equalizer drawn in two dimensional (for QAM signaling) space. We refer to an "open eye pattern" when threshold decoding can easily differentiate between neighbouring states.

4. The residual intersymbol interference at the output of the equalizer that is defined as $ISI = \frac{\sum_k |f(k)*u(k)|^2}{max\{|f(k)*u(k)|^2\}} - 1$.

5. The signal to noise power ratio (SNR) measured at the output of the channel (input to equalizer).

We must mention that the objective of most existing blind equalization algorithms is to open the eye pattern of the signal to a degree where decisions based on threshold decoding are reliable enough to switch to the simple and fast decision-directed (DD) algorithm [2,3]. Recall that the DD algorithm is one of the simplest Bussgang type equalizers. However, despite its blind properties, the DD algorithm in most cases fails to equalize channels that introduce severe distortion to completely close the eye pattern of the signal [2,3,33]. Thus, more sophisticated blind equalization algorithms need to be applied prior to the DD equalization.

Figure 7 demonstrates the ability of the TEA, [21], to track the minimum and maximum phase cepstrum parameters of the channel. The TEA was implemented as explained in section IV(B), with $p = q = 6$ and $\delta(n) = \frac{1}{trace[\mathbf{P}^H(n)\mathbf{P}(n)]}$, and converged to the true parameters at around 500 iterations.

Figures 8, 9, and 10 demonstrate and compare the convergence of the TEA and the convergence of the Stop-and-Go (SG), [33], and Benveniste-Goursat (BG), [32], linear equalizers with the channel examples of figure 6. The MSE and SER were calculated by ensemble averaging over 10 experiments with independent signal and noise realizations and by time averaging over 100 samples for each realization. Also, the eye pattern at iteration (n) was obtained by drawing the equalizer output for a specific number of samples around (n) for all 10 realizations. In figure 8 the theoretical mean-square-error lower bounds MSE_{out} (for the TEA) and MMSE (for the SG) are provided as well. The $MMSE < MSE_{out}$ because the TEA is a zero forcing (ZF) equalizer while the two Bussgang algorithms are minimum mean- square- error equalizers. The simulation results show that the TEA has faster initial convergence and therefore, opens the eye pattern of the signal faster than the two Bussgang algorithms. This behaviour is typical among most polyspectra algorithms. However, the convergence speed of the TEA slows down after initial convergence. On the other hand, the SG and BG algorithms have slow initial convergence and speed up later when the eye pattern of the signal starts to open. This is attributed to the structure of these algorithms that transforms automatically to the DD algorithm after the eye pattern opens (see Table I).

In figure 11, the performance of four decision- feedback- equalization (DFE) algorithms is depicted with the channel example 1 of figure 6 and 64-QAM signaling, [81]. The decision feedback predictor (PRED), [41], is based exclusively on classical linear prediction principles that utilize second-order statistics and thus, fails to equalize the nonminimum phase channel. The decision feedback tricepstrum equalization algorithm (TEA), [21], and the extended tricepstrum equalization algorithm (ETEA), [24], utilize exclusively the fourth- order cumulants of the channel output. Finally, the decision feedback polycepstra and prediction equalization algorithm (POPREA), [81], is basically a combination of the PRED and ETEA algorithms. Thus, the TEA, ETEA and POPREA belong to polyspectra family of blind equalizers. The ETEA and POPREA utilize a different definition of fourth- order cumulants[10] than the TEA and achieve faster convergence. Furthermore, the fourth-order cumulants utilized by ETEA and POPREA allow them to operate efficiently even in the presence of low to moderate carrier frequency offset (FO) which is often present in communication signals due to imperfect demodulation [2,24,33]. This is clearly demonstrated in the equalized eye patterns of POPREA in figure 11.

It has been mentioned in section III that the Bussgang equalizers might converge to local equilibria of their non-convex cost function. This has been sufficiently investigated and clearly demonstrated in the literature [54-59]. This type of convergence behaviour is illustrated for the Godard (p=2) algorithm in figure 12[11]. The communication channel has a transfer function $F(z) = \frac{1}{1+0.6z^{-1}}$[12] and the input signal is 2-PAM. Obviously the ideal linear equalizer (in absence of noise) is of length 2 with transfer function $U(z) = 1 + 0.6z^{-1}$. The Godard cost function with respect to the two unknown parameters of the equalizer exhibits two local and two global equilibria (minima) as it is shown at the top of figure 12. The two global minima differ by a sign reversal and therefore cannot be distinguished by blind deconvolution methods. Five different initialization settings were considered for the Godard equalizer: (1.3, -1), (-1.9, 1), (-1.7, -0.2), and (0.1, 0.3). We observe that for the first and fifth settings the Godard algorithm converged to the undesired local equilibria. This problem is also demonstrated by the corresponding MSE convergence.

---

[10] Different definitions of the fourth-order cumulants of complex signals can be obtained by conjugating one or more signal terms

[11] From: "Analysis of Bussgang Equalization Schemes", by Jacob, Wah-Hing Leung, B.A.Sc. Thesis, ECE Dept., University of Toronto, April 1992.

[12] The same channel has been considered earlier in [55]

To overcome the problem of convergence to local equilibria associated with Bussgang equalizers three general directions have been followed in the literature. The first tries to determine conditions that will allow efficient initialization of the algorithm [2]. However, this is not possible without having partial knowledge of the channel properties. The second direction recognizes that the problem of local convergence is due to the utilization of the LMS algorithm in the minimization of a non-convex cost function. Therefore, a different type of minimization algorithms should be applied. In [61,62], a simulated annealing algorithm was proposed for minimizing the Godard cost function. Simulated annealing is a random search procedure capable of escaping from local solutions. One such example is illustrated in figure 13 where, assuming the same initial equalizer settings, the Godard algorithm with LMS adaptation locks into a local solution, however, the Godard cost function with simulated annealing converges to the correct solution. Nevertheless this has been accomplished at the expense of high computational complexity. The third direction looks towards designing cost functions for blind equalization that are convex with respect to the equalizer parameters and thus easy to minimize by means of simple LMS type algorithms. A block minimization algorithm with a convex corresponding cost function has been proposed in [60]. The ability of this algorithm to open the eye pattern of the signal after approximately 300 iterations is shown in figure 14. However, this algorithm requires the storage of a large number of samples before adaptation begins and is more complex than the classical LMS. Finally, another algorithm, the criterion with memory nonlinearity (CRIMNO) algorithm, utilizes an augmented Godard (p=2) function with additional terms that include the weighted autocorrelation lags of the equalizer output [63,110]. This transforms the memoryless nonlinear Godard cost function into a nonlinear cost function with memory. The inclusion of memory into the cost function results in faster convergence of the CRIMNO algorithm as it is clearly demonstrated in figure 15. Furthermore, by properly adjusting the weights of the additional terms the CRIMNO algorithm is able to avoid convergence to local equilibria.

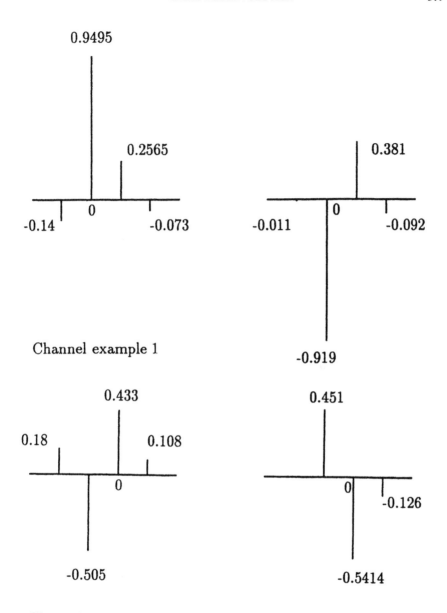

Channel example 1

Channel example 2

Figure 6: Complex Discrete Channel Examples. (From Hatzinakos, [81], with permission of Elsevier Science Publishers)

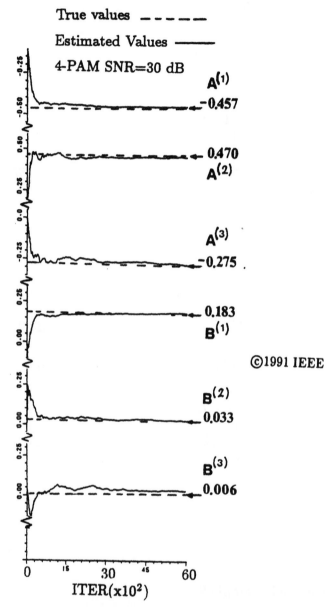

Figure 7: Convergence of the first three minimum phase and maximum phase cepstrum parameters of the channel with transfer function $F(z) = (1 - 0.183z)(1 - 0.1987z^{-1})(1 + 0.656z^{-1})$. (From Hatzinakos and Nikias, [21], with permission of IEEE)

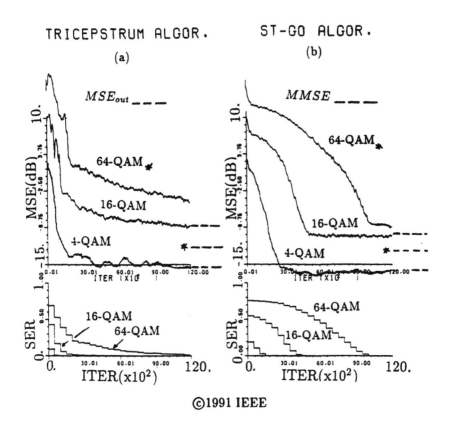

©1991 IEEE

Figure 8: Mean square error convergence (top) and symbol error rate (bottom) vs number of iterations for a $N = 31$ tap linear equalizer with the channel example 1. $SNR = 20$ dB for 4-QAM and 16-QAM, $SNR = 30$ dB for 64-QAM. a) TEA with $p = q = 6$ and $\delta(n) = \frac{1}{trace[\mathbf{P}^H(n)\mathbf{P}(n)]}$. b) Stop-and-Go with $\gamma = 10^{-3}$, $\alpha = 1$ for 4-QAM, $\gamma = 4 \cdot 10^{-4}$, $\alpha = 2.5$ for 16-QAM, and $\gamma = 10^{-4}$, $\alpha = 6$ for 64-QAM. (From Hatzinakos and Nikias, [21], with permission of IEEE)

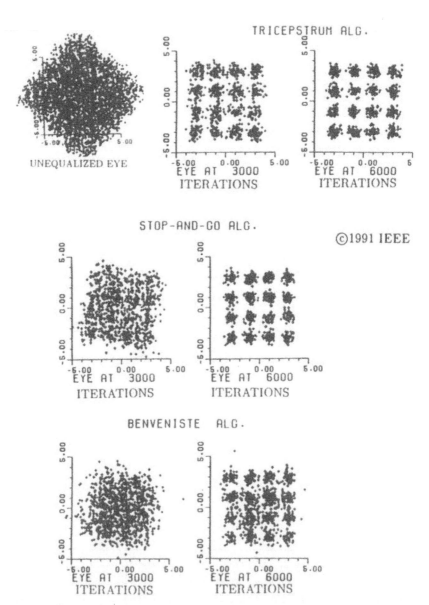

Figure 9: Distorted eye pattern (channel example 1) and equalized eye patterns of a 16-QAM signal constellation with various blind equalization algorithms. $SNR = 20$ dB. Parameters for TEA and Stop-and-Go are as in figure 8. For Benveniste-Goursat, $\gamma = 10^{-4}$, $\alpha = 2.5$, $k_1 = 3$ and $k_2 = 1$. (From Hatzinakos and Nikias, [21], with permission of IEEE)

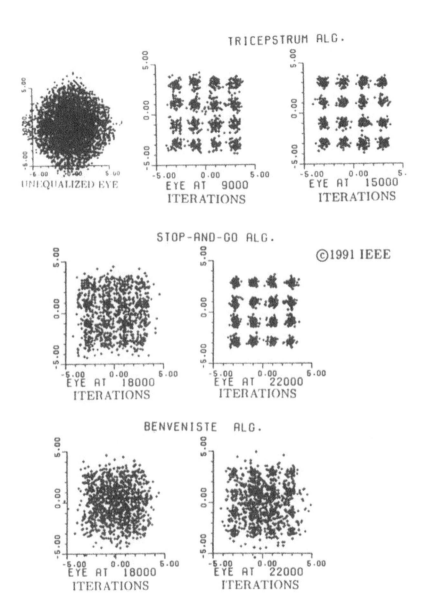

Figure 10: Distorted eye pattern (channel example 2) and equalized eye patterns of a 16-QAM signal constellation with various blind equalization algorithms. $SNR = 25$ dB. Parameters are as in figure 9. (From Hatzinakos and Nikias, [21], with permission of IEEE)

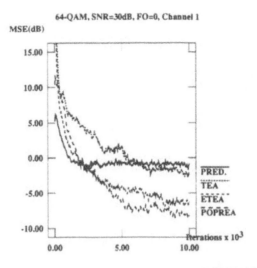

©1992 ELSEVIER

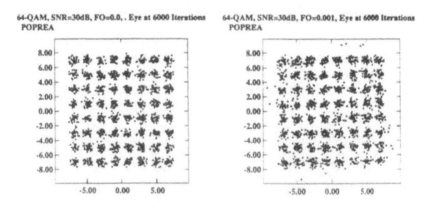

Figure 11: Mean square error performance and eye patterns of different blind decision feedback algorithms with 64-QAM and channel example 1.(From Hatzinakos, [81], with permission of Elsevier Science Publishers)

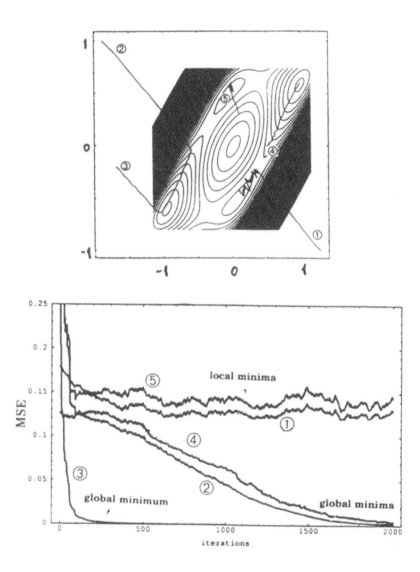

Figure 12: Contour diagram of the Godard cost function and mean square error convergence under different initial equalizer settings.

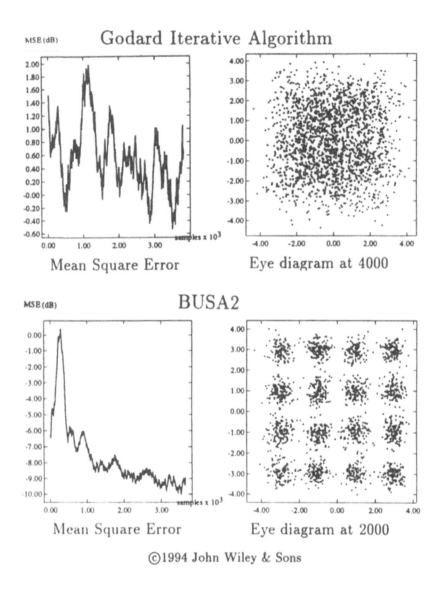

Figure 13: Mean square error convergence and eye diagram (16-QAM) of the Godard cost function with LMS optimization (top) and simulated annealing optimization (bottom). (From Ilow, Hatzinakos and Venetsanopoulos, [62], with permission of John Wiley & Sons, Ltd.)

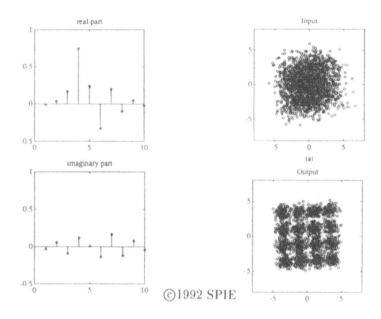

©1992 SPIE

Figure 14: Impulse response of complex channel (left) and eye diagram prior and after blind equalization (right) with the block minimization algorithm after 300 iterations with a block of 2000 samples. (From Kennedy and Ding, [60], with permission of the authors and of SPIE Publications)

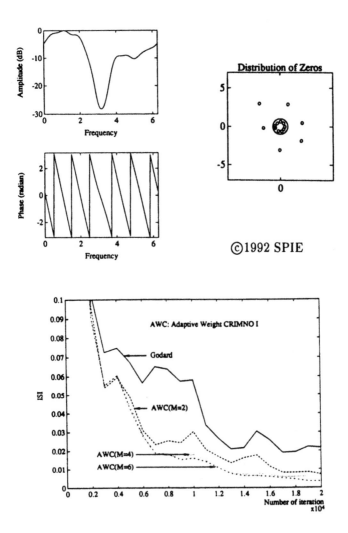

Figure 15: Channel characteristics (top) and residual intersymbol interference at the output of the CRIMNO equalizer of memory size M. (From Chen, Nikias and Proakis, [63], with permission of the authors and of SPIE Publications)

IX   Summary

An overview of the major approaches to the problem of blind deconvolution is given. Without loss of generality, the treatment of the problem focused on the blind identification and equalization of digital communication channels. In Table II, an attempt is made to classify the general characteristic of the five approaches presented in this chapter. It must be emphasized that this classification provides a statement of the average and not the individual behaviour of the algorithms in each approach. For example, it has been found experimentally that polyspectra algorithms in general converge faster than the Bussgang algorithms. Nevertheless, the convergence behaviour of each algorithm is highly dependent on the channel characteristics, choice of parameters, and equalizer initialization among others.

Table II: General characteristics of blind deconvolution methods

	Bussgang	Polyspectra	Probabilistic	Minimum Entropy	Cyclic Spectrum
Complexity	Low	High	High	Low to Moderate	moderate
Relative Convergence Speed From 1 (low) to 3 (high	1-2	1.5-2.5	2-3	1-2	
Robustness to additive noise		Gaussian noise			Stationary noise
Restrictions to input data distribution	non-Gaussian	non-Gaussian		non-Gaussian	
Possibility of ill-convergence	Yes			Yes	
Channel Identifiability					Restricted

There are other blind deconvolution approaches in the literature that cannot be easily classified under the examined five approaches, such as references [111-114] among others. Nevertheless, a common characteristic of all existing blind deconvolution algorithms is their slow convergence compared to trained algorithms. Thus, a lot of research effort is directed currently towards the development of faster blind deconvolution algorithms.

X   References

1. Y. Sato, " A Method for Self-Recovering Equalization for Multilevel Amplitude-Modulation Systems ", *IEEE Trans. on Communications*, Vol. 23, pp. 679-682, (1975).

2. D. N. Godard, " Self Recovering Equalization and Carrier Tracking in two-Dimensional Data communication Systems ", *IEEE Trans. on Communications*, Vol. 28, pp. 1867-1875, Nov. (1980).

3. A. Benveniste, M. Goursat, and G. Ruget, " Robust Identification of a Nonminimum Phase System: Blind Adjustment of a Linear Equalizer in Data Communications ", *IEEE Trans. on Automatic Control*, Vol. 25(3), pp. 385-398, (1980).

4. G. J. Foschini, " Equalizing Without Altering or Detecting Data ", *AT&T Techn. J.*, Vol.64(8), pp. 1885-1911, (1885).

5. G. R. Ayers and J. C. Dainty, " Iterative Blind Deconvolution Method and its Applications ", *Opt. Letters*, Vol. 13(7), pp. 547-549, (1988).

6. A. P. Petropulu and C. L. Nikias, " Blind Deconvolution Using Signal Reconstruction from Partial Higher Order Cepstral Information ", *IEEE Trans. on Signal Processing* ", Vol. 41(6), pp. 2088-2095, (1993).

7. J. L. Yen, "Image Reconstruction in Synthesis Radiotelescope Arrays", in *Array Signal Processing*, ed. S. Haykin, Prentice Hall, Englewood Cliffs, NJ, pp. 293-350, (1985)

8. S. S. Kuo and R. J. Mammone, "An Iterative Projection Technique for Blind Image Restoration", *Journal of Visual Communication and Image Representation*, Vol. 4(1), pp. 92-101, (1993).

9. R. A. Wiggins, " Minimum entropy Deconvolution", *Geoexploration*, Vol.16, pp. 21-35, (1978).

10. L. C. Wood and S. Treitel, "Seismic Signal Processing", *Proc. of IEEE*, Vol. 63(4), pp. 649-661, (1975).

11. A. V. Oppenheim and R. W. Schafer, *Discrete-Time Signal Processing*, Prentice Hall, Englewood Cliffs, NJ, (1989)

12. R. W. Schafer and L. Rabiner, "Digital Representations of Speech Signals *Proc. of the IEEE*, Vol. 63(4), pp. 662-677, (1975).

13. T. G. Stockham, Jr., T. M. Cannon, and R. B. Ingebretsen, " Blind Deconvolution Through Digital Signal Processing ", *Proc. of the IEEE*, Vol. 63(4), pp. 678-692, (1975).

14. *Blind Deconvolution*, ed. S. Haykin, Prentice Hall, Englewood Cliffs, NJ, (1994).

15. J. G. Proakis, *Digital Communications*, McGraw Hill, (1983).

16. S. Haykin, *Adaptive Filter Theory*, Prentice Hall, Englewood Cliffs, NJ, (1991).

17. D. Hatzinakos and C. L. Nikias, " Estimation of Multipath Channel Response in Frequency Selective Channels ", *IEEE J. on Sel. Ar. in Com.*, Vol. 7(1), pp. 12-19, (1989).

18. A. P. Petropulu and C. L. Nikias, "Blind Deconvolution of Coloured Signals Based on Higher-Order Cepstra and Data Fusion", *IEE Proceedings-F*, Vol. 140(6), pp. 356-361, (1993).

19. J. Makhoul, "Linear Prediction: A Tutorial Review", *Proc. of the IEEE*, Vol. 63(4), pp. 561-580, (1975).

20. C. L. Nikias and A. P. Petropulu, *Higher-Order Spectral Analysis: A Digital Signal Processing Framework*, Prentice Hall, Englewood Cliffs, NJ, (1983).

21. D. Hatzinakos and C. L. Nikias, " Blind Equalization Using a Tricepstrum Based Algorithm ", *IEEE Trans. on Communications*, Vol. 39(5), pp. 669-682, (1991).

22. J. G. Proakis, and C. L. Nikias, "Blind Equalization", *Proceedings of SPIE*, Vol. 1565, pp. 76-87, (1991).

23. N. K. Jablon, " Joint Blind Equalization, Carrier Recovery, and Timing Recovery for High-Order QAM Signal Constellations ", *IEEE Trans. on Signal Processing*, Vol. 40(6), pp. 1383-1398, (1992).

24. D. Hatzinakos, " Carrier Phase Tracking and Tricepstrum- Based Blind Equalization," *Computers & Electrical Eng.*, Vol. 18(2), pp. 109-118, (1992).

25. R. Godfrey and F. Rocca, " Zero Memory Nonlinear Equalization ", *Geophysical Prospecting*, Vol. 29, pp. 189-228, (1981).

26. S. Bellini, " Blind Equalization and Deconvolution", *Proceedings of SPIE*, Vol. 1565, pp. 88-101, (1991).

27. S. Bellini and F. Rocca, " Blind Deconvolution: Polyspectra or Bussgang Techniques ", in *Digital Communications*, eds E. Biglieri and G. Prati, North-Holland, pp. 251-262 (1986).

28. V. Weerackody and S. A. Kassam, " A Computationally Efficient Algorithm for Blind Equalization ", *Proc. of the 22nd Conf. on Information Sciences and Systems*, pp. 416-419, Princeton, (1988).

29. V. Weerackody, S. A. Kassam, and K. R. Laker, "A Reduced Complexity Adaptive Filter Algorithm for Blind Equalization ", *Proc. of the 23nd Conf. on Information Sciences and Systems*, pp. 73-77, Valtimore, (1989)

30. J. R. Treichler and B. G. Agee, " A New Approach to Multipath Correction of Constant Modulus signals ", *IEEE Trans. on Acoustics, Speech, and Signal Processing*, Vol. 31(2), pp. 459-471, (1983).

31. J. R. Treichler and M. G. Larimore, " New Processing Techniques Based on the Constant Modulus Algorithm ", *IEEE Trans. on Acoust., Speech and Signal Processing*, Vol. 33(2), pp. 420-431, (1985).

32. A. Benveniste and M. Goursat, " Blind Equalizers ", *IEEE Trans. on Communications*, Vol. COM-32, pp. 871-883, (1984).

33. G. Picchi and G. Prati, " Blind Equalization and Carrier Recovery Using a Stop-and-Go Decision-Directed Algorithm", *IEEE Trans. on Communications*, Vol. 35(9), pp. 877-887, (1987).

34. G. Picchi and G. Prati, " A Blind SAG-SO-DFD-FS Equalizer ", *Proc. of the IEEE Clobecom*, pp. 30.4.1-30.4.5, (1986).

35. D. Hatzinakos, " Blind Equalization Using Stop-and-Go Adaptation Rules ", *Optical Enginnering*, Vol. 31(6), pp. 1181-1188, (1992).

36. I. Cha and S. A. Kassam, " Blind Equalization Using Radial Basis Function Networks ", *Proc. of the Canadian Conf. on Electrical and Computer Eng.*, pp. TM3,13,1-13.4, Toronto, Ont. (1992).

37. D. N. Godard and P. E. Tririon, " Method and Device for Training an Adaptive Equalizer by Means of an Unknown Data Signal in a QAM Transmission System ", United States Patent No. 4,227,152, (1980).

38. R. Gooch, M. Ready, and J. Svoboda, " A Lattice-Based Constant Modulus Adaptive Filter ", *Proc. of the 20-th Asilomar Conference,* pp. 282-286, (1987).

39. S. D. Gray, " Multipath Reduction Using Constant Modulus Conjugate Gradient Techniques ", *IEEE J. on Sel. Ar. in Comm.* , Vol. 10(8), pp. 1300-1305, (1992).

40. O. W. Kwon, C. K. Un, and J. C. Lee, " Performance of Constant Modulus Adaptive Digital Filters for Interference Cancellation ", *Signal Processing,* (Elsevier, Pbl.), Vol. 26, pp. 185-196, (1992).

41. O. Macchi and A. Hachicha, " Self-Adaptive Equalization based on a Prediction Principle ", *Proc. of the IEEE Globecom,* pp. 46.2.1-46.2.5, (1986).

42. M. J. Ready and R. P. Gooch, " Blind Equalization Based on Radius Directed Adaptation ", *Proc. of the IEEE ICASSP,* pp. 1699-1702, (1990).

43. F. J. Ross and D. P. Taylor, " An Enhancement to Blind Equalization Algorithms ", *IEEE Trans. on Communications,* Vol. 39(5), pp. 636-639, (1991).

44. W. Sethares, G. A. Rey, and C. R. Johnson, Jr., " Approaches to Blind Equalization of Signals with Multiple Modulus ", *Proc. of the IEEE ICASSP,* pp. 972-975, (1989).

45. S. J. Nowlan and G. E. Hinton, " A Soft Decision-Directed LMS Algorithm for Blind Equalization ", *IEEE Trans. on Communications,* Vol. 41(2), pp. 275-279, (1993).

46. J. Karaoguz, " An unsupervised Gaussian Cluster Formation Technique as a Bussgang Blind Deconvolution Algorithm ", Proc. of the IEEE ISCAS, pp. 719-722, Chicago, IL, (1993).

47. J. J. Shynk, R. P. Gooch, G Krishnamurthy, and C. C. Chan, "A Comparative Study of Several Blind Equalization Algorithms ", Proceedings of SPIE, Vol. 1565, pp. 102-117, (1991).

48. O. Macchi, and E. Eweda, " Convergence Analysis of Self Adaptive Equalizers ", *IEEE Trans. Inform. Theory,* Vol. IT-30, pp. 161-176, (1984).

49. J. E. Mazo, " Analysis of Decision-Directed Equalizer Convergence ", *Bell Syst. Tech. J.,* Vol. 59, pp. 1857-1876, (1980).

50. V. Weerackody, S. A. Kassam, and K. R. Laker, " Convergence Analysis of an Algorithm for Blind Equalization ", *IEEE Trans. on Communications*, COM-39(6), pp. 856-865, (1991). .

51. R. A. Kennedy, G. Pulford, B. D. O. Anderson, and R. R. Bitmead, " When Has a Decision-Directed Equalizer Converged ", *IEEE Trans. on Communications*, COM-37(8), pp. 879-884, (1989).

52. R. Kumar, " Convergence of a Decision-Directed Adaptive Equalizer ", *Proc. Conf. Decision and Control*, pp. 1319-1324, (1983).

53. J. J. Bussgang, " Crosscorrelation Functions of Amplitude-Distorted Gaussian Signals", *M.I.T. Technical Report*, No. 216, (1952).

54. C. K. Chan, and J. J. Shynk, " Stationary Points of the Constant Modulus Algorithm for Real Gaussian Signals ", *IEEE Trans. on Acoustics, Speech, and Signal processing*, Vol. 38(12), pp. 2176-2181, (1990).

55. Z. Ding, R. A. Kennedy, B. D. O Anderson, and C. R. Johnson, " Ill-convergence of Godard Blind Equalizers in Data Communication Systems ", *IEEE Trans. on Communications*, Vol. 39(9), pp. 1313-1327, (1991).

56. Z. Ding, R. A. Kennedy, B. D. O Anderson, and C. R. Johnson, " Local Convergence of the Sato Blind equalizer and Generalizations Under Practical Constraints ", *IEEE Trans. on Information Theory*, Vol. 39(1), pp. 129-144, (1993).

57. Z. Ding, and C. R. Johnson, " On the Nonvanishing Stability of Undesirable Equilibria for FIR Godard Blind Equalizers ", *IEEE Trans. on Signal Processing*, Vol. 41(5), pp. 1940-1944, (1993).

58. S. Verdu, " On the Selection of Memoryless Adaptive Laws for Blind Equalization in Binary Communications ",*Proc. 6-th Intl. Conf. on Analysis and Optimization of Systems*, pp. 239-249, Nice, France, (1984)

59. C. R. Johnson, Jr., S. Dasgupta, and W. A. Sethares, " Averaging Analysis of Local Stability of a Real constant Modulud Algorithm Adaptive Filter ", *IEEE Trans. on Acoustics, Speech,and Signal Processing*, Vol. 36(6), pp. 900-910, (1988).

60. R. A. Kennedy and Z. Ding, " Blind Adaptive Equalizers for Quadrature Amplitude Modulated Communication Systems Based on Convex cost Functions ", *Optical Engineering*, Vol. 31(6), pp. 1189-1199, (1992).

61. J. Ilow, D. Hatzinakos, and A. N. Venetsanopoulos, " Simulated Annealing Optimization in Blind Equalization ", *Proc. of the IEEE MILCOM*, pp. 216-220, San Diego, (1992).

62. J. Ilow, D. Hatzinakos, and A. N. Venetsanopoulos, " Blind Equalizers with Simulated Annealing Optimization for Digital Communication Systems ", *Int. Jour. on Adaptive Control and Signal Processing*, Vol. 8, (1994).

63. Y. Chen, C. L. Nikias, and J. G. Proakis, " Blind Equalization with Criterion with Memory Nonlinearity ", *Optical Engineering*, Vol. 31(6), pp. 1200-1210, (1992).

64. C. L. Nikias and M. R. Raghuveer, " Bispectrum Estimation: A Digital Signal Processing Framework ", *Proceedings of IEEE*, Vol. 75(7), pp. 869-891, (1987).

65. J. M. Mendel, " Tutorial on Higher-Order Statistics (Spectra) in Signal Processing and System Theory: Theoretical Results and Some Applications ", *Proc. IEEE*, Vol. 79, pp. 278-305, (1991).

66. D. Hatzinakos and C. L. Nikias, "Blind Equalization Based on Higher-Order Statistics" in *Blind Deconvolution*, ed. S. Haykin, Prentice Hall, Englewood Cliffs, NJ, pp. 181-258, (1994).

67. H. H. Chiang and C. L. Nikias, " Adaptive Deconvolution and Identification of Nonminimum phase FIR Systems Based on Cumulants ", *IEEE Trans. on Automatic Control*, Vol. 35(1), pp.36-47, (1990).

68. C. L. Nikias and H. H. Chiang, " Higher-Order Spectrum Estimation via Noncausal Autoregressive Modeling and Deconvolution ", *IEEE Trans. Acoustics, Speech, and Signal Processing*, Vol. 36(12), pp. 1911-1914, (1988).

69. G. B. Giannakis and J. M. Mendel, "Identification of Nonminimum Phase Systems Using Higher-Order Statistics", *IEEE Trans. Acoustics, Speech, and Signal Processing*, Vol. 37(3), pp. 360-377, (1989).

70. F. C. Zheng, S. McLaughlin, and B. Mulgrew, " Blind Equalisation of Multilevel PAM Data for Nonminimum Phase Channels via Second-

and Fourth-Order Cumulants ", *Signal Processing*, Elsevier Pbl., Vol. 31, pp. 313-327, (1993).

71. J. K. Tugnait " A Globally Convergent Adaptive Blind Equalizer Based On Second-and Fourth-Order Statistics ", *Proc. of the IEEE ICC*, pp. 1508-1512, (1992).

72. J. K. Tugnait " Adaptive Filters and Blind Equalizers for Mixed Phase Channels ", *Proc. of the SPIE*, Vol. 1565, pp. 209-220, (1991).

73. K. S. Lii and M. Rosenblatt, " Deconvolution and Estimation of Transfer Function Phase and Coefficients for Non-Gaussian Linear Processes ", *Ann. Statist*, Vol. 10, pp. 1195-1208, (1982).

74. B. Porat and B. Friedlander, " Blind Equalization of Digital Communication Channels Using High-Order Moments ", *IEEE Trans. Acoustics, Speech, and Signal Processing*, Vol. 39, pp. 522-526, Feb. (1991).

75. F. C. Zheng, S. McLaughlin, and B. Mulgrew, " Blind Equalization of Nonminimum Phase Channels: Higher Order Cumulant Based Algorithm ", *IEEE Trans. on Signal Processing*, Vol. 41(2), pp. 681-691, (1993).

76. J. K. Tugnait, " Blind Estimation of Digital Communication Channel Impulse Response", *IEEE Trans. Communications*, Vol. 42(2/3/4), pp. 1606-1616, (1994).

77. J. Ilow, D. Hatzinakos, and A. N. Venetsanopoulos, " Blind Deconvolution Based on Cumulant Fitting and Simulated Annealing Optimization ", *Proc. of the 26-th Asilomar Conf. on Sign., Syst., and Comp.*, Pacific Grove, CA, (1992).

78. R. Pan and C. L. Nikias, " The Complex Cepstrum of Higher-Order Cumulants and Nonminimum Phase System Identification ", *IEEE Trans. Acoustics, Speech, and Signal Processing*, Vol. 36, pp. 186-205, (1988).

79. A. G. Bessios and C. L. Nikias, " Blind Equalization Based on Cepstra of the Power Cepstrum and Tricoherence ", *Proc. of the SPIE*, Vol. 1565, pp. 166-177, (1991).

80. D. H. Brooks and C. L. Nikias, " Multichannel Adaptive Blind Deconvolution Using the Complex Cepstrum of Higher-Order Cross-Spectra ", *IEEE Trans. on Signal Processing* , Vol. 41(9), pp. 2928-2934, (1993).

81. D. Hatzinakos, " Blind Equalization Using Decision Feedback Prediction and Tricepstrum Principles ", *Signal Processing*, Elsevier Pbl., Vol. 36, pp. 261-276, (1994).

82. D. Hatzinakos, " Blind Equalization Based on Prediction and Polycepstra Principles ", *IEEE Trans. on Communications*, in print, (1994).

83. D. Donoho, "On Minimum Entropy Deconvolution", in *Applied Time Series Analysis II*, Academic Press, pp. 565-608, (1981).

84. O. Salvi and E. Weinstein, " New Criteria For Blind Deconvolution of Nonminimum Phase Systems (Channels) ", *IEEE Trans. Inform. Theory*, Vol. 36, pp. 312-321, March (1990).

85. J. K. Tugnait, " Comments on 'New Criteria For Blind Deconvolution of Nonminimum Phase Systems (Channels)"', *IEEE Trans. on Information Theory*, Vol. 38(1), pp. 210-213, (1992).

86. M. Boumahdi and J. L. Lacoume, " Blind Identification of Non-Minimum Phase FIR Systems Using the Kurtosis", *7-th SP Workshop on Statistical Signal & Array processing*, pp. 191-194, Quebec City, (1994).

87. J. G. Proakis, "Adaptive Algorithms for Blind Channel Estimation", *Proc. of the Canadian Conf. on Electrical and Computer Eng.*, pp. TA2.18.1-TA2.18.5, Toronto, (1992).

88. G. Kawas and R. Vallet, " Joint Detection and Estimation for Transmission over Unknown Channels ", *Proc. of the Douzienne Colloque GRETSI*, Juan-les-Pins, France, (1989).

89. N. Sheshadri, " Joint Data and Channel Estimation Using Blind Trellis Search Techniques ", in *Blind Deconvolution*, ed. S. Haykin, Prentice Hall, Englewood Cliffs, NJ, pp. 259-286 (1994).

90. N. Sheshadri, " Joint Data and Channel Estimation Using Blind Trellis Search Techniques ", *IEEE Trans. on Communications*, Vol. 42(2/3/4), pp. 1000-1011, (1994).

91. M. Ghosh and C. L. Weber, " Maximum Likelihood Blind Equalization ", *Optical Engineering*, Vol. 31(6), pp. 1224-1228, (1992).

92. E. Zervas, J. Proakis and V. Eyuboglu, " A Quantized Channel Approach to Blind Equalization ", *Proc. of the IEEE ICC*, pp. 1539-1543, Chicagi, ILL, (1992).

93. E. Zervas and J. Proakis, " A Sequential Algorithm for Blind Equalization ", *Proc. of the IEEE MILCOM*, pp. 231-235, San Diego, CA, (1992).

94. E. Zervas, J. Proakis and V. Eyuboglu, " Effects of Constellation Shaping on Blind Equalization ", *Proc. of the SPIE*, Vol. 1565, pp. 178-187, (1991).

95. M. Feder and J. A. Catipovic, " Algorithms for Joint Channel Estimation and Data Recovery - Application to Equalization in Underwater Communications ", *L. of Optical Engineering*, Jan. (1991).

96. A. Neri, G. Scarano, and G. Jacovitti, " Bayesian Iterative Method for Blind Deconvolution ", *Proc. of the SPIE*, Vol. 1565, pp. 196-208, (1991).

97. K. Giridhar, J. J. Shynk, and R. A. Iltis, " Bayesian/Decision-Feedback Algorithm for Blind Adaptive Equalization ", *Optical Engineering*, Vol. 31(6), pp. 1211-1223, (1992).

98. R. A. Iltis, J. J. Shynk, and K. Giridhar, " Bayesian Algorithms for Blind Equalization Using Parallel Adaptive Filtering", *IEEE Trans. on Communications*, Vol. 42(2/3/4), pp. 1017-1032, (1994).

99. W. A. Gardner, *Statistical Spectral Analysis; A Non-Probabilistic Theory*, Prentice Hall, Englewood Cliffs, NJ, (1988).

100. W. A. Gardner, " A New Method of Channel Identification ", *IEEE Trans. Communications*, Vol. 39, pp. 813-817, (1991).

101. L. Tong, G. Xu and T. Kailath, " A new Approach to Blind Identification and Equalization of Multipath Channels ", *Proc. of the 25-th Asilomar Conference*, pp. 856-860, (1991).

102. L. Tong, G. Xu and T. Kailath, " Blind Identification and Equalization of Multipath Channels ", *Proc. the IEEE ICC*, Chicago, IL, (1992).

103. Y. Chen and C. L. Nikias, " Blind Identifiability of A Band-Limited Nonminimum Phase System from its Output Autocorrelation ", *Proc of IEEE ICASSP*, Vol. IV, pp. 444-447, Minneapolis, MN, (1993).

104. Z. Ding and Y. Li "On Channel Identification Based on Second Order Cyclic Spectra", *IEEE Trans. on Signal Processing*, Vol. 42(5), pp. 1260-1264, (1994).

105. Y. Li and Z. Ding, " Blind Channel Identification based on Second-Order Cyclostationary Statistics ", *Proc. of the IEEE ICASSP*, pp. IV-81-84, Minneapolis, MN, (1993).

106. D. Hatzinakos, "Nonminimum Phase Channel Deconvolution Using the Complex Cepstrum of the Cyclic Autocorrelation" *IEEE Trans. on Signal Processing*, to appear in November (1994).

107. D. Hatzinakos, " Blind System Identification Based on the Complex Cepstrum of the Cyclic Autocorrelation ", *Proc. of the IEEE ISCAS*, pp. 726-729, Chicago, IL, (1993).

108. J. Ilow and D. Hatzinakos, " Recusrive Least Squares Algorithm for Blind Deconvolution of Channels with Cyclostationary Inputs ", *Proc. of the IEEE MILCOM*, Vol 1, pp. 123-127, Boston, MA, (1993).

109. M. K. Tsatsanis and G. B. Giannakis, " Blind Equalization of Rapidly Fading Channels Via Exploitation of Cyclostationarity and Higher-Order Statistics ", *Proc. of the IEEE ICASSP*, pp. IV-85-88, Minessota, MN, (1993).

110. A. G. Bessios and C. L. Nikias, " Multichannel Adaptive Blind Equalization with CRIMNO-MSE Technique ", *Proc. of the IEEE MILCOM*, pp. 236-240, San Diego, CA, (1992).

111. T. Li, " Blind Identification and Deconvolution of Linear Systems Driven by Binary Random Sequences ", *IEEE Trans. on Information Theory*, Vol. 38(1), pp. 26-38, (1992).

112. O. Salvi and E. Weinstein, " Super-Exponential Methods for Blind Deconvolution ", *IEEE Trans. on Information Theory*, Vol. 39(2), pp. 504-519, (1993).

113. S. Verdu, B. D. O. Anderson and R. A. Kennedy, " Blind Equaliztion Without Gain Identification ", *IEEE Trans. on Information Theory*, Vol. 39(1), pp. 292-297, (1993).

114. S. Bellini and F. Rocca, " Asymptotically Efficient Blind Deconvolution ", *Signal Processing*, (Elsevier Pbl.), Vol. 20, pp. 193-209, (1990).

Time-Varying System Identification and Channel Equalization Using Wavelets and Higher-Order Statistics

Michail K. Tsatsanis

University of Virginia
Charlottesville, VA 22903-2442

I. INTRODUCTION

Parametric modeling of signals and systems provides a compact description of the underlying process and facilitates further processing of the data
(e.g., in deconvolution or filtering problems). Most of the work in parametric system identification however, relies upon the *stationarity* assumption
for the observed signal, or equivalently, on the *time-invariance* (TI) of the
underlying system. This assumption, although mathematically convenient,
is not always valid for various signals encountered in several applications.

Time-varying (TV) systems arise naturally in a variety of situations
including speech analysis [1] (due to the constantly changing vocal tract),
seismic processing [2] (due to earth's space-varying absorption) and array

Copyright © 1995 by Academic Press, Inc.
All rights of reproduction in any form reserved.

processing (due to moving sources). Other examples include time-delay estimation, echo cancellation, radar and sonar problems and many more applications of system identification. The growing interest in time-frequency representations and TV spectral analysis (e.g., [3]) indicates the importance of nonstationary signal analysis.

A major application of system identification and deconvolution appears in digital transmission through channels with multipath effects or bandwidth constraints. Intersymbol Interference (ISI) is present in this case, due to delayed copies of the transmitted signal arriving through the multiple paths, or due to the transmitter and receiver filters [4]. ISI is a major impeding factor in high-speed digital transmission and its effects can be significantly more severe compared with those of additive noise. Thus, the use of some *channel equalization* procedure is essential for the recovery and detection of the transmitted symbols.

It is common practice in communication applications to assume that the intersymbol interference does not change throughout the transmission period, i.e., the channel is *time-invariant* (TI). In many cases however, ISI is induced by multipath effects from a changing environment, thus a *time-varying* channel has to be considered. Examples of TV channels (called *frequency-selective fading links*) include over the horizon communications [4] (due to random changes in the ionosphere), the underwater acoustic channel [5] (due to local changes in the temperature and salinity of the ocean layers) and microwave links [6]. An equally important application appears in radio transmission to a mobile receiver, as for example in cellular telephony. In this case, the multipath effect from reflections at nearby buildings is constantly changing as the vehicle moves. In order to equalize these fading links, identification and deconvolution of TV systems and channels should be considered. This is the general topic of this work.

The most popular approach for TV channel estimation and equalization has been to employ an adaptive algorithm, in order to track the chanel's changing parameters [4, Ch. 6,7], [7]. Typically, a training sequence (known to the receiver) is transmitted at the beginning of the session so that the equalizer can adapt its parameters. After the training period, the equalizer usually switches to a decision-directed mode. In this mode, the previously detected symbols are assumed to be correct and are fed back to the adaptive algorithm, to update the parameter estimates. In this way, the algorithm can follow the time variations, provided they are sufficiently slow in comparison to the algorithm's convergence time.

Despite their popularity and simplicity, adaptive algorithms are derived under the *stationarity* assumption and do not take *explicitly* into account the TV nature of the channel. Thus, they can only be used for slowly changing channels and systems. Moreover, in the decision feedback (DF) mode they suffer runaway effects and divergence problems, whenever a deep fading or rapid change occurs. For this reason they require periodic retraining.

In order to overcome these problems, further modeling of the channel's variations needs to be incorporated into the equalization procedure. A second, probabilistic approach would be to regard each TV system coefficient as a stochastic process. In this framework, the TV identification problem is equivalent to estimating these "hidden" processes. If the statistics of these processes are a priori *known*, Kalman filtering techniques can be employed to estimate the TV coefficients from input/output data [5]. It is not clear however, how to estimate those statistics since the TV coefficients are not directly observed. Moreover, this as well as simpler random walk models rely on the *random coefficient* assumption, which is reasonable only when there are many, randomly moving reflectors in a multipath channel (e.g., ionospheric channel). It will not be valid for different setups, e.g., channels with occasional jumps or periodic variations.

A third approach, on which we will focus, is based on the expansion of each TV coefficient onto a set of basis sequences. If a combination of a small number of basis sequences can well approximate each coefficient's time-variation, then the identification task is equivalent to the estimation of the parameters in this expansion, which do not depend on time. This approach transforms the problem into a time-invariant one and has been used for the estimation of TV-AR models in the context of speech analysis [1], [8]. However, the performance of these methods depends crucially on the wise choice of a basis set, which can capture the dynamics of the channel's variations in a parsimonious way. Several polynomial [9], [10], and prolate spheroidal sequences [1] have been proposed in the past, although accompanied by no quantitative justification.

Here, we defer the discussion on the choice of the basis sequences for Section V, where the wavelet basis is advocated for the general case. We motivate the basis expansion approach however in Section II, where we show that the mobile radio, multipath channel can be described by a periodically varying model. Each TV coefficient is given as a combination of some complex exponentials. Thus, the use of an exponential basis in this framework, proves the usefulness and applicability of the basis expansion approach.

Basis expansion ideas provide a valuable tool for extending RLS and LMS type adaptive algorithms to the rapidly varying systems case. Moreover, they offer a framework into which the more challenging problem of *blind* or output only identification of the TV channel can be addressed.

Blind or *self recovering* equalization procedures use output only information and therefore do not require a training period. Thus, they are useful in applications where no training sequence is available [11], [12], [13], [14], [15]. Examples include broadcasting to many receivers (e.g., HDTV broadcasting), where the transmitter cannot be interrupted to initiate new training sessions, and multipoint data networks, where the cost of training each individual terminal point is prohibitive in terms of network management [16]. Blind methods (in the TI case) typically involve the minimization of criteria based on the signal's statistics in the place of the mean square error [11], [15]. Thus, they do not lend themselves easily to TV extensions, since the statistics in this case vary with time and cannot be easily estimated.

In Section IV basis expansion ideas are employed to address the blind equalization problem for rapidly fading channels. Second- and fourth-order nonstationary moments and cumulants are used to recover the TV channel coefficients. Identifiability of the channel from these output statistics is shown and novel linear and nonlinear algorithms are proposed based on instantaneous approximations of the TV moments. The performance of these methods is studied and strong convergence of the proposed algorithm is shown.

In an effort to keep the presentation as general as possible, we do not refer to any specific basis throughout these derivations. However, the choice of an appropriate basis set is crucial for the success of this approach. While for certain cases, the choice of the basis sequences is clearly dictated by the channel dynamics (e.g., mobile radio channel), for the general case it is not a trivial problem [8].

Motivated by the success of *multiresolution* methods in signal and image compression, [17], [18], in Section V we study the applicability of the wavelet basis for the parsimonious description of the TV system coefficients. Wavelet expansions offer a time-scale analysis of the signal and provide information about global as well as local behavior at different resolution depths. The promise of multiresolution expansions of the TV coefficients is that most of their energy will be concentrated into the low-resolution approximation, and hence the detail signals can be discarded without affecting the quality of the approximation. In this way a parsimonious approximation to the channel's variations is obtained.

While this approach can provide an acceptable overall approximation to the system's trajectory, it will not be able to track rapid changes or transient fadings which usually manifest themselves in the detail signal. Thus, some important parts of the detail signal have to be kept as well, similarly to image coding procedures. We should be able to locally "zoom into" the details when necessary (e.g., in an abrupt change or transition) or in other words, select the appropriate resolution depth locally, depending on the variability of the system's coefficients.

In Section V we formulate this problem as a model selection problem and use information theoretic criteria [19] or hypothesis testing procedures [20] to automatically select the appropriate resolution depth. The proposed algorithm incorporates maximum likelihood, or simpler blind methods, and provides a general framework for the estimation of TV systems, where no specific a priori knowledge on the nature of the time variations is assumed.

II. BASIS EXPANSIONS AND TV CHANNELS

Let us consider a general TV-ARMAX system described by the model

$$y(n) = \sum_{k=1}^{p} a(n;k)y(n-k) + \sum_{k=0}^{q} b(n;k)w(n-k) + \sum_{k=0}^{r} c(n;k)v(n-k) \ , \ (1)$$

where $y(n)$ is the system's output, $w(n)$ is the input and the third term of the RHS represents observation noise. Notice the explicit dependence of the parameters $a(n;k)$, $b(n;k)$, $c(n;k)$, on time, since the system is TV. In a communications framework, $y(n)$ represents the received signal, $w(n)$ the transmitted symbols and $v(n)$ the channel noise. While eq. (1) is a rather general parametric description of the fading channel, simpler models are sufficient for many applications. For example, in most cases the additive noise is assumed to be white and Gaussian, hence the third term in the

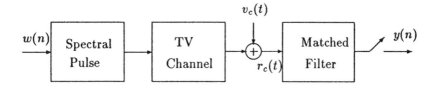

Figure 1: Actual Setup

RHS of eq. (1) reduces to $v(n)$. Moreover, in multipath channels with no reverberation, the first term may also be absent, resulting in an FIR model

$$y(n) = \sum_{k=0}^{q} b(n;k)w(n-k) + v(n) \ . \tag{2}$$

In this work we focus on the equalization problem and some results are specialized for the model of eq. (2). However, we frequently discuss extensions to the general case of eq. (1).

Equation (2) is indeed a general discrete-time equivalent model for a fading channel as discussed next. Consider the setup of Fig. 1 for digital transmission through a fading link and let $h_c(t; \tau)$ be the overall impulse response of the the TV channel in cascade with the transmitter's spectral shaping filter [1]. The received signal is then

$$r_c(t) = \sum_{k=0}^{\infty} w(k)h_c(t; t - kT) + v_c(t) \ , \tag{3}$$

where $w(k)$ is the discrete-time symbol stream, T is the symbol period, and $v_c(t)$ is additive Gaussian noise. Throughout this chapter, we will consider the information symbols $w(n)$ to be i.i.d., equiprobable, drawn from a QAM constellation with variance σ_w^2; e.g., for the 4-QAM case, $w(n)$ can take the values $\sqrt{\sigma_w^2/2} \ (\alpha + j\beta)$, where $\alpha, \beta = \pm 1$; In general, $w(n)$ can take a

---

[1] We use the subscript c to denote continuous-time signals.

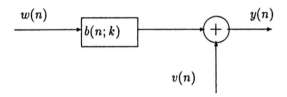

Figure 2: Equivalent Discrete-Time Model

finite number of distinct values $w^{(m)}$, $m = 1, \ldots, M$, on the complex plain, depending on the chosen constellation.

After the receiver filter and sampler, we obtain the discrete samples

$$y_c(nT + \tau_0) = \sum_{k=0}^{\infty} w(k)b_c(nT + \tau_0; nT - kT + \tau_0) + v_c(t) , \qquad (4)$$

where τ_0 is the transmission delay and $b_c(t; \tau)$ is the convolution of $h_c(t; \tau)$ with the impulse response of the receiver filter. The sampler has period T, since we deal only with symbol spaced equalizers here.

The impulse response $b_c(t; \tau)$ is infinite in general. However, it is common practice in communication literature, to truncate it at some order q (FIR approximation) and thus arrive at the equivalent discrete-time model of eq. (2), (see also Fig. 2).

We should notice here, that both TV models of eq. (1) and (2) have too many degrees of freedom to be practically useful. In other words, the estimation of the TV parameters in eq. (1) or (2) without any further constraint in their variation, is an ill posed problem. To understand why, consider the simpler model of eq. (2) and assume that both input and output data are given ($w(n)$ and $y(n)$ respectively). Then, for every new data point $\{y(n), w(n)\}$ received, one more equation is obtained with $q + 1$ unknowns $\{b(n; k)\}_{k=0}^{q}$. Therefore, the system of equations generated by (2) is underdetermined and does not lead us to a meaningful solution.

Thus, further modeling of the system's time-history is required in order to obtain a more constrained and parsimonious description, in order to facilitate the identification task. In this work, we propose to expand each TV system coefficient onto some basis sequences $f_l(n)$, $l = 1, \ldots, L$

$$b(n; k) = \sum_{l=1}^{L} \theta_{kl} \, f_l(n) \ . \tag{5}$$

In this way, the system is only parametrized by the expansion coefficients θ_{kl}, $k = 0, \ldots, q$, $l = 1, \ldots, L$ and a drastic reduction in the number of unknown parameters is achieved. However, one might question the quality of approximation to the actual coefficients provided by (5) and hence, the applicability of this approach. Indeed, the wise choice of the sequences $f_l(n)$ is crucial for the success of this method. Of equal importance is the choice of the appropriate expansion order L [21]. Although here, we defer the discussion on the basis selection for Section V, we would like to motivate our approach by studying the mobile radio channel in more detail. It can be shown that this channel is naturally modeled by equations (2), (5), where the basis sequences are complex exponentials. This serves as a motivation for the usefulness and applicability of the basis expansion procedure.

A. THE MOBILE RADIO CHANNEL

Let us consider a radio channel with multipath, where the receiver is on a constantly moving platform while the transmitter and the reflectors are fixed (see Fig. 3). The transmitted signal is

$$s_c(t) = \mathrm{Re}[\sum_{k} w(k)g_c(t - kT)e^{j2\pi f_c t}] \ , \tag{6}$$

where f_c is the carrier frequency and $g_c(t)$ is the spectral shaping pulse.

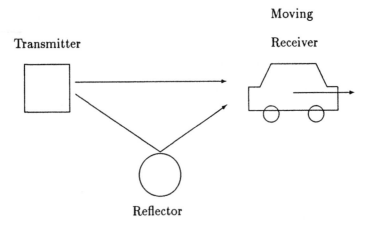

Figure 3: Multipath in Mobile Telephony

Due to multipath, several delayed copies of $s_c(t)$ arrive at the receiver

$$r_c(t) = \sum_{k=1}^{L} A_l(t)s_c(t - \tau_l(t)) + v_c(t) , \qquad (7)$$

where $\tau_l(t)$ and $A_l(t)$ are each path's delay and gain, respectively (notice that they change with time, since the receiver is moving). Substituting (6) into (7) we obtain

$$r_c(t) = \text{Re}[y_c(t)e^{j 2\pi f_c t}] + v_c(t) , \qquad (8)$$

$$y_c(t) = \sum_{l=1}^{L} A_l(t)e^{j 2\pi f_c \tau_l(t)} \sum_{k} w(k)g_c(t - kT - \tau_l(t)) . \qquad (9)$$

where $y_c(t)$ represents the complex envelope. At the receiver, the signal passes through a filter matched to the transmitter's spectral pulse and is then sampled every T seconds. The discrete time output after the matched filter and the sampler is given by

$$y(n) = \int y_c(t)g_c(t - nT)dt . \qquad (10)$$

Substituting (9) into (10) and assuming that $A_k(t) \approx A_{ln}$, $\tau_l(t) \approx \tau_{ln}$ are approximately constant for a symbol period (a piecewise constant approximation), we obtain

$$y(n) = \sum_{l=1}^{L} A_{ln} e^{j 2\pi f_c \tau_{ln}} \sum_{k} r_g([n-k]T - \tau_{ln}) w(n) + v(n) \; , \qquad (11)$$

where $r_g(t)$ is the (deterministic) correlation of the spectral shaping pulse. If we further approximate the path delay variation with a linear function of time, i.e., $\tau_{ln} = \lambda_l n + \lambda_l^0$ (first degree approximation), and truncate $r_g(t)$, for $|t| > MT$, and some $M > 0$, we obtain

$$y(n) = \sum_{k=n-q_M}^{n+q_M} \sum_{l=1}^{L} [r_g(kT - \tau_{ln}) A_{kn}] e^{j 2\pi f_c (\lambda_l n + \lambda_l^0)} w(n-k) + v(n) \; , \quad (12)$$

for an order q_M. Furthermore, if we observe that the term in square brackets is approximately constant with n (compared with the rate of change of the exponential), we arrive at the following equivalent *discrete-time* model:

$$y(n) = \sum_{n=0}^{q} b(n; k) w(n-k) + v(n) \qquad (13)$$

$$b(n; k) = \sum_{l=1}^{L} \theta_{kl} \, e^{j \alpha_l n} \; , \qquad (14)$$

for some complex constants θ_{nk}, some order q and frequencies $\alpha_l = 2\pi f_c \lambda_l$. The model of equations (13), (14) is identical to that of (2), (5), with the exponentials used as basis, i.e., $f_l(n) = e^{j \alpha_l n}$.

Thus, the mobile radio channel, due to the high frequency modulation inherent in the transmitted signal, can be considered as a discrete-time, linear, periodically varying channel. Each TV coefficient is given as a linear combination of exponentials as shown in (14). This derivation reveals an interesting application which fits rather nicely into the basis expansion approach used in this work. This analysis however, is valid only when the multipath effect is caused by a small number ($q = 1, 2$, or 3) of distinct reflectors (as in cellular telephony). It does not cover channels with many randomly moving reflectors (random channel), or channels with abrupt jumps, or even smooth but non-periodic variations. A more general multiresolution approach is discussed in Section V, to deal with these cases. In the meanwhile, we consider $f_l(n)$ to be given and focus on the estimation of the expansion parameters θ_{kl}.

Notice that the decomposition of eq. (5) transforms the problem into a time-invariant one. Indeed, the coefficient's time-variations are captured by the sequences $f_l(n)$ while the parameters θ_{kl} do not depend on time. Adaptive and blind methods for the estimation of θ_{kl} are proposed in the sequel, by exploiting the general framework of equations (2), (5). For the case of the mobile radio channel, the frequencies α_l have to be estimated as well, in order to completely characterize the channel. We will not deal with this issue here; the interested reader is referred to [22], [23], where cyclostationarity tests are used [24], to address this problem.

III. ADAPTIVE ESTIMATION / EQUALIZATION

Let us consider the following ARX system

$$y(n) = \sum_{k=1}^{p} a(n;k)y(n-k) + \sum_{k=0}^{q} b(n;k)w(n-k) + v(n) \ , \qquad (15)$$

where the TV coefficients $a(n;k)$, $b(n;k)$ are given by the expansion

$$a(n;k) = \sum_{l=1}^{L} \theta_{kl}^{(a)} f_l(n) \ , \quad b(n;k) = \sum_{l=1}^{L} \theta_{kl}^{(b)} f_l(n) \ . \qquad (16)$$

Then by substituting (16) into (15), we obtain the following input/output relationship,

$$y(n) \quad = \quad \sum_{k=0}^{q} \sum_{l=1}^{L} \theta_{kl}^{(a)} [f_l(n)y(n-k)]$$

$$+ \sum_{k=0}^{q} \sum_{l=1}^{L} \theta_{kl}^{(b)} [f_l(n)w(n-k)] + v(n) \quad , \quad n = 1, \dots, N \ . \quad (17)$$

Typically, at the beginning of each session a known sequence is transmitted to serve as a training pattern for the receiver's equalizer. Thus, during the training mode, $w(n)$ and $y(n)$ are given and eq. (17) represents a linear model w.r.t. the unknown parameters $\theta_{kl}^{(a)}$, $\theta_{kl}^{(b)}$. It can be written in vector form as

$$y(n) = \mathbf{f}_n^T \underline{\theta} + v(n) \quad , \quad n = 1, \dots, N \quad (18)$$

where $\underline{\theta} = [\theta_{11}^{(a)} \dots \theta_{pL}^{(a)} \ \theta_{11}^{(b)} \dots \theta_{qL}^{(b)}]$ is the parameter vector and $\mathbf{f}_n^T \triangleq [f_1(n)y(n-1) \dots f_L(n)y(n-p) \ f_1(n)w(n) \dots f_L(n)w(n-q)]$ is the data vector (^{T} stands for transpose). The formulation of eq. (18) transforms the TV channel estimation task into a classical linear regression problem. Then, the optimum estimator for $\underline{\theta}$ (in the mean-square sense) is provided by minimizing the modeling mean-square error,

$$\hat{\underline{\theta}} = \ \arg\min_{\underline{\theta}} \ E\{|y(n) - \mathbf{f}_n^T \underline{\theta}|^2\} \ . \quad (19)$$

Following standard procedures (e.g., [25]), the minimizer of eq. (19) can be estimated adaptively by combining gradient descent methods and stochastic approximations. The resulting "LMS type" adaptive algorithm is given by the simple iteration

$$\hat{\underline{\theta}}(n + 1) = \hat{\underline{\theta}}(n) + \mu e(n) \mathbf{f}_n^* \ , \quad (20)$$

where the error is

$$e(n) = y(n) - \mathbf{f}_n^T \hat{\underline{\theta}}(n) \ . \quad (21)$$

The algorithm should be initialized with $\hat{\underline{\theta}} = 0$, and the stepsize μ should be a small positive number; μ represents a compromise between fast convergence and increased "excess error" in the steady state.

An adaptive solution with faster convergence and no excess error is possible, based on Recursive Least Squares (RLS) methods (see for example [25]). In this formulation, eq. (18) is written in matrix form as

$$\mathbf{y} = \mathbf{F}\underline{\theta} + \mathbf{v} \quad , \tag{22}$$

where $\mathbf{F} \triangleq [\mathbf{f}_1 \ldots \mathbf{f}_N]^T$, $\mathbf{y} \triangleq [y(1) \ldots y(N)]^T$ $\mathbf{v} \triangleq [v(1) \ldots v(N)]^T$, and $\underline{\theta}$ can be estimated by minimizing a Least squares criterion,

$$\hat{\underline{\theta}} = \arg\min_{\underline{\theta}} [\frac{1}{N}\mathbf{v}^{T*}\mathbf{v}] \quad . \tag{23}$$

Under the white Gaussian assumption for $v(n)$, eq. (23) coincides with the maximum likelihood estimate of $\underline{\theta}$, and the procedure is efficient. The solution is given in closed form by

$$\hat{\underline{\theta}} = (\mathbf{F}^{*T}\mathbf{F})^{\dagger}\mathbf{F}^{*T}\mathbf{y} \tag{24}$$

where $*$ stands for complex conjugation and $\dagger$ denotes the pseudoinverse. As commonly done in equalization problems, Recursive LS (RLS) can be applied to (22) to reduce computations as well as memory requirements [4]. In Table 1, we provide the RLS algorithm for the TV model of eq. (17).

The proposed algorithms adaptively estimate the expansion parameters $\theta_{kl}^{(a)}$, $\theta_{kl}^{(b)}$. Although they are applied to a TV environment, the formulation of eq. (18) is very similar to the general linear TI approach. Thus, questions regarding the performance and asymptotic behavior of these algorithms can be answered following standard techniques. For example, consistency of the LS estimate can be shown under certain conditions as discussed in [26], [27].

In the TI case, adaptive equalizers usually switch to a decision feedback (DF) mode, after the training period is over. Thus, it would be of interest to examine whether a decision directed approach is also applicable to the

Initialization: $\mathbf{P}(0) = \delta\mathbf{I}, \quad \hat{\underline{\theta}}(0) = 0, \quad 0 < \lambda < 1$

Recursion: $\mathbf{K}(n) = \dfrac{\lambda^{-1}\mathbf{P}(n-1)\mathbf{f}_n}{1+\lambda^{-1}\mathbf{f}_n^{*T}\mathbf{P}(n-1)\mathbf{f}_n}$

$\alpha(n) = y(n) - \hat{\underline{\theta}}^T(n-1)\mathbf{f}_n$

$\hat{\underline{\theta}}(n) = \hat{\underline{\theta}}(n-1) + \alpha(n)\mathbf{K}^*(n)$

$\mathbf{P}(n) = \lambda^{-1}\mathbf{P}(n-1) - \lambda^{-1}\mathbf{K}(n)\mathbf{f}_n^{*T}\mathbf{P}(n-1)$

Table 1: RLS adaptive equalizer

TV case. In this way, the equalizer will be able to even track slow variations in the expansion parameters $\theta_{kl}^{(a)}$, $\theta_{kl}^{(b)}$.

During normal transmission, both in the TV and TI case the symbols $w(n)$ are not known to the receiver, to be used in the algorithm's iteration. To overcome this problem, the decoded symbols $\hat{w}(n)$ are used in place of the true ones. If the probability of decoding errors is sufficiently small, the algorithm will still converge to the true parameters.

In the current setup, the goal is to decode $\hat{w}(n)$ and update the estimate $\hat{\underline{\theta}}(n)$, given the previously decoded symbols $\hat{w}(n)$, $\hat{w}(n-1), \ldots, \hat{w}(n-q)$; to this end every possible value for $\hat{w}(n)$, is hypothesized, and the one which minimizes the error between the estimated and received signal $y(n)$ is selected. Then, $\hat{\underline{\theta}}(n)$ is computed through the LMS or RLS algorithm using the data $y(n)$ and $\hat{w}(n)$. The algorithm is summarized in Table 2.

goal:	given $\hat{w}(n-1), \ldots, \hat{w}(n-q), \hat{\underline{\theta}}(n-1)$ estimate $\hat{w}(n), \hat{\underline{\theta}}(n)$.

step 1:	for $m = 1, \ldots, M$ hypothesize: $\hat{w}(n) = w^{(m)}$, and compute: $$\begin{aligned} e^{(m)}(n) &= y(n) - \sum_{l=1}^{L} \hat{\theta}_{0l}(n-1)[f_l(n)w^{(m)}] \\ &- \sum_{k=1}^{q}\sum_{l=1}^{L} \hat{\theta}_{kl}(n-1)[f_l(n)\hat{w}(n-k)] \end{aligned}$$
step 2:	Pick m_0 which minimizes the error $|e^{(m)}(n)|^2$ and set $\hat{w}(n) = w^{(m_0)}$.
step 3:	Update $\hat{\underline{\theta}}(n)$ through the recursion of Table 1 or Equation 21, using $\hat{w}(n), \ldots, \hat{w}(n-q), \hat{\underline{\theta}}(n-1)$.

Table 2: Decision-Feedback Equalizer

This algorithm provides a simple "nearest neighbor" rule to detect the transmitted symbols, on top of estimating the parameter vector $\underline{\theta}$. In this way, the actual equalization of the channel is combined with the estimation procedure. However, this rule is not efficient and possesses no optimality. Superior results should be expected if maximum likelihood detection procedures are employed.

Under the Gaussianity assumption for the additive, white noise $v(n)$, the negative log-likelihood of the data (after dropping unnecessary terms) can be written as

$$L(\underline{\theta}) = \sum_{n=1}^{N} |y(n) - \mathbf{f}_n^T \underline{\theta}|^2 \ . \tag{25}$$

Thus, if some estimate $\hat{\underline{\theta}}$ of the channel's parameters is provided, the ML estimate for the input vector $\mathbf{w} \triangleq [w(1), \ldots, w(N)]^T$ is given by

$$\hat{\mathbf{w}}_{ML} = \arg\min_{\mathbf{w}} \sum_{n=1}^{N} |y(n) - \mathbf{f}_n^T \underline{\theta}|^2 \quad , \tag{26}$$

where the elements of vector $\mathbf{w}$ are allowed to take all possible QAM values. By exploiting the fact that the cost function in (25) is cumulative, dynamic programming methods (such as the Viterbi algorithm) can be applied, and the optimal $\hat{\mathbf{w}}$ can be recursively computed [4].

Similarly to the TI case, the Viterbi decoder can be coupled with an adaptive channel estimator (e.g. Table 1) which should operate on the data with a delay D equal to the Viterbi decoding delay. In this way, both output and (decoded) input data will be available for the algorithm's operation, (see [4]).

Both the Viterbi approach and the algorithm of Table 2, similarly to all decision directed schemes, are nonlinear procedures and will not converge to the true parameters unless a good initialization point is provided. Thus, either initial training or some self-recovering procedure is always needed. For the cases where training is too costly, or simply not available due to the network's architecture, it would be of interest to examine the applicability of blind equalization methods in this fading environment. Successful blind recovery of TV channels could potentially enable the integration of fading links and mobile terminal points in multipoint or broadcast data networks. This topic is discussed next.

IV. BLIND CHANNEL ESTIMATION

While the importance and usefulness of blind deconvolution methods is widely recognized in the TI case, and self-recovering equalizers are becoming increasingly popular, no such procedures have been developed (to the best of our knowledge) for rapidly varying systems. In this work we attempt to fill this gap and propose blind equalization algorithms for fading channels.

In the TI case, blind methods typically exploit relationships and properties of output statistics in order to recover the unknown channel's impulse response or derive the equalizing filter [16], [13]. They solve the problem in the statistics' domain, after estimating them from the given data. Extension of this general philosophy to the fading case faces a number of difficulties, associated with the TV nature of the channel.

The first problem involves the choice of the appropriate statistics which contain sufficient information to recover the system's TV impulse response. In other words, the statistics used should guarantee unique identifiability of the TV channel. The second difficulty arises when trying to estimate these TV statistics. Apparently, since they change with time, they cannot be estimated through conventional time-averaging.

The identifiability problem is studied next. We show that higher-order moments are needed to uniquely characterize the channel, while the estimation question is discussed later. In this part of the work, we concentrate on the FIR channel of equations (2), (5). Extensions to more general ARMA systems presents an interesting future problem.

A. TV MOMENTS AND CUMULANTS

Similarly to the TI case, we will use second- and higher-order moments and cumulants of the received signal. Let us define the $(k + L)$th moment of $y(n)$ as

$$m_{kly}(n; \tau_1, \ldots, \tau_{k+l-1}) \stackrel{\triangle}{=} E\{\underbrace{[y(n) \cdots y(n + \tau_{k-1})]}_{k}\}^*$$

$$\times\underbrace{[y(n+\tau_k)\cdots y(n+\tau_{k+l-1})]\}}_{l} \quad (27)$$

where the subscripts k and l denote the conjugated and unconjugated parts in the expectation. Since some modulation schemes are better described by complex constellations, we develop the proposed methods for complex channels and signals. Notice the explicit dependence of the moments on n, since $y(n)$ is nonstationary. According to (27), we can define two different second-order correlations of $y(n)$, namely

$$m_{02y}(n;\tau) \stackrel{\triangle}{=} E\{y(n)y(n+\tau)\} \ , \quad (28)$$

$$m_{11y}(n;\tau) \stackrel{\triangle}{=} E\{y^*(n)y(n+\tau)\} \ , \quad (29)$$

depending on the number of conjugated factors in the product. In a communications setup, the moment defined in (28) is rather useless, since it is identically zero for most signal constellations. For example, for the popular 4-QAM or 16-QAM constellation, one can easily check that $m_{02y}(n;\tau) \equiv m_{20y}(n;\tau) \equiv 0$ due to symmetry. Thus, we concentrate on the moment defined by (29) in the sequel.

Using the definition (29) as well as the modeling equation (2), it can be shown that the TV correlation is related with the channel's impulse response in the following way,

$$m_{11y}(n;\tau) = \gamma_{11w}\sum_{k=0}^{q} b^*(n;k)b(n+\tau;k+\tau) + \sigma_v^2\delta(\tau) \ , \quad (30)$$

where $\gamma_{11w}\stackrel{\triangle}{=}E\{|w(n)|^2\} = \sigma_w^2$, provided that the input $w(n)$ is i.i.d. and zero-mean. Equation (30) indicates that $c_{11y}(n;\tau)$ contains information about the TV channel. In the next subsection however, it will become clear that this information is not sufficient to recover the TV impulse response. Thus, higher-order statistics of $y(n)$ have to be used.

Most signal constellations are symmetrically distributed around zero, hence third order moments (and in fact all odd-order moments) are identically zero. For this reason we resort to fourth-order moments defined as

$$m_{22y}(n; \tau_1, \tau_2, \tau_3) \triangleq E\{y^*(n)y^*(n + \tau_1)y(n + \tau_2)y(n + \tau_3)\} \ . \qquad (31)$$

Notice that conjugation is not always necessary in the fourth order case. For example,

$$m_{04y}(n; \tau_1, \tau_2, \tau_3) \triangleq E\{y(n)y(n + \tau_1)y(n + \tau_2)y(n + \tau_3)\} \ , \qquad (32)$$

will be non-zero for many cases (e.g., for the 4-QAM constellation we can check that $E\{w^4(n)\} = -\gamma_{11w} \neq 0$). For cases where eq. (32) is zero however, we have to use the moment of eq. (31).

Higher than second-order moments are not related to the impulse response in a straightforward manner as in (30). For this reason, different statistics called cumulants have been used in system identification problems. While the second-order cumulant of a zero-mean process coincides with the autocorrelation,

$$c_{11y}(n; \tau) \equiv m_{11y}(n; \tau) \ , \qquad (33)$$

higher-order cumulants are defined as combinations of moment products. For the fourth-order case, cumulants are defined as

$$
\begin{aligned}
c_{04y}(n; \tau_1, \tau_2, \tau_3) \ \triangleq \ & m_{04y}(n; \tau_1, \tau_2, \tau_3) - m_{02y}(n + \tau_2; \tau_3 - \tau_2)m_{02y}(n; \tau_1) \\
& - \ m_{02y}(n; \tau_2)m_{02y}(n + \tau_1; \tau_3 - \tau_1) \\
& - \ m_{02y}(n; \tau_3)m_{02y}(n + \tau_1; \tau_2 - \tau_1) \ , \qquad (34)
\end{aligned}
$$

or

$$c_{22y}(n; \tau_1, \tau_2, \tau_3) \ \triangleq \ m_{22y}(n; \tau_1, \tau_2, \tau_3) - m_{02y}(n + \tau_2; \tau_3 - \tau_2)m_{20y}(n; \tau_1)$$

$$- \quad m_{11y}(n; \tau_2) m_{11y}(n + \tau_1; \tau_3 - \tau_1)$$

$$- \quad m_{11y}(n; \tau_3) m_{11y}(n + \tau_1; \tau_2 - \tau_1) \,, \tag{35}$$

depending on the use of conjugation. More general definitions and properties of cumulants can be found in [28]; a recent tutorial is [29]. Notice that for many constellations only the first term in the RHS of (35) is nonzero and the cumulant coincides with the corresponding moment

$$c_{04y}(n; \tau_1, \tau_2, \tau_3) = m_{04y}(n; \tau_1, \tau_2, \tau_3) \,. \tag{36}$$

Similarly to the second-order case, one can show using the cumulant definitions and (2) that

$$
\begin{aligned}
c_{04y}(n; \tau_1, \tau_2, \tau_3) \quad=\quad & \gamma_{04w} \sum_{k=0}^{q} b(n; k) b(n + \tau_1; k + \tau_1) \\
\times \quad & b(n + \tau_2; k + \tau_2) b(n + \tau_3; k + \tau_3) \,,
\end{aligned} \tag{37}
$$

$$
\begin{aligned}
c_{22y}(n; \tau_1, \tau_2, \tau_3) \quad=\quad & \gamma_{22w} \sum_{k=0}^{q} b^*(n; k) b^*(n + \tau_1; k + \tau_1) \\
\times \quad & b(n + \tau_2; k + \tau_2) b(n + \tau_3; k + \tau_3) \,,
\end{aligned} \tag{38}
$$

under the i.i.d. assumption for the input; γ_{22w} and γ_{04w} represent the cumulants of the input and are given by $\gamma_{04w} \triangleq c_{04w}(0,0,0) = E\{w^4(n)\} - 3E\{w^2(n)\}^2$ and $\gamma_{22w} \triangleq c_{22w}(0,0,0) = E\{|w(n)|^4\} - 2\sigma_w^4 - E\{w^2(n)\}^2$. Notice that the additive Gaussian noise does not affect the cumulant of the received signal, because Gaussian processes have all cumulants identically zero for any order greater than two.

Equations (30), (37) and (38) relate the output statistics with the system's parameters and form the basis of many blind estimation techniques. In the TI case, the output statistics are typically estimated from sample averages; then the system's parameters are selected so that the *theoretical* cumulants computed by eq. (37), or (38) best match the ones estimated from the data. Such *cumulant matching* procedures have been successfully applied to output-only ARMA identification [30], while optimal variations have been reported in [31], [32]. In order to employ cumulant matching approaches in the TV case, we have to guarantee that the TV coefficients are uniquely identified from these statistics. Otherwise, the minimization of such criteria would not lead to a meaningful parameter estimate. This issue is discussed next.

B. IDENTIFIABILITY OF TV CHANNELS

Let us consider the channel of eq. (2) where the TV coefficients are given by (5) for some *known* sequences $f_l(n)$, $l = 1, \ldots, L$. We wish to examine whether second-order statistics are sufficient to uniquely characterize the parameter vector $\underline{\theta}$; i.e., whether there is an one-to-one relationship between $\underline{\theta}$ and the output correlations $c_{11y}(n; \tau | \theta)$. The answer is negative, as explained in the next proposition.

Proposition 1 : *Given the model of (2), (5), with i.i.d. input $w(n)$, there exist parameter vectors $\underline{\theta}^{(1)} \neq \underline{\theta}^{(0)}$, such that*

$$c_{11y}(n; \tau | \underline{\theta}^{(0)}) \equiv c_{11y}(n; \tau | \underline{\theta}^{(1)}) \quad \forall n, \tau \tag{39}$$

$$\square .$$

As a simple counterexample, which proves the validity of proposition 1, consider a system $\underline{\theta}^{(0)}$ with two TV coefficients ($q = 1$), given by two basis sequences ($L = 2$) as follows,

$$b^{(0)}(n; 0) = f_1(n) + f_2(n) \ , \quad b^{(0)}(n; 1) = 0.5 f_1(n) + 0.5 f_2(n) \ . \tag{40}$$

One can show that this system has identical autocorrelation with the system $\underline{\theta}^{(1)}$ given by

$$b^{(1)}(n; 0) = 0.5 f_1(n) + 0.5 f_2(n) \ , \quad b^{(1)}(n; 1) = f_1(n) + f_2(n) \ . \tag{41}$$

Indeed, from eq. (30) it is true that

$$
\begin{aligned}
c_{11y}(n; 0|\underline{\theta}^{(0)}) &= \gamma_{11w}[|b^{(0)}(n; 0)|^2 + |b^{(0)}(n; 1)|^2] \\
&= \gamma_{11w}[|b^{(1)}(n; 0)|^2 + |b^{(1)}(n; 1)|^2] = c_{11y}(n; 0|\underline{\theta}^{(1)})(42)
\end{aligned}
$$

and

$$
\begin{aligned}
c_{11y}(n; 1|\underline{\theta}^{(0)}) &= \gamma_{11w}[b^{(0)}(n; 0)]^* b^{(0)}(n + 1; 1) \\
&= \gamma_{11w}[b^{(1)}(n; 0)]^* b^{(1)}(n + 1; 1) = c_{11y}(n; 1|\underline{\theta}^{(1)}) \quad (43)
\end{aligned}
$$

resulting in identical second-order statistics for every time-point. In order to further identify the cause of the identifiability problem, let us express output correlations as a function of the parameters $\underline{\theta}$, using (30), (2) and (5),

$$
\begin{aligned}
c_{11y}(n; \tau|\underline{\theta}) &= \gamma_{11w} \sum_{k=0}^{q} b^*(n; k) b(n + \tau; k + \tau) + \gamma_{11v}\delta(\tau) \\
&= \gamma_{11w} \sum_{k=0}^{q} \sum_{l_1,l_2=1}^{L} [\theta_{kl_1}^* \theta_{k+\tau,l_2}] [f_{l_1}^*(n) f_{l_2}(n + \tau)] + \gamma_{11v}\delta(\tau) \\
&= \gamma_{11w} \sum_{l_1,l_2=1}^{L} c_{11\theta}^{(\tau)}(l_1, l_2) [f_{l_1}^*(n) f_{l_2}(n + \tau)] + \gamma_{11v}\delta(\tau) \quad (44)
\end{aligned}
$$

where $c_{11\theta}^{(\tau)}(l_1, l_2) \triangleq \sum_{k=0}^{q} [\theta_{kl_1}^* \theta_{k+\tau,l_2}]$. Notice that $c_{11y}(n; \tau|\underline{\theta})$ depends on the parameters only through the quantity $c_{11\theta}^{(\tau)}(l_1, l_2)$. If we think of θ_{kl} as the impulse response (indexed by k) of an FIR vector multichannel system with $l = 1, \ldots, L$ channels, then $c_{11\theta}^{(\tau)}(l_1, l_2)$ corresponds to the cross-correlation between channels l_1 and l_2. Thus, similarly to the the multichannel case, identifiability of $\underline{\theta}$ cannot be guaranteed from output correlations (see e.g., [33]). For example, if all L channels have a common zero, then all auto- and cross-spectra contain this zero, along with its mirror image (w.r.t. the unit circle). Hence, this zero can never be resolved. In conclusion, Proposition 1 shows that second-order statistics are inadequate for estimating the

TV channel and thus motivates the use of higher-order moments. It would be an interesting research problem to study certain fading channels and examine whether natural constraints appear on the position os their zeros or spectral nulls. Under certain assumptions, identifiability may be possible from autocorrelation information and the use of higher-order cumulants can be avoided.

In an effort to address the most general case here, we use second- and fourth-order information to recover the TV system. Thus, we wish to raise the identifiability question in the context of fourth-order cumulants. The following proposition shows that under some mild conditions, fourth-order cumulants are sufficient to uniquely characterize the channel up to a scalar phase ambiguity.

Proposition 2 : *Given the model of (2), (5), with i.i.d. input drawn from a known constellation, assume that:*

(AS1) *$f_l(n)$ are linearly independent sequences of n with finite power, i.e.,*

$$0 < \lim_{N \to \infty} \frac{1}{N} \sum_{n=1}^{N} |f_l(n)|^2 < \infty,$$

(AS2) *for every fixed τ_1, τ_2, τ_3, the product sequences $f_{l_1}(n) f_{l_2}(n + \tau_1)$ $f_{l_3}(n + \tau_2) f_{l_4}(n + \tau_3)$ are linearly independent and of finite power for $1 \leq l_1 \leq l_2 \leq l_3 \leq l_4 \leq L$.*

Then, for $\underline{\varrho}^{(1)} \neq \underline{\varrho}^{(0)} e^{j\phi}$ and $\phi = k\pi/2$, it holds that:

(i) *$\exists$ some $n, \tau_1, \tau_2, \tau_3$ such that*

$$c_{04y}(n; \tau_1, \tau_2, \tau_3 | \underline{\varrho}^{(1)}) - c_{04y}(n; \tau_1, \tau_2, \tau_3 | \underline{\varrho}^{(0)}) \neq 0 \ , \qquad (45)$$

(ii) *and,*

$$\lim_{N \to \infty} \frac{1}{N} \sum_{n=q+1}^{N} \sum_{\tau_1, \tau_2, \tau_3 = -q}^{q} |c_{04y}(n; \tau_1, \tau_2, \tau_3 | \underline{\varrho}^{(1)}) - c_{04y}(n; \tau_1, \tau_2, \tau_3 | \underline{\varrho}^{(0)})|^2 \neq 0$$

$$\square \ (46)$$

Before proceeding to the proof of Proposition 2, let us mention that amplitude and scalar phase ambiguities are inherent in all blind methods since there is no access to the input. Usually, in practice, phase ambiguities are handled by encoding information in the symbol's phase differences, instead of the phases themselves (e.g., differential PSK modulation). Within this phase ambiguity however, Proposition 2 proves identifiability of the channel in part (i), and in part (ii) shows that the cumulant difference is present throughout the data record, i.e., it is a sustained sequence of n, of finite power. (AS2) on the linear independence of the basis sequences holds for most practical problems (see [22] for equivalent conditions on periodically varying systems).

Proof: We follow the philosophy of the proof of Proposition 1 and establish the equivalence with a multichannel identification problem. Similarly to (44), we can express the output cumulants as functions of the parameters θ_{kl} as follows

$$
\begin{aligned}
c_{04y}(n;\underline{\tau}|\underline{\theta}) &= \gamma_{04w} \sum_{l_1,l_2,l_3,l_4=1}^{L} c_{04\theta}^{(\tau)}(l_1,l_2,l_3,l_4) \\
&\times [f_{l_1}(n)f_{l_2}(n+\tau_1)f_{l_3}(n+\tau_2)f_{l_4}(n+\tau_3)]
\end{aligned}
$$

$$(47)$$

where $c_{04\theta}^{(\tau)}(l_1,l_2,l_3,l_4) = \sum_{k=0}^{q}\theta_{kl_1}\theta_{k+\tau_1,l_2}\theta_{k+\tau_2,l_3}\theta_{k+\tau_3,l_4}$. If we consider again θ_{kl} as a multichannel impulse response, we identify $c_{04\theta}^{(\tau)}(l_1,l_2,l_3,l_4)$ as the fourth-order cross-cumulant of an FIR multichannel system. From (47) we see that for $\underline{\theta}^{(1)} \neq \underline{\theta}^{(0)}e^{j\phi}$, we have

$$
\begin{aligned}
&c_{04y}(n;\underline{\tau}|\underline{\theta}^{(1)}) - c_{04y}(n;\underline{\tau}|\underline{\theta}^{(0)}) \\
&= \gamma_{04w} \sum_{l_1,l_2,l_3,l_4=1}^{L} [c_{04\underline{\theta}^{(1)}}^{(\tau)}(l_1,l_2,l_3,l_4) - c_{04\underline{\theta}^{(0)}}^{(\tau)}(l_1,l_2,l_3,l_4)]
\end{aligned}
$$

$$\times [f_{l_1}^*(n) f_{l_2}^*(n + \tau_1) f_{l_3}(n + \tau_2) f_{l_4}(n + \tau_3)] . \tag{48}$$

By appealing to the cumulant based identifiability results of multichannel systems [33], it follows that for some $\underline{\tau}$, at least one of the differences in the RHS of (48) must be nonzero (otherwise $\underline{\theta}^{(1)} \equiv \underline{\theta}^{(0)}$). This, together with the linear independence assumption proves part (i).

Under the additional assumption (AS2), the RHS of (48) has finite power; then, so does the LHS, which completes the proof of the proposition.   □

Similar identifiability results can be derived using the statistics $c_{22y}(n; \underline{\tau}|\underline{\theta})$ in place of $c_{04y}(n; \underline{\tau}|\underline{\theta})$. However, in many cases $c_{04y}(n; \underline{\tau}|\underline{\theta})$ is computed more easily because of eq. (36), and is thus preferred. More results on identifiability from the moments $m_{22y}(n; \underline{\tau}|\underline{\theta})$ can be found in [34].

Proposition 2 suggests that the minimization of a cumulant matching criterion can identify the correct model, i.e., if $y(n)$ is generated by some unknown model $\underline{\theta}^{(0)}$, and the cumulants $c_{04y}(n; \underline{\tau}|\underline{\theta})$ are given, then $\underline{\theta}^{(0)}$ can be found as the minimizer over all $\underline{\theta}$ of the cost function

$$\frac{1}{N} \sum_{n=q+1}^{N} \sum_{\tau_1, \tau_2, \tau_3 = -q}^{q} |c_{04y}(n; \tau_1, \tau_2, \tau_3|\underline{\theta}) - c_{04y}(n; \tau_1, \tau_2, \tau_3|\underline{\theta}^{(0)})|^2 .$$

A major obstacle to this approach however, is the difficulty of obtaining estimates of the statistics $c_{04y}(n; \tau_1, \tau_2, \tau_3|\underline{\theta}^{(0)})$ from the data. Since the statistics are TV, time-averaging approaches are not applicable. This problem is addressed next.

C. MOMENT MATCHING APPROACH

In this section we propose moment matching criteria for the estimation of the TV channel. Let us consider the general cost function of the form

$$
J_N(\underline{\theta}) = \frac{\lambda}{N} \sum_{n=q+1}^{N} \sum_{\tau_1, \tau_2, \tau_3 = -q}^{q} |m_{04y}(n; \tau_1, \tau_2, \tau_3) - m_{04y}(n; \tau_1, \tau_2, \tau_3|\underline{\theta})|^2
$$
$$
+ \frac{(1-\lambda)}{N} \sum_{n=q+1}^{N} \sum_{\tau=-q}^{q} |m_{11y}(n; \tau) - m_{11y}(n; \tau|\underline{\theta})|^2 \qquad (49)
$$

for some $\lambda \in (0, 1]$. The reason for incorporating second-order statistics as well in (49) is to improve performance, as in the TI case (e.g., see [30]). A strong additive noise component in the received signal however, will exhibit considerable bias in the second order statistics of the signal. In this case we should chose $\lambda = 1$. The moments $m_{04y}(n; \tau_1, \tau_2, \tau_3|\underline{\theta})|$ and $m_{11y}(n; \tau|\underline{\theta})$ in eq. (49) can be computed from (36), (44) and (49).

In case $m_{04y}(n; \tau_1, \tau_2, \tau_3|\underline{\theta})|$ is zero, the moment $m_{22y}(n; \tau_1, \tau_2, \tau_3|\underline{\theta})|$ can be used resulting in a cost function

$$
J_N(\underline{\theta}) = \frac{\lambda}{N} \sum_{n=q+1}^{N} \sum_{\tau_1, \tau_2, \tau_3 = -q}^{q} |m_{22y}(n; \tau_1, \tau_2, \tau_3) - m_{22y}(n; \tau_1, \tau_2, \tau_3|\underline{\theta})|^2
$$
$$
+ \frac{(1-\lambda)}{N} \sum_{n=q+1}^{N} \sum_{\tau=-q}^{q} |m_{11y}(n; \tau) - m_{11y}(n; \tau|\underline{\theta})|^2 \qquad (50)
$$

The moment $m_{22y}(n; \tau_1, \tau_2, \tau_3|\underline{\theta})|$ in this case is computed from the corresponding cumulant as (see also eq. (35))

$$
m_{22y}(n; \tau_1, \tau_2, \tau_3) \overset{\Delta}{=} c_{22y}(n; \tau_1, \tau_2, \tau_3) + m_{02y}(n + \tau_2; \tau_3 - \tau_2)m_{20y}(n; \tau_1)
$$
$$
+ m_{11y}(n; \tau_2)m_{11y}(n + \tau_1; \tau_3 - \tau_1) + m_{11y}(n; \tau_3)m_{11y}(n + \tau_1; \tau_2 - \tau_1),
$$
$$
\qquad (51)
$$

after all lower-order moments have been computed; the cumulant $c_{22y}(n; \tau_1, \tau_2, \tau_3)$ is given by (38), (5). In both cost functions (49) and (50), we resort to instantaneous approximations to estimate the signal's TV statistics. Thus, we use the trivial estimators $y_{04}(n; \tau_1, \tau_2, \tau_3) \overset{\Delta}{=} y(n)y(n + \tau_1)y(n + \tau_2)y(n + \tau_3)$ and $y_{11}(n; \tau_1, \tau_2, \tau_3) \overset{\Delta}{=} y^*(n)y(n + \tau)$ in place of the ensemble statistics

$m_{04y}(n; \tau_1, \tau_2, \tau_3)$ and $m_{11y}(n; \tau)$ respectively. With these substitutions the proposed cost function becomes

$$
\begin{aligned}
\hat{J}_N(\underline{\theta}) \;=\; & \frac{\lambda}{N} \sum_{n=q+1}^{N} \sum_{\tau_1, \tau_2, \tau_3 = -q}^{q} |y_{04}(n; \tau_1, \tau_2, \tau_3) - m_{04y}(n; \tau_1, \tau_2, \tau_3 | \underline{\theta})|^2 \\
& + \frac{(1-\lambda)}{N} \sum_{n=q+1}^{N} \sum_{\tau=-q}^{q} |y_{11}(n; \tau) - m_{11y}(n; \tau | \underline{\theta})|^2
\end{aligned}
\tag{52}
$$

while (50) can be modified in a similar way. The proposed estimation algorithm is based on the minimization of (52). The rational of this approach is that instantaneous moment estimators, although rather noisy, are at least unbiased and hence can be written as

$$
y_{04}(n; \tau_1, \tau_2, \tau_3) = m_{04y}(n; \tau_1, \tau_2, \tau_3 | \underline{\theta}) + e_{04}(n; \tau_1, \tau_2, \tau_3) \;,\;\; \forall n, \tau_1, \tau_2, \tau_3
\tag{53}
$$

$$
y_{11}(n; \tau) = m_{11y}(n; \tau | \underline{\theta}) + e_{11}(n; \tau) \;,\;\; \forall \tau \,,
\tag{54}
$$

with zero-mean error terms. Equations (53), (54) represent a nonlinear regression problem which can be solved by the minimization of the energy of the error terms as suggested by (52). Despite the fact that instantaneous approximations are rather inaccurate, the minimization yields surprisingly reliable results because a large number of them is collected in (52). Notice that equations of the form (53), (54), are gathered for every time point in (52).

The proposed algorithm consists of the following more specific steps:

step 1: Compute $y_{04}(n; \tau_1, \tau_2, \tau_3)$, for all $n = q+1, \ldots, N$, $\tau_1, \tau_2, \tau_3 = -q, \ldots, q$ and $y_{11}(n; \tau)$ for $n = q+1, \ldots N$, $\tau = -q, \ldots, q$.

step 2: Compute $\hat{\underline{\theta}}_N$ to be the minimizer of the cost function (52) over $\underline{\theta}$, using some nonlinear optimization procedure. □

Next we study the asymptotic behavior of this algorithm, in order to assess its performance and quantify earlier claims about its reliability. The next proposition shows strong consistency of the algorithm under some assumptions.

Proposition 3 *If $y(n)$ is given by the model of (2), (5) for some parameter vector $\underline{\theta}^{(0)} \in \Theta$, where Θ is a compact parameter set, then under the assumptions of Proposition 2 and the additional assumption*

(AS4) $|f_l(n)| < C < \infty, \forall n, l,$

the minimizer of (52), $\hat{\underline{\theta}}_N$, is a strongly consistent estimator of $\underline{\theta}^{(0)}$ within a complex phase ambiguity $\phi = k\pi/2$, i.e.,

$$\hat{\underline{\theta}}_N \xrightarrow{w.p.1} \underline{\theta}^{(0)} e^{jk\pi/2} \ , \ k \in \mathbf{Z}, \text{ as } N \to \infty \ . \tag{55}$$

□

Proof: In order to show strong convergence of the minimizer $\hat{\underline{\theta}}_N$ to $\underline{\theta}^{(0)} e^{j\phi}$, we have to show that the cost function $\hat{J}_N(\underline{\theta})$ converges to some limit $J(\underline{\theta})$, uniformly in $\underline{\theta}$, as $N \to \infty$, and that $\underline{\theta}^{(0)} e^{j\phi}$ is the minimizer of $J(\underline{\theta})$. Then, it follows that the minimizer of $\hat{J}_N(\underline{\theta})$ will converge w.p.1 to the minimizer of $J(\underline{\theta})$. To this end, we add and subtract $m_{04}(n; \underline{\tau}|\underline{\theta}^{(0)})$ to the first term of (52) as follows

$$\frac{\lambda}{N} \sum_{n=q+1}^{N} \sum_{\underline{\tau}=-q}^{q} |y_{04}(n; \underline{\tau}) - m_{04y}(n; \underline{\tau}|\underline{\theta}^{(0)})$$

$$+ m_{04y}(n; \underline{\tau}|\underline{\theta}^{(0)}) - m_{04y}(n; \underline{\tau}|\underline{\theta})|^2$$

$$= \frac{\lambda}{N} \sum_{n=q+1}^{N} \sum_{\underline{\tau}=-q}^{q} |y_{04}(n; \underline{\tau}) - m_{04y}(n; \underline{\tau}|\underline{\theta}^{(0)})|^2$$

$$+ \frac{\lambda}{N} \sum_{n=q+1}^{N} \sum_{\underline{\tau}=-q}^{q} |m_{04y}(n; \underline{\tau}|\underline{\theta}^{(0)}) - m_{04y}(n; \underline{\tau}|\underline{\theta})|^2$$

$$+ \frac{2\lambda}{N} \sum_{n=q+1}^{N} \sum_{\underline{\tau}=-q}^{q} \text{Re}\{[y_{04}(n; \underline{\tau}) - m_{04y}(n; \underline{\tau}|\underline{\theta}^{(0)})]$$

$$\times [m_{04y}(n; \underline{\tau}|\underline{\theta}^{(0)}) - m_{04y}(n; \underline{\tau}|\underline{\theta})]^*\}. \tag{56}$$

The first term in the RHS of (56) does not depend on θ and is irrelevant to the minimization. The third term, as $N \to \infty$, can be shown to tend w.p.1 to

$$\lim_{N \to \infty} \frac{2\lambda}{N} \sum_{n=q+1}^{N} \sum_{\underline{\tau}=-q}^{q} \mathrm{Re}\{E\{y_{04}(n;\underline{\tau}) - m_{04y}(n;\underline{\tau}|\underline{\theta}^{(0)})\}$$
$$\times \; [m_{04y}(n;\underline{\tau}|\underline{\theta}^{(0)}) - m_{04y}(n;\underline{\tau}|\underline{\theta})]^*\}, \quad (57)$$

uniformly in $\underline{\theta} \in \Theta$; this is true due to (AS4) and the strong convergence results of [35]. But the limit in (57) is zero because the expectation is zero. Thus, the only term that survives in (56) is the second one. Following a similar procedure for the second term in (52) we conclude that, as $N \to \infty$, $J_N(\underline{\theta})$ tends to the equivalent limit

$$J(\underline{\theta}) \;=\; \lim_{N \to \infty} \frac{1}{N} \sum_{n=q+1}^{N} [\sum_{\tau_1,\tau_2,\tau_3=-q}^{q} |m_{04y}(n;\tau_1,\tau_2,\tau_3|\underline{\theta}^{(0)})$$
$$-m_{04y}(n;\tau_1,\tau_2,\tau_3|\underline{\theta})|^2$$
$$+ \sum_{\tau=-q}^{q} |m_{11y}(n;\tau|\underline{\theta}^{(0)}) - m_{11y}(n;\tau|\underline{\theta})|^2] \neq 0 \; . \; (58)$$

Comparing (58) with (49) and using Proposition 2, [cf. 46)] we see that $\underline{\theta}^{(0)}e^{j\phi}$ (for $\phi = k\pi/2$) minimizes (58). Hence, the minimizer of $J_N(\underline{\theta})$ will converge to $\underline{\theta}^{(0)}e^{j\phi}$ w.p.1 as $N \to \infty$.
□

Similar convergence properties can be shown if we use the moment matching criterion of eq. (50), instead of (49), (see [34]). More discussion on the optimization procedure, as well as stochastic gradient versions of this algorithm can be found in [36].

We close this discussion with the remark that the proposed algorithm is a nonlinear one, hence it is sensitive to the the initial choice of the parameter estimates. No global convergence is guaranteed from an arbitrary initial point, as the algorithm may be trapped in a local minimum. For this reason, alternative linear methods are needed to provide a satisfactory initialization. This motivates the derivations of the next section.

D. LINEAR BLIND METHODS

The combination of instantaneous approximations and basis expansions can provide linear solutions, if we allow some overparametrization in the description of the TV statistics. For example, second-order statistics can be linearly estimated if we use instantaneous approximations in the LHS of the expansion of eq. (44),

$$y^*(n)y(n+\tau) = \gamma_{11w} \sum_{l_1,l_2=1}^{L} c_{11\theta}^{(\tau)}(l_1,l_2)[f_{l_1}^*(n)f_{l_2}(n+\tau)] + \gamma_{11v}\delta(\tau) + e_{11}(n;\tau),$$

(59)

where $e_{11}(n;\tau)$ is the error term. By considering $c_{11\theta}^{(\tau)}(l_1,l_2)$ as unknowns (overparametrization) in eq. (59), we obtain a linear LS problem. Once eq. (59) is solved for $c_{11\theta}^{(\tau)}(l_1,l_2)$, the TV correlations $c_{11y}(n;\tau)$ can be reconstructed from (44).

In order to linearly estimate $\underline{\theta}$ we need to combine second- and fourth-order information. In particular, we will use the q-slice of the of the fourth-order cumulants (i.e., $c_{22y}(n;0,k,q)$) similarly to the q-slice algorithm available for the TI case [37]. From (38), the q-slice can be expressed as

$$c_{22y}(n;0,k,q) = \gamma_{22w}b^*(n;0)b^*(n;0)b(n+k;k)b(n+q;q) .$$

(60)

while the q lag of the autocorrelation is

$$c_{11y}(n;q) = \sigma_w^2 b^*(n;k)b(n+q;q) .$$

(61)

Combining (60), (61) and (2) we can write

$$c_{22y}(n;0,k,q) = \frac{\gamma_{22w}}{\sigma_w^2}b^*(n;0)b(n+k;k)c_{11y}(n;q)$$

(62)

Step 1:   Collect eq. (59) for $n = q + 1, \ldots, N$, and solve for $\hat{c}^{(\tau)}_{11\theta}(l_1, l_2)$ using LS. Repeat for every $\tau = -q, \ldots, q$.

Step 2:   Estimate $\hat{c}_{11y}(n; \tau)$ from $\hat{c}^{(\tau)}_{11\theta}(l_1, l_2)$ and (44).

Step 3:   Combine (62) and (64) for $n = q + 1, \ldots, N$, and solve for $\hat{\alpha}(k; l_1, l_2)$ using LS.

Step 4:   Compute $\hat{\theta}_{kl}$ from (65).

Table 3: Linear blind equalizer

$$= \frac{\gamma_{22w}}{\sigma_w^2} \sum_{l_1, l_2 = 1}^{L} \alpha(k; l_1, l_2)[f_{l_1}^*(n) f_{l_2}(n + k) c_{11y}(n; q)] \ ,$$

where

$$\alpha(k; l_1, l_2) = \theta_{0l_1}^* \theta_{kl_2} \qquad (63)$$

The q-slice of the fourth-order moment is given from the corresponding cumulant from (51). Using instantaneous approximations for the LHS of (51) we obtain

$$y^*(n)y^*(n + k)y(n + q) = c_{22y}(n; 0, k, q) + c_{11y}^2(n; k)c_{11y}^2(n; q)$$
$$+ \ e_{22}(n; 0, k, q) \ . \qquad (64)$$

Then substituting (62) into (64) and treating the α's as unknowns, we obtain a system of linear equations which can be solved using LS (where second-order statistics in eq. (64) have been already linearly estimated using eq. (59)).

Finally, given the estimates of $\alpha(k; l_1, l_2)$, the parameters θ_{kl} can be estimated from (63). To resolve the complex phase ambiguity we set w.l.o.g. $arg(\theta_{00}) \equiv 0$. Hence, $\theta_{00} = |\theta_{00}| = \sqrt{\alpha(0; 0, 0)}$ and θ_{kl} is estimated as

$$\theta_{kl} = \frac{\alpha(k; 0, l)}{\sqrt{\alpha(0; 0, 0)}} . \tag{65}$$

The algorithm is summarized in Table 3. The application of recursive solutions to these LS systems in order to provide adaptive versions of the linear blind algorithm, is an interesting future problem.

We close this discussion with a remark that throughout the blind approach of this section, as well as the adaptive one proposed earlier, the basis sequences $f_l(n)$ are assumed to be a priori known. Moreover, even the expansion order L has to be fixed and given. Apparently, these assumptions limit the general applicability of these methods. In the next section multiresolution ideas are employed to relax these assumptions.

V. MULTIRESOLUTION DESCRIPTIONS

Wavelet expansions offer compact representations of 1-D and 2-D signals and have been successfully applied to speech and image coding problems [17], [18]. They provide successive approximations of the original signal at different resolutions. In this way, most of the signal's energy is compressed into the coarse resolution component, resulting in a more compact description. At the same time, local details (e.g., occasional edges or transitions) are preserved at some parts of the fine resolution (or detail) components. Thus, these parts of the detail signal, together with the low resolution approximation offer a compressed version of the signal's information.

Prompted by the success and general applicability of wavelet methods in signal analysis, we propose, in this parametric framework, to use multiresolution descriptions for the TV coefficients $a(n; k)$, $b(n; k)$. We expand each

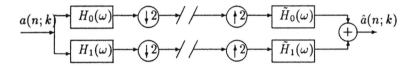

Figure 4: Dyadic Perfect Reconstruction Filter Bank

TV coefficient onto a wavelet basis and keep the most significant expansion parameters.

These coefficients (for fixed k) can be thought of as sequences of n, and the basis expansion approach of (16), as an attempt for a parsimonious description of these signals $a(n; k)$, $b(n; k)$. The basis expansion approach in TV modeling is thus analogous to transform coding in signal compression problems. In both cases, an orthogonal transform is desired, such that most parameters in the transform domain will be (close to) zero, while only few of them carry all the signal's information.

From this analogy, and given the popularity of wavelets in signal analysis, we expect the proposed multiresolution methods to provide a satisfactory approximation to most system trajectories, with smooth variations and even occasional jumps or transient changes. The advantages of wavelet coding, compared with more classical DCT and Fourier methods, are exploited in this section, in a parametric setup.

Let us define the problem more explicitly. Consider the channel description of eq. (15) where $w(n)$ is the transmitted signal and $v(n)$ the additive noise. While up to now we considered general expansions of the form (16) to model the TV coefficients, we now focus on the wavelet basis.

In discrete-time, multiresolution expansions are computed through multirate filter banks [17]. Let us, for a fixed k, analyze the TV coefficient $a(n; k)$ (or similarly $b(n; k)$) through the dyadic filter bank of Fig. 4. If the filters $h_0(n)$, $h_1(n)$ fulfill certain conditions then perfect reconstruction at the output is possible (e.g., [38]). Then, the TV coefficient is given by the synthesis equation

$$a(n;k) = \sum_m \zeta_{1,m}^{(a_k)} \tilde{h}_0(n - 2m) + \sum_m \xi_{1,m}^{(a_k)} \tilde{h}_1(n - 2m) , \qquad (66)$$

where $\tilde{h}_l(n) = h_l(-n)$, $l = 1, 2$; the first term in the RHS of (66) corresponds to the low-resolution part of $a(n;k)$ while the second to the detail signal. Equation (66) represents an expansion of $a(n;k)$ onto a basis generated by translations of $\tilde{h}_0(n)$, $\tilde{h}_1(n)$. The expansion parameters $\zeta_{1,m}^{(a_k)}$, $\xi_{1,m}^{(a_k)}$ are given by the analysis equations (see Fig. 4)

$$\zeta_{1,m}^{(a_k)} = \sum_l h_0(l) a(2m - l; k) , \qquad (67)$$

$$\xi_{1,m}^{(a_k)} = \sum_l h_1(l) a(2m - l; k) . \qquad (68)$$

We use subscript 1 in (67), (68) to denote the resolution depth (we arbitrarily assign $a(n;k)$ to be the zero resolution depth [17]). Even coarser resolutions may be obtained by the repeated application of (66). In filtering terms, we further decompose the lower resolution branch of the filter bank in Fig. 4 as in Fig. 5. In the general case, the coefficients at resolution j can always be reconstructed from $j + 1$ similarly to eq. (66) (see also Fig. 5)

$$\zeta_{j,n}^{(a_k)} = \sum_m \zeta_{j+1,m}^{(a_k)} \tilde{h}_0(n - 2m) + \sum_m \xi_{j+1,m}^{(a_k)} \tilde{h}_1(n - 2m) . \qquad (69)$$

We wish to expand the channel's coefficients up to a depth J_{max}. In order to write a closed form expression, we repeatedly back-substitute $\zeta_{j,n}^{(a_k)}$ for $j = 1, \ldots, J_{max} - 1$ in (69) and (66). It can be shown that, following this procedure, $a(n;k)$ is expanded as follows,

$$a(n;k) = \sum_m \zeta_{J_{max},m}^{(a_k)} \tilde{h}_0^{(J_{max})}(n - 2^{J_{max}}m) + \sum_{j=1}^{J_{max}} \sum_m \xi_{j,m}^{(a_k)} \tilde{h}_1^{(j)}(n - 2^j m) ,$$

$$(70)$$

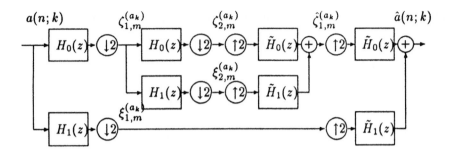

Figure 5: Multiresolution Analysis to a Depth of $J = 2$

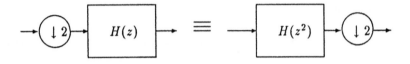

Figure 6: The Noble Identity

where the first term in the RHS represents the approximation at resolution J_{max}, while the second term the summation of J_{max} detail signals; $\tilde{h}_0^{(J_{max})}(n)$, $\tilde{h}_1^{(j)}(n)$ are the inverse Z-transforms of the product transfer functions $H_0^{(J_{max})}(z)$ and $H_1^{(j)}(z)$ defined as

$$H_0^{(J_{max})}(z) = H_0(z)H_0(z^2)\cdots H_0(z^{2^{J_{max}-1}}) \; , \tag{71}$$

$$H_1^{(j)}(z) = H_0(z)H_0(z^2)\cdots H_0(z^{2^{j-2}})H_1(z^{2^{j-1}}) \; , \; j = 1,\ldots,J_{max} \; . \tag{72}$$

The derivation of (70) is better understood in the frequency domain. It exploits a basic property of multirate systems called the *Noble identity* (e.g., see [38]). This identity allows the interchange of filtering and subsampling in multirate processing, and is depicted in Fig. 6. It states that post-processing (after subsampling) by a filter $H(z)$ is equivalent to pre-processing by $H(z^2)$. Using this property, the equivalence between Figures

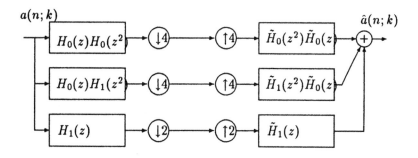

Figure 7: Equivalent Structure

5 and 7 can be easily established.

In the general case of depth J_{max} the repeated application of the noble identity results in an equivalent structure with $J_{max} + 1$ branches with product transfer functions at each branch given by (71), (72). By expressing this filter bank operation in the time-domain we obtain (70).

While the multiresolution nature of this analysis is rather appealing, we should keep in mind that (70) represents nothing more than an orthogonal expansion of $a(n; k)$ onto a basis defined by translations of $\tilde{h}_0^{(J_{max})}(n)$, $\tilde{h}_1^{(j)}(n)$. Moreover, since no approximation was involved in this transformation, the number of expansion parameters in the wavelet domain should equal the number of samples in the time-domain, so that no immediate benefit from this approach is yet evident. The promise is however, like in every other transform based signal description, that fewer significant parameters will exist in the transform domain, while most of them will be close to zero. Indeed, in the present framework, it is common practice to select the filters $h_0(n)$ and $h_1(n)$ as low-pass and high-pass respectively, so that most of the signal's energy is preserved in the low-resolution component (provided that the signal to be analyzed is generally a low-pass one).

In the same way, the TV coefficient's expansion in (70) should compress most information of the system's time-variation in the low-resolution part (provided that these variations have a low-pass nature). For special cases, where the time-variations have a band-pass or nonstationary nature, optimal analysis filters can still be designed as discussed in [39]. However, for most cases encountered in practice, a general low-pass/high-pass decomposition will be sufficient.

By keeping the low-resolution approximation only, a parsimonious description of the TV coefficients is obtained. A more challenging problem however, is related to the accurate description of sharp transitions in the coefficients' time history. Unfortunately, this information is preserved in some parameters of the detail signals. Thus, in order to accurately model the global channel behavior, as well as the local events, the significant parts of the detail signal have to be retained as well. This is not a trivial task however, since the TV coefficients are not immediately observed. Model selection techniques are employed in the next section to address this issue.

A. SELECTION OF RESOLUTION DEPTH

Let the received data $y(n)$, $n = 0, \ldots, N - 1$ be described by the model of (15), (70). In order to simplify the notation, let us gather all the parameters $\zeta^{(b_k)}_{J_{max},m}$, $\xi^{(b_k)}_{j,m}$, $\zeta^{(a_k)}_{J_{max},m}$, $\xi^{(a_k)}_{j,m}$ in a parameter vector $\underline{\theta} = [\theta_1, \ldots, \theta_{max}]^T$, and all transmitted symbols in a vector $\mathbf{w} = [w(0), \ldots, w(N - 1)]^T$. Then, if we substitute (70) into (15) we obtain the modeling equation

$$y(n) = y(n|\underline{\theta}, \mathbf{w}) + v(n) , \tag{73}$$

where

$$
\begin{aligned}
y(n|\underline{\theta}, \mathbf{w}) = {} & \sum_{k=1}^{p} \sum_{m} \zeta^{(a_k)}_{J_{max},m} [\tilde{h}_0^{(J_{max})}(n - 2^{J_{max}} m) y(n - k)] \\
& + \sum_{k=1}^{p} \sum_{j=J_{min}}^{J_{max}} \sum_{m} \xi^{(a_k)}_{j,m} [\tilde{h}_1^j(n - 2^j m) y(n - k)]
\end{aligned}
$$

$$+ \sum_{k=0}^{q} \sum_{m} \zeta_{J_{max},m}^{(b_k)} [\tilde{h}_0^{(J_{max})}(n - 2^{J_{max}} m) w(n - k)]$$

$$+ \sum_{k=0}^{q} \sum_{j=J_{min}}^{J_{max}} \sum_{m} \xi_{j,m}^{(b_k)} [\tilde{h}_1^j(n - 2^j m) w(n - k)] \ . \qquad (74)$$

Eq. (74) shows that the selection of the significant expansion parameters is equivalent to a model selection problem. To understand why, assume for a moment that $\mathbf{w}$ is given (as in the training mode). Then, equations (73), (74) represent a linear regression and the choice of the significant terms of the RHS is precisely a regressor selection problem. Notice also that the expansion in (74) is limited to depths $j = J_{min}, \ldots, J_{max}$ for some J_{min} assuming that $\xi_{j,m}^{(b_k)} = 0$ for $0 < j < J_{min}$. In this way the problem is simplified by exploiting prior information about the low-pass nature of the channel variations. To further eliminate unnecessary terms in (74) we use model selection procedures.

By a model $\mu = \{m_1, m_2, \ldots, m_{d(\mu)}\}$ we mean that some θ's are *a priori* set to zero,

$$\underline{\theta}_\mu = [0, \ldots, \theta_{m_1}, 0, \ldots, \theta_{m_2}, 0, \ldots, \theta_{m_{d(\mu)}}, \ldots] \ , \qquad (75)$$

where $d(\mu)$ stands for the number of the non-zero θ's and hence the dimensionality of the model μ. Thus, model μ describes the data as

$$y(n) = y(n|\underline{\theta}_\mu, \mathbf{w}) + v_\mu(n) \qquad (76)$$

where $y(n|\underline{\theta}_\mu, \mathbf{w})$is given by (74) after deleting all terms corresponding to zero elements of $\underline{\theta}_\mu$.

In this setup, we seek a model selection procedure to determine the appropriate $\underline{\theta}_\mu$, or equivalently keep only the important terms in (74).

B. MAXIMUM LIKELIHOOD METHODS

The appropriate model structure can be selected by testing the likelihood of the data under each candidate model. The most widely used such procedure, derived from information theoretic considerations, is the minimization of Akaike's AIC criterion [19]. Several consistent and optimal versions of this criterion have been studied in the literature [40], [41]. The AIC criterion is given by

$$AIC(\mu_i) = -2\frac{1}{N}\log(L_{\mu_i}) + 2\frac{1}{N}d(\mu_i) \ , \tag{77}$$

where L_{μ_i} is the maximized likelihood of the data under model μ_i. In the current setup, we have to distinguish two cases. If both $y(n)$ and $w(n)$ are known (training period), then L_{μ_i} is given by

$$L_{\mu_i} = \max_{\underline{\theta}_{\mu_i}} L\{y(n),\ w(n),\ n = 0,\ldots N-1\} \ . \tag{78}$$

In the more general case, where $w(n)$ is unknown, L_{μ_i} is

$$L_{\mu_i} = \max_{\mathbf{w},\underline{\theta}_{\mu_i}} L\{y(n),\ n = 0,\ldots N-1\} \ , \tag{79}$$

where $\mathbf{w}$ can assume only a finite number of values depending on the signal constellation. We continue this discussion, using (79) since (78) represents a special case of the former.

Under the white, complex Gaussian assumption for the additive noise, with uncorrelated real and imaginary parts of equal variance, the maximum log-likelihood is

$$\log(L_{\mu_i}) = \max_{\mathbf{w},\underline{\theta}_{\mu_i}} \{-\sum_{n=0}^{N-1} \frac{|\hat{v}(n|\mathbf{w},\underline{\theta}_{\mu_i})|^2}{\hat{\sigma}_v^2(\mathbf{w},\underline{\theta}_{\mu_i})} - N\log\hat{\sigma}_v^2(\mathbf{w},\underline{\theta}_{\mu_i}) - N\log\pi\} \ , \tag{80}$$

where $\hat{v}(n|\mathbf{w},\underline{\theta}_{\mu_i}) = y(n) - y(n|\mathbf{w},\underline{\theta}_{\mu_i})$ represents the estimated residuals given $\mathbf{w}$, $\underline{\theta}_{\mu_i}$; $\hat{\sigma}_v^2(\mathbf{w},\underline{\theta}_{\mu_i})$ is the ML estimate of σ_v^2, given by

$$\hat{\sigma}_v^2(\mathbf{w},\underline{\theta}_{\mu_i}) = \frac{1}{N-d(\mu)} \sum_{n=0}^{N-1} |\hat{v}(\mathbf{w},\underline{\theta}_{\mu_i})|^2 \ . \tag{81}$$

Substituting (81) and (80) into (77) we obtain

$$AIC(\mu_i) = 2 + 2\log \pi + 2\min_{\mathbf{w},\underline{\theta}_{\mu_i}} \{\log \hat{\sigma}_v^2(\mathbf{w},\underline{\theta}_{\mu_i})\} + \frac{2d(\mu_i)}{N} \quad , \tag{82}$$

or after deleting constant terms

$$AIC(\mu_i) = 2\min_{\mathbf{w},\underline{\theta}_{\mu_i}} \{\log \hat{\sigma}_v^2(\mathbf{w},\underline{\theta}_{\mu_i})\} + \frac{2d(\mu_i)}{N} \quad , \tag{83}$$

The statistic of eq. (83) has to be computed for every candidate model, and the minimum has to be selected. This minimization is possible, at least in pronciple, since for a fixed $\mathbf{w}$, the problem is linear and the minimizing $\underline{\theta}_{\mu_i}$ is given by the solution of a LS system defined by (73), (74). Thus, a brute force method would involve an exhaustive search through all possible values of $\mathbf{w}$, with a LS system solved, and $\log \hat{\sigma}_v^2(\mathbf{w},\underline{\theta}_{\mu_i}^{(LS)})$ computed at each value. Then the AIC is given by

$$AIC(\mu_i) = 2\min_{\mathbf{w}} \{\log \hat{\sigma}_v^2(\mathbf{w},\hat{\underline{\theta}}_{\mu_i}^{(LS)})\} + \frac{2d(\mu_i)}{N} \quad . \tag{84}$$

Although (84) requires a prohibitive amount of computations, it is a well defined index for each candidate model and, at least in principle, provides a solution to the problem. Thus, before exploring simplifications of this maximum likelihood approach, we would like to continue using (84) and complete the discussion on the resolution selection.

The proposed algorithm starts with a minimal model, containing only the terms of (74) corresponding to the low-resolution signal, and gradually adds on terms of the detail signal, which are indicated as significant according to the AIC test. The algorithm is described in more detail by the following steps:

Step 1:

    a) Based on prior information and computational constraints select the depths J_{min}, J_{max}, and the order q.

b) Initialize the model μ_0 using basis sequences from depth J_{max} only (low-resolution signal).

c) Compute $AIC(\mu_0)$

Step 2:   For every depth $j = J_{max} + 1, \ldots, J_{min}$ repeat step 3.

Step 3:   For every basis sequence at depth j:

a) Formulate an alternative model μ_{i+1} by adding this candidate sequence (and $q + 1$ more parameters) to model μ_i.

b) Decide between the two candidate models

$$H_0 \quad : \quad \mu_{true} = \mu_i$$

$$H_1 \quad : \quad \mu_{true} = \mu_{i+1}$$

and accept the new basis sequence if

$$AIC(\mu_{i+1}) < AIC(\mu_i); . \tag{85}$$

We should mention here that the proposed algorithm is not exclusively tied to the AIC criterion used in step 3. Hypothesis testing procedures have also been applied to the model selection problem [42], [43] and can be used in this framework. Some researchers argue that hypothesis testing procedures offer the flexibility of a threshold selection and are more versatile [20]. Moreover, these approaches hold asymptotically even for non-Gaussian data.

Hypothesis testing procedures are based on a statistic formed by the ratio of the residual energies under the candidate models μ_1 and μ_2,

$$s[d(\mu_1), d(\mu_2)] = N \frac{\hat{\sigma}_v^2(\mu_1) - \hat{\sigma}_v^2(\mu_2)}{\hat{\sigma}_v^2(\mu_1)} \quad , \tag{86}$$

It can be shown that this statistic is asymptotically $\chi^2[d(\mu_2) - d(\mu_1)]$ distributed if $\mu_2 \supset \mu_1$ [26, pg. 422]. Based on this observation, the statistical significance of (86) can be tested, using an appropriate threshold for the desired confidence level. Thus, (86) can be used in place of the AIC criterion in step 3 of the algorithm.

In both cases, the proposed algorithm exploits model selection techniques to locally select the appropriate resolution for the wavelet expansion. It increases the resolution until a satisfactory approximation is achieved, as indicated by the statistical test. In this way, the multiresolution description is adapted to the local variability of the TV coefficients. Moreover, this algorithm provides a maximum likelihood (ML) estimate of the transmitted symbol stream, and thus simultaneously performs the equalization and detection task. The exhaustive search however, implied by the ML approach of eq. (84), makes this algorithm computationaly too demanding. Therefore, it is of interest to examine special cases or suboptimal solutions which reduce the amount of computations required.

C. SUBOPTIMAL METHODS

As mentioned earlier, during the training period where $w(n)$ is given, eq. (74) represents a linear model and the computation of the AIC for each candidate model only amounts to the solution of a LS problem. Thus, in this case the problem is drastically simplified. Therefore, for cases where the significant basis sequences can be determined during the training period, the ML method is readily applicable.

This approach may fit well in the context of periodically varying channels discussed earlier, but unfortunately cannot be applied to the current multiresolution framework. The reason is that the wavelet expansion of eq. (70) involves *locally concentrated* basis sequences. Thus, not all basis sequences can be tested using training data, since some of them contribute only to later parts of the data record, i.e., they are zero (or close to zero) throughout the training part.

An interesting case where a linear solution is possible in this multiresolution framework, is when the channel can be approximated by a causal, totally stable, all pole model,

$$y(n) = \sum_{k=1}^{p} a(n;k)y(n-k) + w(n) + v(n) \ . \tag{87}$$

By expanding $a(n;k)$ on a multiresolution basis using (16) and identifying $e(n) = w(n) + v(n)$ as an error term, we obtain a linear regression with i.i.d. non-Gaussian residuals, which is not limited to the training period. Thus, the proposed algorithm with the test of (86) can be implemented by solving only a linear problem for each candidate model.

Apart from this special case, the application of the ML approach to general ARMA or MA channels involves a prohibitive amount of computations. Some computational savings can be achieved if the iterative procedure, proposed in [44] (for the TI case) is used. At each iteration of this procedure, $\mathbf{w}$ was fixed and the optimal $\underline{\theta}_{\mu_i}$ was computed using LS and then $\underline{\theta}_{\mu_i}$ was fixed and a more accurate estimate for $\mathbf{w}$ was computed through the Vitterbi algorithm. However, no global convergence of this algorithm can be guaranteed.

Next, we would like to discuss blind solutions to the resolution selection problem. Surprisingly enough, these methods although suboptimal, provide a simple and linear solution and are well matched with the blind estimation techniques discussed earlier. In order to grasp the main idea of this approach, let us concentrate on the zero lag of the signal's TV autocorrelation $m_{11y}(n;0) = E\{|y(n)|^2\}$. Let us consider a general expansion as in (5) and, similarly to (73), express $m_{11y}(n;0)$ as

$$\begin{aligned} m_{11y}(n;0|\underline{\theta}_\mu) &= \sigma_w^2 \sum_{k=0}^{q} |b(n;k)|^2 + \sigma_v^2 \\ &= \sigma_w^2 \sum_{l_1,l_2=1}^{L} [\sum_{k=0}^{q} \theta_{kl_1}^* \theta_{kl_2}] f_{l_1}^*(n) f_{l_2}(n) + \sigma_v^2 \end{aligned} \tag{88}$$

using (30) and (5). If we rearrange terms, (88) can be written more explic-

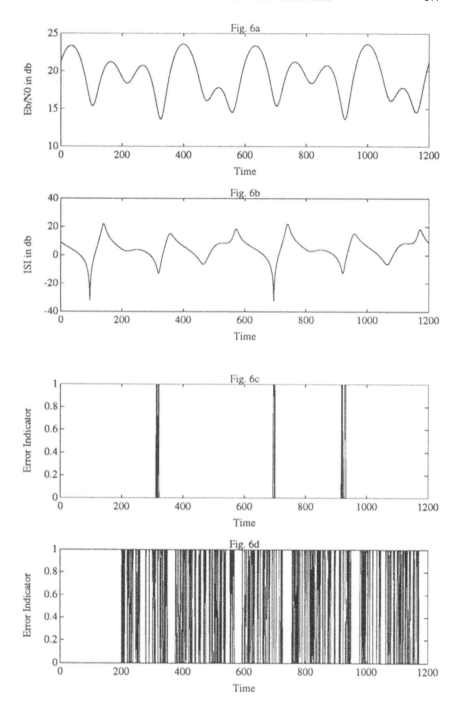

Figure 8: Decision-feedback equalization of a TV channel

itly as

$$m_{11y}(n;0|\underline{\theta}_\mu) = \sigma_w^2 \sum_{1 \le l_1 < l_2 \le L} 2Re\{[\sum_{k=0}^{q} \theta_{kl_1}^* \theta_{kl_2}] f_{l_1}^*(n) f_{l_2}(n)\}$$

$$+ \sigma_w^2 \sum_{l=1}^{L} [\sum_{k=0}^{q} |\theta_{kl}|^2] |f_l|^2 + \sigma_v^2 . \tag{89}$$

Using instantaneous approximations for the LHS of (89) as in Section IV, we obtain

$$|y(n)|^2 = m_{11y}(n;0|\underline{\theta}_\mu) + e(n|\underline{\theta}_\mu) \tag{90}$$

where $e(n|\underline{\theta}_\mu)$ is a zero-mean error term (under model μ). Equations (89), (90) define a linear regression of $|y(n)|^2$ onto the basis sequences $f_{l_1}^*(n) f_{l_2}(n)$ for $n = 0, \ldots, N-1$. They form the basis for a linear blind model selection algorithm. Each new candidate model in Step 3 of the depth selection algorithm with one more basis sequence, would involve $L+1$ more regressors in (89). Then the χ^2 test of eq. (86) can be used in this linear setup, to test the significance of the new model.

The regression of (89), (90) can be extended to more lags of the auto-correlation or higher moments and is well tied with the blind procedures of Section IV. Notice that no estimates of the channel parameters are provided at this step; hence, once the significant basis sequences have been determined, the blind methods of Section IV have to be applied. This procedure possesses no optimality but offers a simple solution to the resolution selection problem.

VI. SIMULATIONS

In this section we present some simulation examples to illustrate the potential and applicability of the proposed methods in rapidly fading environments. Some comparisons with more traditional adaptive methods are also presented and the superiority of the new approach is shown. How-

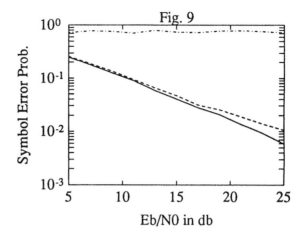

Figure 9: Error-probability curves

ever, these examples do not constitute a complete simulations analysis of
the proposed methods, and do not study the effects of different coding or
modulation schemes on the performance of the proposed equalizers.

In the first test case we simulated a mobile radio channel and applied the
adaptive and blind solution, proposed in Sections III and IV, to estimate
and equalize the TV channel. We used a two tap channel given by (2) with
$q = 1$, while the TV coefficients were given by a linear combination of three
basis sequences ($L = 3$ in (5)), corresponding to a direct path plus two
reflectors

$$f_1(n) = 1 + j \ , \quad f_2(n) = e^{j\frac{2\pi}{T_2}n} \ , \quad f_3(n) = e^{j\frac{2\pi}{T_3}n} \ . \tag{91}$$

The periods were chosen to be $T_2 = 120$ and $T_3 = 200$ samples. These
numbers are not far from reality for the mobile radio channel, for a carrier
frequency of 900 MHz, bit rate around 20 Kbit/sec, and a vehicle moving
at 100 Km/h.

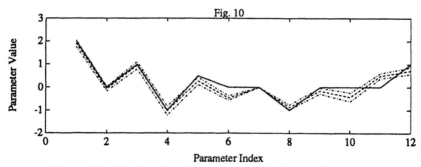

Figure 10: Blind estimation: True and estimated parameters

The RLS based algorithm of Tables 1 and 2 was used. The data frame consisted of 1200 samples where the first 200 were used for training. 4-QAM modulation was used and AWGN was added at the output. Figures 8a and 8b show how the SNR and ISI change with time in this periodically varying channel. SNR and ISI are plotted (in db) as functions of time. Figure 8c illustrates the performance of the proposed adaptive method. It shows the locations of decoding errors in the data frame. The indicator function (which becomes one, when an error occurs) is plotted versus time. In comparison, Fig. 8d shows the error pattern when the conventional RLS decision-feedback equalizer is used. Figure 8d clearly shows that the conventional approach loses track of the TV channel soon after the training period, and performs very poorly. Thus, this channel presents an example of a rapidly fading environment, where the time-variations are too fast for the conventional adaptive algorithms to follow.

In order to get a more general impression of the proposed algorithm's performance, we have plotted the symbol error probabilities versus SNR in Fig 9, obtained through Monte-Carlo simulations (50 iterations per SNR point, 2 db increments). The solid line represents demodulation with a priori known channel, the dashed line represents the estimated channel case (using the proposed decision-feedback method), and the dashed-dotted line

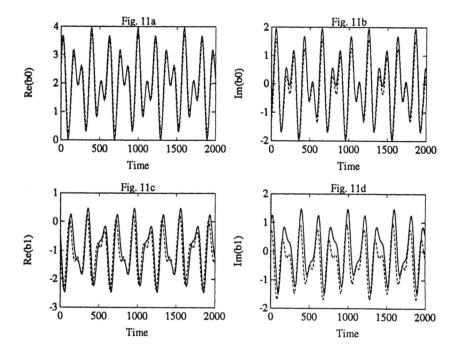

Figure 11: True and estimated TV tap coefficients

shows the poorly performing conventional decision-feedback method. No error correcting coding was used.

The blind method of Section IV was also applied to the same channel. The data length was 2000 samples and the average SNR was 5 db. The proposed minimization procedure with linear initialization was used to estimate the expansion coefficients. Figure 10 shows the true expansion coefficients θ_{kl} (solid line) as well as the estimated ones, using 10 Monte-Carlo runs (dashed: mean, dashed-dotted: $\pm$ standard deviation); θ_{kl} is complex here, so we plot the real part (index 1-6) followed by the imaginary part (index 7-12). In Fig. 11 we show the true (solid line) and the reconstructed (dashed line) periodically varying channel coefficients. We have used (5) and the blind estimates of θ_{kl} to reconstruct $b(n; k)$. We plot the real and

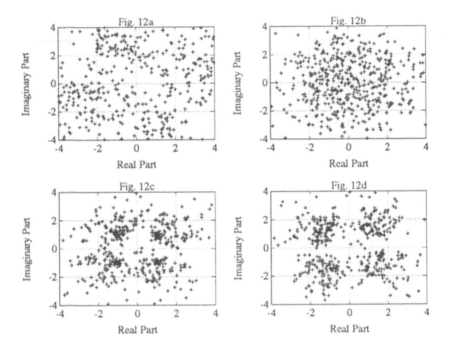

Figure 12: Received symbols: Before and after equalization

imaginary parts of $b(n; k)$ in separate subplots. Figure 11 shows that the proposed algorithm can follow the rapid variations of the channel fairly well.

Finally, in Fig. 12 we plot the received symbols (before and after equalization) on a constellation graph. Figure 12a shows the unequalized symbols which suffer severe ISI. Figure 12b shows these symbols after equalization using the conventional DF approach, Fig. 12c after the proposed adaptive method and Fig. 12d after the novel blind method. In Figures 12b-12d the ISI due to previous symbols is subtracted from the current symbol, using the channel estimate. The figure shows that the novel methods manage to remove the ISI in situations where the conventional methods fail.

In the second test case, we examined a TV-AR channel described by

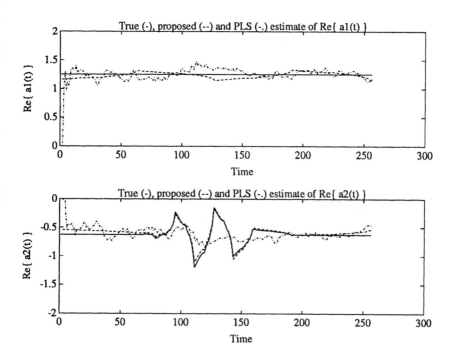

Figure 13: True and estimated TV tap coefficients

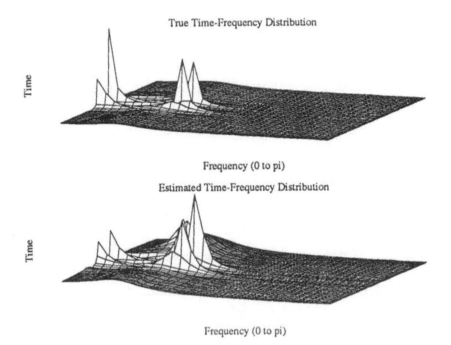

True Time-Frequency Distribution

Frequency (0 to pi)

Estimated Time-Frequency Distribution

Frequency (0 to pi)

Figure 14: True and estimated TV frequency response

(87). The two TV coefficients are shown in Fig. 13 (solid line); the first coefficient remains constant throughout the data record, while the second exhibits a transient change at the center of the record. This corresponds to a transient change in the frequency response of this generally low-pass channel as shown in Fig. 14. The noise was additive, white, Gaussian of 12db SNR. The resolution selection algorithm of Section V.C was applied using a standard wavelet basis taken from [45]. The estimated coefficients are shown in dashed lines while the RLS estimates are shown in dot-dashed lines. Notice that the transient jumps are too fast for the RLS algorithm to follow.

In order to illustrate how the algorithm proceeds, we have plotted in Fig. 15 the χ^2-statistic value (+ symbol) for each iteration as well as the

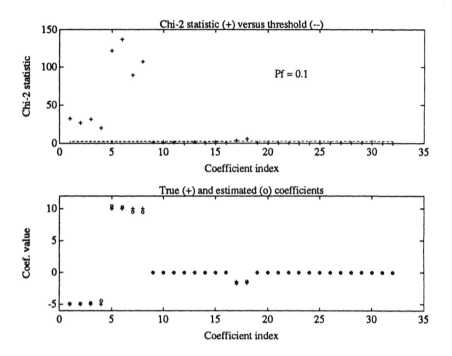

Figure 15: χ^2 test at each iteration

corresponding threshold (dashed line). Each iteration determines the significance of one added regressor. In this figure we visualize how the algorithm decided which regressors to keep. In the same figure, we also show the true and estimated expansion parameters corresponding to these regressors. We see that most parameters are zero and are correctly identified as such by the algorithm.

In the third case, we tested a TV-MA channel as in (2) of order $q = 1$. Its TV coefficients exhibit abrupt changes (solid line in Fig. 16). These coefficients correspond to a channel with an in band spectral null, whose frequency drifts at the center of the data record as shown in Fig. 17. The Haar wavelet basis was used in this example, which is best suited for describing discontinuities and jumps.

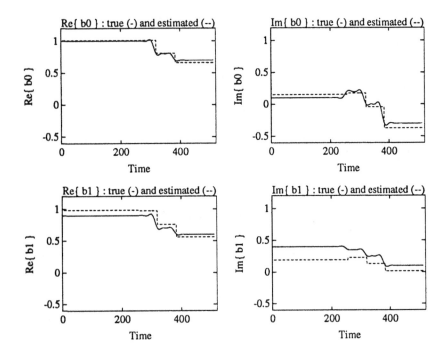

Figure 16: Blind method: True and estimated TV tap coefficients

TV Frequency Response

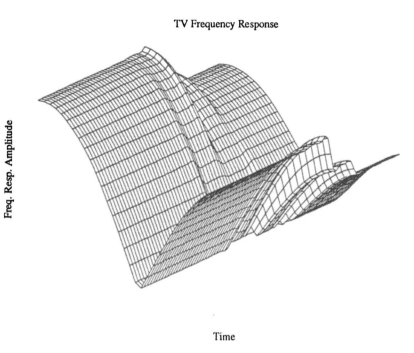

Freq. Resp. Amplitude

Time

Figure 17: Channel's TV frequency response

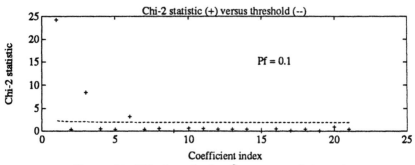

Figure 18: Blind method: χ^2 test at each iteration

The regression of (89), (90) was used to blindly determine the resolution depth. The significant regressors (i.e., model parameters) were determined using the χ^2-test. The iterations of the algorithm are visualized in Fig. 18, where only three components of the detail signal are shown to be significant. After the significant regressors have been determined, the blind method of Section IV was employed to estimate the channel. The estimated TV coefficients are shown in Fig. 16 with dashed lines. They represent a piecewise constant approximation to the actual trajectory.

VII. CONCLUSIONS

Throughout this chapter, basis expansion ideas are employed to identify TV systems and equalize rapidly fading channels. This work illustrates the potential of basis expansion tools for addressing challenging questions regarding adaptive and blind estimation of these TV channels. It also provides a general framework, into which multiresolution descriptions fit nicely, and cumulant based methods can be extended to cover the TV case.

Several more questions regarding the performance and optimality of the proposed methods remain open. Also, recursive and more computational-ly efficient implementations need to be studied. Finally, the usefulness of these procedures in other deconvolution problems, different than the communications setup, should be examined.

ACKNOWLEDGEMENTS

It is my pleasure to thank Prof. G. B. Giannakis and Prof. S. G. Wilson for several insightful discussions on various topics related to this work.

References

[1] Y. Grenier, "Time-dependent ARMA modeling of nonstationary signals," *IEEE Trans. on ASSP*, Vol. **31**, No. 4, pp. 899-911, Aug. 1983.

[2] L. A. McCarley, "An autoregressive Filter Model for Constant Q Attenuation", *Geophysics*, vol. 50, no. 5, pp. 749-758, May 1985.

[3] F. Hlawatsch, and G. F. Boudreaux-Bartels, "Linear and Quadratic Time-Frequency Signal Representations", *IEEE Signal Proc. Magazine*, pp. 21-67, April 1992.

[4] J. Proakis, *Digital Communications*, Mc Graw Hill, 1989.

[5] R. A. Iltis, and A. W. Fuxjaeger, "A Digital DS Spread-Spectrum Receiver with Joint Channel and Doppler Shift Estimation," *IEEE Trans. on Communications* vol. 39, no. 8, pp. 1255-1267, August 1991.

[6] L. Greenstein and B. Czekaj, "Modeling Multipath Fading Responses Using Multitone Probing Signals and Polynomial Approximation," *Bell System Tech. Journal*, vol 60, pp. 193-214, 1981.

[7] J. G. Proakis, "Adaptive equalization techniques for acoustic telemetry channels," *IEEE Journal of Oceanic Engr.*, vol. 16, pp. 21-31, January 1991.

[8] L. A. Liporace, "Linear Estimation of Nonstationary Signals," *J. of Acoust. Soc. of Am.*, vol. 58, no. 6, pp. 1288-1295, 1975.

[9] T. Subba Rao, "The Fitting of Nonstationary Time-Series Models with Time-Dependent Parameters", *J. Royal Statistical Soc. Series B*, vol 32, no 2, pp. 312-322, 1970.

[10] M. J. Hinich, and R. Roll, "Measuring Nonstationarity in the Parameters of the Market Model," *Research in Finance (a research annual)*, ed. H. Levy, vol 3, 1981.

[11] A. Benveniste, and M. Goursat, "Blind Equalizers," *IEEE Trans. on Communications*, vol. 32, no. 8, pp. 871-883, August 1984.

[12] D. Hatzinakos and C.L. Nikias, "Blind equalization using a tricepstrum based algorithm," *IEEE Trans. on Communications*, vol. 38, pp. 669-682, 1991.

[13] B. Porat, and B. Friedlander, "Blind Equalization of Digital Communication Channels Using Higher-Order Statistics," *IEEE Trans. on Signal Processing*, vol. 39, no. 2, pp. 522-526, February 1991.

[14] O. Shalvi and E. Weinstein, "New Criteria for Blind Deconvolution of Nonminimum Phase Systems (Channels)," *IEEE Trans. on Information Theory* vo.l 36, no. 2, pp. 312-321, March 1990.

[15] J. K. Tugnait, "Blind channel estimation and adaptive blind equalizer initialization," *Proc. Intl. Conf. on Communications*, Denver, CO, June 1991.

[16] D. N. Godard, "Self-Recovering Equalization and Carrier-Tracking in Two-Dimensional Data Communication Systems," *IEEE Trans. on Communications*, vol. 28, no 11, pp. 1867-1875, November 1980.

[17] S. Mallat, "A Theory of Multiresolution Signal Decomposition: the Wavelet Representation," *IEEE Trans. on Pattern Anal. and Machine Intell.*, vol. 11, no. 7, pp. 674-693, July 1989.

[18] N. Moayeri, I. Daubechies, Q. Song, and H. S. Wang, "Wavelet Transform Image Coding Using Trellis Coded Vector Quantization", *ICAS-SP Proc.*, vol. IV, pp. 405-408, San Francisco, CA, March 23-26, 1992.

[19] H. Akaike, "Information Theory and an Extension of the Maximum Likelihood Principle", in *Proc. 2nd Int. Symp. Information Theory*, B. N. Petrov and F. Csaki, eds., Akademiai Kiado, Budapest, pp. 267-281, 1973.

[20] T. Söderström, "On Model Structure Testing in System Identification", *Int. Journal of Control*, vol 26, no. 1, pp. 1-18, 1977.

[21] F. Kozin, and F. Nakajima, "The Order Determination Problem for Linear Time-Varying AR Models", *IEEE Trans. on Auto. Control*, vol. AC-25, no. 2, pp. 250-257, April 1980.

[22] M. K. Tsatsanis and G. B. Giannakis, "Equalization of Rapidly Fading Channels, Part I: Adaptive Methods", *IEEE Trans. on Communications*, Dec. 1992 (submitted).

[23] G. B. Giannakis, G. Zhou, and M. K. Tsatsanis, "On blind channel estimation with misses and equalization of periodically varying channels," *Proc. 26th Asilomar Conf. on Signals, Systems, and Computers*, pp. 531-535, Pacific Grove, CA, Oct. 26-28, 1992.

[24] A.V. Dandawate and G.B.Giannakis, "Detection and Classification of Cyclostationary Signals Using Cyclic-HOS: A Unifying Approach", *Proc. of SPIE Conf., Advanced Sig. Proc. Alg., Arch. and Implem.*, San Diego, CA, July 1992.

[25] S. Haykin, "Adaptive Filter Theory," *Prentice Hall*, 1986.

[26] L. Ljung, *System Identification: Theory for the user*, Prentice-Hall inc., Englewood Cliffs, N.J., 1987.

[27] U. Grenander, "On the Estimation of Regression Coefficients in the Case of an Autocorrelated Disturbance," *The Annals of Mathematical Statistics*, vol 25, no 2, pp. 252-272, June 1954.

[28] D. R. Brillinger, *Time series: data analysis and theory*, Holden-day Inc., San Francisco, 1981.

[29] J. M. Mendel, "Tutorial on Higher-Order Statistics (Spectra) in Signal Processing and System Theory: Theoretical Results and Some Applications," *Proceedings of the IEEE*, vol. 79, no 3, pp. 278-304, 1991.

[30] J.K. Tugnait, "Identification of Linear Stochastic Systems via Second- and Fourth-Order Cumulant Matching," *IEEE Trans. on Information Theory*, vol. 33, pp. 393-407, May 1987.

[31] B. Friedlander and B. Porat, "Asymptotically Optimal Estimation of MA and ARMA Parameters of Non-Gaussian Processes from Higher-Order Moments," *IEEE Trans. on Automatic Control*, vol. 35, pp. 27-35, Jan 1990.

[32] G. B. Giannakis, and M. K. Tsatsanis, "A unifying maximum-likelihood view of cumulant and polyspectral measures for non-Gaussian signal classification and estimation", *IEEE Trans. on Information Theory*, vol. 38, pp. 386-406, March 1992.

[33] A. Swami, G. B. Giannakis, and J. Mendel, "A unified approach to modeling multichannel ARMA processes," *Proc. of Intl. Conf. on ASSP, (ICASSP'89)*, pp. 2182-2185, Glascow, Scotland, May 1989; see also *IEEE Trans. on Signal Processing*, (to appear).

[34] M. K. Tsatsanis, "On Wavelets and Time-Varying System Identification", *Ph.D. Thesis*, University of Virginia, September 1992.

[35] A.V. Dandawate and G.B.Giannakis, "Ergodic results for non-stationary processes," *Proc. of 25th Conf. on Info. Sciences and Systems*, pp. 976-983, The Johns Hopkins Univ., Baltimore, March 20-22, 1991.

[36] M. K. Tsatsanis and G. B. Giannakis, "Equalization of Rapidly Fading Channels, Part III: Multiresolution Methods", *IEEE Trans. on Communications*, (submitted).

[37] G. B. Giannakis, "Cumulants: A powerful tool in signal processing," *Proceedings of IEEE*, (Letters), pp. 1333-1334, September 1987.

[38] P.P. Vaidyanathan, "Multirate Digital Filters, Filter Banks, Polyphase Networks, and Applications," *Proceedings of the IEEE*, vol. 78, no. 1, pp. 56-93, Jan. 1990.

[39] M. K. Tsatsanis and G. B. Giannakis, "On the optimum wavelet for statistical multiresolution analysis," *Proc. of 26th Conf. on Info. Sciences and Systems (CISS'92)*, pp. 12-14, Princeton Univ., NJ, March 18-20, 1992. (also submitted to IEEE Trans. on SP).

[40] J. Rissanen, "Modeling by Shortest Data Description", *Automatica*, vol 14, pp. 465-471, 1978.

[41] R. Shibata, "An Optimal Selection of Regression Variables," *Biometrika*, vol 68, pp. 45-54, (correction vol. 69, p. 492), 1981.

[42] K. J. Åström, and T. Bohlin, "Numerical Identification of Linear Dynamic Systems from Normal Operating Records", in *Theory of Self-Adaptive Control Systems; Proc. of Second IFAC Symposium on the Theory of Self-Adaptive Control Systems*, P. H. Hammond, ed., pp. 96-111, Plenum Press, New York, 1966.

[43] K. J. Åström, "Lectures on the Identification Problem - The Least Squares Method", *Report 6806*, Lund Institute of Technology, Sweden, 1968.

[44] M. Feider and J. A. Catipovic, "Algorithms for Joint Channel Estimation and Data Recovery - Application to Equalization in Underwater Communications", *IEEE Journal of Oceanic Engineering*, vol. 16, no. 1, pp. 42-55, Jan. 1991.

[45] I. Daubechies, "Orthogonal Bases of Compactly Supported Wavelets," *Commun. on Pure and Applied Mathematics*, vol. 41, no. 7, pp. 909-996, 1988.

INDEX

W

X